ANTARCTIC RESEARCH SERIES

American Geophysical Union

ANTARCTIC
RESEARCH
SERIES

American Geophysical Union

Volume 1
BIOLOGY OF THE ANTARCTIC SEAS
Milton O. Lee, *Editor*

Volume 2
ANTARCTIC SNOW AND ICE STUDIES
Malcom Mellor, *Editor*

Volume 3
POLYCHAETA ERRANTIA OF ANTARCTICA
Olga Hartman

Volume 4
GEOMAGNETISM AND AERONOMY
A. H. Waynick, *Editor*

Volume 5
BIOLOGY OF THE ANTARCTIC SEAS II
George A. Llano, *Editor*

Volume 6
GEOLOGY AND PALEONTOLOGY OF THE ANTARCTIC
Jarvis B. Hadley, *Editor*

Volume 7
POLYCHAETA MYZOSTOMIDAE AND SEDENTARIA OF ANTARCTICA
Olga Hartman

Volume 8
ANTARCTIC SOILS AND SOIL FORMING PROCESSES
J. C. F. Tedrow, *Editor*

Volume 9
STUDIES IN ANTARCTIC METEOROLOGY
Morton J. Rubin, *Editor*

Volume 10
ENTOMOLOGY OF ANTARCTICA
J. Linsley Gressitt, *Editor*

Volume 11
BIOLOGY OF THE ANTARCTIC SEAS III
Waldo L. Schmitt and George A. Llano, *Editors*

Volume 12
ANTARCTIC BIRD STUDIES
Oliver L. Austin, Jr., *Editor*

Volume 12

ANTARCTIC
RESEARCH
SERIES

Antarctic Bird Studies

Oliver L. Austin, Jr., *Editor*

Published with the aid of a grant from the National Science Foundation

PUBLISHER
AMERICAN GEOPHYSICAL UNION
OF THE
National Academy of Sciences—National Research Council

Publication No. 1686
1968

ANTARCTIC
Volume 12 RESEARCH
SERIES

ANTARCTIC BIRD STUDIES

Oliver L. Austin, Jr., *Editor*

Library of Congress Catalogue Card No. 68-61438

List Price, $16.50

Printed by
The Horn-Shafer Company
DIVISION OF
Geo. W. King Printing Co.
Baltimore, Maryland

THE ANTARCTIC RESEARCH SERIES

The Antarctic Research Series is designed to provide a medium for presenting authoritative reports on the extensive and detailed scientific research work being carried out in Antarctica. The series has been successful in eliciting contributions from leading research scientists engaged in antarctic investigations; it seeks to maintain high scientific and publication standards. The scientific editor for each volume is chosen from among recognized authorities in the discipline or theme it represents, as are the reviewers on whom the editor relies for advice.

Beginning with the scientific investigations carried out during the International Geophysical Year, reports of research results appearing in this series represent original contributions too lengthy or otherwise inappropriate for publication in the standard journals. In some cases an entire volume is devoted to a monograph. The material published is directed not only to specialists actively engaged in the work but to graduate students, to scientists in closely related fields, and to interested laymen versed in the biological and the physical sciences. Many of the earlier volumes are cohesive collections of papers grouped around a central theme. Future volumes may concern themselves with regional as well as disciplinary aspects, or with a comparison of antarctic phenomena with those of other regions of the globe. But the central theme of Antarctica will dominate.

In a sense the series continues the tradition dating from the earliest days of geographic exploration and scientific expeditions—the tradition of the expeditionary volumes which set forth in detail everything that was seen and studied. This tradition is not necessarily outmoded, but in much of the present scientific work one expedition blends into the next, and it is no longer scientifically meaningful to separate them arbitrarily. Antarctic research has a large degree of coherence; it deserves the modern counterpart of the expeditionary volumes of past decades and centuries which the Antarctic Research Series provides.

With the aid of a grant from the National Science Foundation in 1962, the American Geophysical Union initiated the Antarctic Research Series and appointed a Board of Associate Editors to implement it. A supplemental grant received in 1966, the income from the sale of volumes in the series, and income from reprints and other sources have enabled the AGU to continue this series. The response of the scientific community and the favorable comments of reviewers cause the Board to look forward with optimism to the continued success of this endeavor.

To represent the broad scientific nature of the series, the members of the Board were chosen from all fields of antarctic research. At the present time they include: Eugene L. Boudette, representing geology and solid Earth geophysics; A. P. Crary, seismology and glaciology; George A. Llano, botany and zoology; Martin A. Pomerantz, aeronomy and geomagnetism; Morton J. Rubin, meteorology; and Waldo L. Schmitt, marine biology and oceanography. Fred G. Alberts, Secretary of the U. S. Advisory Committee on Antarctic Names, gives valuable assistance in verifying place names, locations, and maps. AGU staff members responsible for the series are: Judith S. McCombs, managing editor, and Marie L. Webner, chief of the editorial office.

Morton J. Rubin
Chairman, Board of Associate Editors
Antarctic Research Series

PREFACE

The birds of Antarctica, and particularly the penguins, have aroused man's interest and his scientific curiosity ever since he first learned of their existence less than two centuries ago. Yet scientific study of them has until recently been only a minor objective of the various expeditions that have visited this most recently discovered and still the least known and least accessible of the continents. The antarctic explorers of the 19th century regarded the birds essentially as a potential source of easily gathered food for men and sled-dogs—and they so used them well into the 20th century. What few bird data and specimens they brought back they acquired largely fortuitously.

Only within the last decade, actually since the inauguration of the International Geophysical Year in 1957, have antarctic expedition complements afforded room for a scientist whose main duties were to investigate the birds. Previously ornithology was pursued by whatever expedition member happened to have the knowledge, ability, interest, and the time to spare for it from his regular duties, traditionally the expedition's doctor. Unquestionably the greatest of these was Edward A. Wilson, medical officer and beloved "Uncle Bill" of the Scott expeditions. Wilson, it will be remembered, was the first to report the amazing breeding of the emperor penguin in the dead of the antarctic winter and to comment on its biological significance. How great was science's loss when he died on the tragic return from the south pole in 1912 is fully evident with last year's publication, under the title *Birds of the Antarctic* (Blandford Press, London), of the detailed bird notes from his diaries and his sensitively accurate field sketches, still vitally alive more than a half-century after he made them.

After Wilson's death comparatively little was added to man's knowledge of antarctic birds until the resurgence of interest and activity in the Antarctic shortly after the end of World War II. There then appeared on the scene a young Englishman, William J. L. Sladen, who, very much in the Wilson tradition, served as medical officer and naturalist on the Falkland Island Dependency Survey expeditions from 1947 to 1951, and who later became my personal friend. Sladen's pioneering studies on the pygoscelid penguins signalled a new era in antarctic ornithology, in which he has played a major role ever since. With the group of exceptional students he attracted to him at Johns Hopkins University, Sladen has stimulated, encouraged, and supervised most of the researches reported on in this 12th volume of the Research Series, and the first volume devoted entirely to antarctic birds.

Each of its eight papers gives striking evidence of the unparalleled opportunities the great antarctic bird rookeries afford for modern biological researches. No new taxa are proposed—it is doubtful that any new birds remain to be discovered in the high latitudes—but each paper presents important new ornithological material. Of singular interest are the unique breeding cycles of the great albatrosses described and explained in the first study. The second, on the ornis of Haswell Island and the only paper dealing essentially with faunistics, also contains important first-hand observations on avian biology, ethology, and demography. The next five papers, detailing various aspects of the behavior and physiology of the Adélie penguin, form a major contribution to our knowledge of this hitherto little-studied species. The final paper describes the banding studies inaugurated by USARP and summarizes the raw data they have already yielded, an invaluable foundation for future studies being planned.

OLIVER L. AUSTIN, JR.

Florida State Museum, Gainesville
March 1968

CONTENTS

The Antarctic Research Series
  Morton J. Rubin v

Preface
  Oliver L. Austin, Jr. vii

Biology of the Great Albatrosses, *Diomedea exulans* and *Diomedea epomophora*
  W. L. N. Tickell 1

The Avifauna of Haswell Island, Antarctica
  Madison E. Pryor 57

Territorial and Social Behavior in the Adélie Penguin
  Richard L. Penney 83

Circadian Rhythms of Activity in the Adélie Penguin, (*Pygoscelis adeliae*) during the Austral Summer
  Dietland Müller-Schwarze 133

Biochemistry of the Adélie Penguin: Studies on Egg and Blood Serum Proteins
  Robert E. Feeney, Richard G. Allison, David T. Osuga, John C. Bigler, and
  Herman T. Miller 151

Salt and Water Metabolism of the Adélie Penguin
  Donald S. Douglas 167

Food Preferences of the Adélie Penguin at Cape Crozier, Ross Island
  William B. Emison 191

The USARP Bird Banding Program, 1958–1965
  W. J. L. Sladen, R. C. Wood, and E. P. Monaghan 213

THE BIOLOGY OF THE GREAT ALBATROSSES, DIOMEDEA EXULANS AND DIOMEDEA EPOMOPHORA

W. L. N. Tickell[1]

Department of Pathobiology, Johns Hopkins University, Baltimore, Maryland

Abstract. The two great albatrosses are separable by size alone from the remaining 11 species in the family Diomedeidae. Except for the early plumage phases in the wandering albatross, morphological differences between the two species are minute. Both breed (if successful) only once every two years. The results of the author's field research on the wandering albatross (*Diomedea exulans*) are compared with the published works of others on the royal albatross (*D. epomophora*).

Many wandering albatrosses breeding at South Georgia travel regularly to New South Wales and back. Royal albatrosses from Campbell Island have been found on both east and west coasts of South America. Apart from differences in the number and location of breeding islands, no differences in oceanic distribution are apparent between the species.

D. exulans shows a small degree of sexual dimorphism. Approximately half the breeding males at South Georgia are in such an advanced state of plumage development that at sea they are indistinguishable from *D. epomophora*. Detailed comparison of all phases of the breeding biology reveals further close similarities between the two species but a significant difference in the length of the fledgling periods. Royal albatross fledglings have plumage similar to their parents, whereas wanderer fledglings are mainly black dorsally. This difference is apparently related to the different fledgling periods. The royal fledglings spend an average of 236 days in or near the nest and leave before the next season's adults arrive. The wanderer chicks take 278 days to fledge and are still present when the next year's adults return to the breeding grounds.

No evidence suggests food resources to be less readily available to wandering albatrosses in some months than others during the winter. Although the wanderer spends its whole first winter in the nest and may fast for as long as 24 days, it is fed an average of 3 times every 10 days. Weight and frequency of feedings decline during the last 50 days in the nest, but the fasts chicks undergo at this time do not exceed those experienced earlier in the fledgling period.

D. exulans pairs that lose their eggs immediately after laying usually return again to breed the following season, but the longer they incubate or feed chicks, the less the chance that they will breed the next season. In all, 20% of pairs breeding in any one year will lay eggs again the following year. No pairs that retained chicks through June laid again the following December. The reproductive physiology of the great albatrosses cycles annually under the influence of extrinsic environmental factors, as in other Procellariiformes, but prolonged care of young evidently causes hormonal or neurological feedback that inhibits gametogenesis.

Annual adult mortality in *D. exulans* at South Georgia is 4.3%, compared with 3% in *D. epomophora* in New Zealand. Modifying Westerskov's theoretical model of the royal albatross population at Campbell Island on the basis of observed mortalities gives a life table for *exulans* indicating that 3.8% of each cohort reach 40 years of age, and 0.4% might theoretically reach 80.

INTRODUCTION

Albatrosses are the largest members of the order Procellariiformes, birds adapted to life in an entirely marine environment. The family Diomedeidae contains 13 species, 11 of which belong to the genus *Diomedea* and 2 to the genus *Phoebetria*. Within the *Diomedea* the two great albatrosses (wingspan approximately 3 meters) are readily separable from all other species (wing span approximately 2 meters).

The field work that produced the main material for this study began at Bird Island, South Georgia (54° 00'S, 38°02'W) in the austral summer of 1958–1959 [*Tickell and Cordall*, 1960]; four more expeditions visited South Georgia between 1958 and 1964 [*Tickell*, 1962a, b; *Tickell et al.*, 1965]. Now available are observations made during six summers and one winter, comprising a total of 26 months of field work.

[1] Present address: The Nature Conservancy, 12, Hope Terrace, Edinburgh, 9, United Kingdom.

Data were obtained for a variety of behavioral and ecological comparisons between different genera and species of albatrosses, but this paper examines principally the two great albatrosses.

The wandering albatross *Diomedea exulans* and the royal albatross *D. epomophora* are similar in size and plumage; they were not recognized as distinct species until 1891 [*Buller*, 1891; *Westerskov*, 1961], and today even experienced workers are frequently unsure of their identity when seen at sea.

The morphological characters distinguishing the two species are as follows [*Murphy*, 1936]:

1. The naricorn plates are more rotund and bulging in *epomophora* and the nostrils more prominent, with the openings circular and directed forward, rather than oval and pointing obliquely upward.

2. The cutting edges of the maxilla in *epomophora* have a black line along the outside; those of *exulans* do not.

3. The eyelids of *epomophora* are black, those of *exulans* white.

4. The chicks of *epomophora* fledge directly into a plumage almost as white as that of the adult. The young *exulans* is black dorsally except for a white face, and the gradual transformation of this plumage over the years goes through many intermediate stages before the predominantly white or so-called "chionoptera" stage is reached.

D. exulans has been the principal subject of my own field work at South Georgia. The studies of *epomophora* by other workers are well documented [see *Richdale*, 1950, 1951; *Sorensen*, 1950; *Westerskov*, 1959, 1960a, b, 1961, 1963]. The circumstances attending the separate studies of the two species have differed considerably, but the two birds are manifestly so closely related that it is both proper and illuminating to analyze the new *exulans* data in terms of what is already known of *epomophora*.

Besides their great size and close morphological similarity, the two species are unusual among sea birds in having a biennial breeding cycle. This dissertation purposes to define the nature of the biennial cycle more precisely and to investigate those specific differences and affinities between the two species that may be of evolutionary significance.

Literature. The literature pertaining to the great albatrosses is not extensive. Because of confusion about the identity of the two species, the reports of naturalists prior to the 1930's are rarely reliable unless they refer to known breeding grounds. *Jameson* [1958] reviews the literature on *exulans* thoroughly and revives interest in the work of *Idrac* [1924] on albatross flight. *Swales* [1965] on a recent visit to Gough Island confirms the observations that Comer [in *Verrill*, 1895]

made on its breeding there. *Murphy* visited the nesting grounds in the Bay of Isles at South Georgia in 1912–1913 and later [1936] suggested that *exulans* in order to breed annually deserts the chicks and leaves them to survive on stored reserves of body fat. *Wilkins* [1923], like Murphy, camped near nesting wandering albatrosses in the Bay of Isles during the last two weeks of December 1921, but in view of what is known of albatrosses today it is difficult to credit some of his observations, which seem to be no more than whaler's hearsay. *Matthews* [1929, 1951] visited the albatross nesting grounds on South Georgia two or three times and predicted that, because *exulans* parents take a whole year to rear their chicks, they breed only in alternate years. *Turbott* [1951] was the first to publish observations of nestling *exulans* being fed in midwinter, and *Rankin's* [1951] voyages about South Georgia are notable for his estimates of the *exulans* breeding population.

Richdale's work on *epomophora* and *bulleri* and his comparative discussions of other albatrosses still form the principal body of knowledge and theories on albatrosses [*Richdale*, 1949, 1950, 1952]. His studies of *epomophora* came about through the rare chance of a pair of these birds attempting to nest at Taiaroa Head, Otago Peninsula, on the mainland of New Zealand in the early 1930's. Persistent human interference at first prevented successful breeding, but strict protection permitted the gradual buildup of a small colony of 17 pairs over 30 years. Richdale had so few (1 to 12) pairs of royals to study that any experimental disturbance of their nesting routine was undesirable. He was able to demonstrate that nestling royal albatrosses were fed throughout the winter, and that parents that reared chicks did not breed the following summer. Since Richdale's studies no further field work on the Taiaroa Head royals has been published.

The largest known breeding population of *epomophora* (circa 4600 pairs) is on Campbell Island, where *Sorensen* [1950] studied them in 1942, 1943, 1945, and 1946. Campbell Island is notable in that among the large number of royals a few wandering albatrosses (some five pairs) have been found nesting [*Oliver*, 1955]. The only other locality where both species breed together is the Auckland Islands, where a few royals nest on an islet adjacent to one where large numbers of wanderings breed (Falla, personal communication; Fleming, personal communication). In recent years *Westerskov* [1959, 1960a, b, 1961, 1963] studied *epomophora* and published a statistical model for the survival of the Campbell Island population.

The work of the New South Wales Albatross Study Group [*Gibson and Sefton,* 1955, 1959, 1960; *Gibson,* 1963, 1967] and my own activities have reinforced each other in demonstrating an unexpected regularity in the movements of *exulans* between Australia and South Georgia [*Tickell and Gibson,* 1967]. Banding at Campbell Island [*Sorensen,* 1954; *Kinsky,* 1959, 1961, 1963, personal communication; *Robertson,* 1964] has indicated similar movements by *epomophora* between the New Zealand island and South America.

Breeding stations. As shown in Figure 1, the royal albatross's breeding range is confined to four of the New Zealand shelf islands between latitudes 44°S and 52°30'S:

South Island, New Zealand		
(Taiaroa Head)	45°45'S	170°45'E
Chatham Islands	44°00'S	176°30'W
Campbell Island	52°30'S	169°15'E
Auckland Islands	50°40'S	166°10'E

This range is much more restricted both in latitude and circumpolar extent than that of the wandering albatross, which breeds both farther north and farther south. The known breeding grounds of *exulans* lie in the southern oceans between latitudes 37°S and 54°S. They are:

Mid-Atlantic Ridge	Tristan da Cunha group	37°20'S	12°45'W
	Gough Island	40°15'S	10°00'W
Kerguelen Plateau	Marion Island	46°50'S	37°40'E
	Îles Crozet	46°25'S	51°40'E
	Îles de Kerguélen	49°30'S	69°30'E
	Amsterdam Island	37°50'S	69°30'E
New Zealand shelf	Antipodes Islands	49°40'S	168°45'E
	Auckland Islands	50°40'S	166°10'E
	Campbell Island	52°30'S	169°15'E
	Macquarie Island	54°40'S	158°50'E
Scotia group	South Georgia	54°00'S	38°05'W

Subspecies. The wandering albatrosses breeding on Gough Island and the Tristan da Cunha group are smaller than those at the more southerly breeding stations. The name *D. exulans dabbenena* Mathews is generally accepted for them. *Murphy* [1936] includes all other wandering albatrosses under the name *D. exulans exulans* Linné, but adds: "It is not unlikely that a race distinct from the antarctic form also nests on some parts of the New Zealand area." As yet too few morphological and anatomical data are available

Fig. 1.   Breeding islands of *Diomedea exulans* (no underline) and *D. epomophora* (solid underline. A broken underline indicates that both species breed.)

from the different breeding stations to determine whether more than the two races, *exulans* and *dabbenena,* are recognizable.

Two races of royal albatrosses have been described, and the narrower breeding distribution of this species gives the trinomina more geographical precision. According to *Falla* [1938] the larger southern form, *D. epomophora epomophora* Lesson, breeds on Campbell Island and on Enderby in the Auckland Islands, and the northern populations nesting on The Sisters and Fortyfours in the Chatham Islands and at Taiaroa Head, New Zealand, are assignable to *D. e. sanfordi* Murphy.

Because breeding stations are so few, the biogeography of the great albatrosses can be discussed in terms of their island populations, on the assumption that these represent breeding isolates of the species. It is possible that isolated populations may differ significantly in their behavior and biology while showing no distinguishable morphological differences.

DEFINITIONS

Wherever possible I have adopted *Richdale*'s [1950, 1952] definitions, but some of these I have redefined in my own terms, and others I have rarely used:

Adult. Any bird, male or female, that is incubating or that is known to have incubated an egg or brooded a chick in a previous season or does so subsequently in the same season. No bird is called adult unless breeding can be proved.

Keeping company. Two birds of the opposite sex staying quietly together either at or away from a nest. The behavior acquires significance only when the same individuals are in each other's presence more than in that of other individuals. I cannot adopt Richdale's definition of keeping company, which combines subadults and failed breeders in the one category "unemployed."

Pair (unqualified). Any male and female that have been observed frequently keeping company and for which there is no evidence of breeding (i.e. an egg) nor record of previous breeding. This category includes birds that have not yet bred because of physiological or behavioral immaturity or both, as well as birds that, having bred previously, have lost their mates and are in the process of establishing a new pair bond.

Mated pair or breeding pair. A male and female that are sharing incubation or care of a chick, or that are known to have shared incubation in a previous season, or do so subsequently in the same season.

Unsuccessful pair or failed breeders. A mated pair that in any given season loses its egg or chick.

Breeding season. The time between the arrival of the first adults (November) and the departure of the last birds that are not feeding chicks (the next May).

Yearling fledgling. An advanced fledgling one year of age about to leave the breeding grounds in December (peak of departure, December 10). Ages of all birds of known age are figured from the time the egg was laid (laying peak, December 24).

Subadult. (*a*) A bird of known age (banded as a chick or fledgling) that has returned to the breeding ground but has not yet bred. (*b*) A bird of unknown age that has been observed on the breeding grounds for at least three consecutive seasons without being known to breed. (Richdale's usage of the term differs in that he includes birds of breeding capacity that have lost their mates.)

Territory. I assign this term to what Richdale calls property rights: "Property rights . . . of a mated albatross may include its nest site, its nest and contents, its chick out of nest, and its mate. If unmated, a bird

may claim property rights over a potential nest site and even over another bird of the opposite sex with which it is apparently attempting to form a mated pair and which seems favorably disposed towards the aspirant, or in other words, a potential sex partner."

Pair bond. The sum of psychological and physiological factors that cause the bird to respond to a given individual of the opposite sex in such a way as to allow breeding.

Widow or widower. A female or male known to have bred previously and present on the breeding grounds for at least one season without its former mate.

Demipopulation. In biennially breeding species that have not suffered catastrophic egg loss for about a decade, slightly more than half the total population, the demipopulation, breed each year.

Fast. An arbitrary interval of five or more days between two consecutive feedings of a chick.

METHODS OF STUDY

I carried out field work on Bird Island, South Georgia (Figure 2), during the summer seasons of 1958–1959, 1960–1961, 1962–1963, and 1963–1964, and also during the winter of 1963. No ornithological expedition visited Bird Island in the 1959–1960 season, but W. N. Bonner, the Falkland Islands Government Sealing Officer, kindly recorded the nest occupants of the study area for me that year. During the summer of 1961–1962 H. Dollman, who had been my field assistant the previous season, carried out a planned program of routine observations with the help of C. F. le Feuvre. I was also assisted in the field at various times by P. A. Cordall and R. Pinder.

When the study started in 1958 the best map of South Georgia available was on a 1:500,000 scale, showing Bird Island less than two centimeters long. Therefore P. A. Cordall plane-tabled Bird Island at 1:12,500 (Figure 3). The place names in heavy type on this map are officially recognized. Recognition has not been requested for names of other minor features shown and used in the text for the study areas established for various phases of the investigations.

Wanderer Valley, because of its proximity to the base camp at the head of Jordan Cove, was the site of most studies of marked individuals. Within Wanderer Valley about 140 pairs nested each season. The most populous part of the valley, containing 80 to 90 nests, was termed the "extended study area." A sector of about one hectare in the middle of the valley

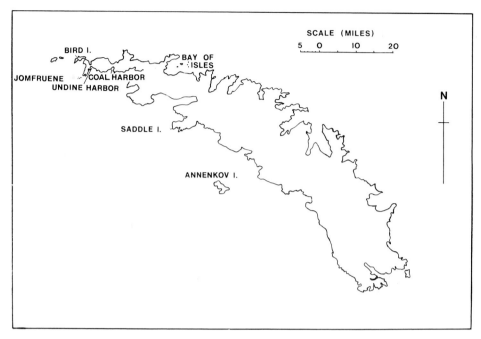

Fig. 2. South Georgia breeding grounds of *D. exulans*.

containing 25 to 30 nests was the "inner study area" where most detailed observations were made.

Wanderer Ridge, adjoining Wanderer Valley, supported a high concentration of some 240 nests and was used for periodic counts to assess egg or chick mortality and fledgling departures. Long Ridge, another sector employed for the same purpose, was visited very infrequently and permitted check-counts less affected by observational disturbances. Gony Ridge was used for counting adult arrivals and for a long-term experiment to determine how rapidly two demipopulations equalize after the removal of all eggs in one season.

All the breeding pairs on the island were censused at the end of January each year. All nests in Wanderer Valley that had eggs or chicks were marked with stout wooden posts 3 centimeters square and 65 centimeters long, with numbered metal plates, painted orange, and driven about 50 centimeters into the ground leaving about 15 centimeters exposed. Substantial markers are a necessity because the albatrosses pull small ones out when gathering nesting material. The marked nests were plotted on a plane-tabled map of the valley at 1:960. Those in the inner study area were plotted annually by tape measure triangulation on a 1:340 scale map on which the nest positions could be determined to within 15 centimeters.

In the inner study area all adults and subadults were marked with numbered metal bands, plastic color spiral bands, or paint spot combinations [*Tickell*, 1960]. This area was visited daily when field workers were on the island. Most adults and many subadults in the extended study area were banded and color-marked. Nests in this area were inspected daily during the breeding season up to the end of brooding and thereafter less frequently. In the remainder of Wanderer Valley most adults were banded but none were color-marked, and visits were made at about weekly intervals.

Capturing wandering albatrosses on the South Georgia breeding grounds presents little difficulty. Unless recent handling has made the bird wary of an approaching observer, wanderers do not normally flee from man. The egg or chick can usually be removed from under an incubating or brooding adult and replaced without the adult leaving the nest. Indeed, if the incubating parent is pushed off the nest, it generally scrambles back on again at once. Birds that do walk or run away can usually be caught readily if chased downwind.

Individual operators favor different approaches, but all methods of handling the birds eventually involve a bold grab at the head or bill, which is very sharp. The tool of the trade at Bird Island, locally called a "trodge hook," is a shorter and stouter replica of the "puffin hook" that *Lockley and Russell* [1953] describe. The easiest (and safest) way for a lone opera-

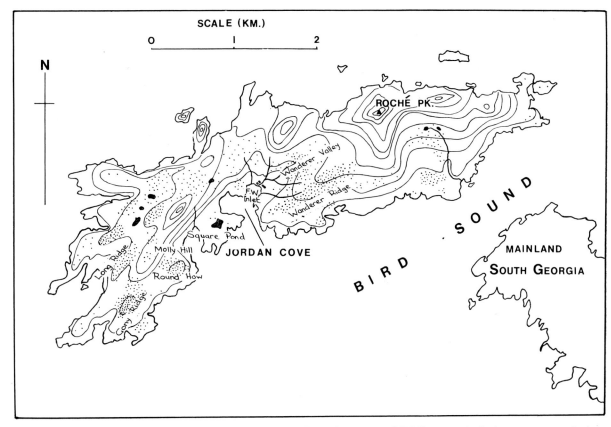

Fig. 3.   Bird Island, South Georgia, showing distribution of *D. exulans* nests. Officially recognized place names are in heavy type, contour lines at 30-meter intervals.

tor to band an albatross is by holding the bird's head snugly between his legs as he works (Figure 4).

Off New South Wales the birds are caught at sea from small motor boats, preferably on calm days when the birds have difficulty taking off. *Gibson and Sefton* [1959] describe how they approach the bird downwind and, as it swims away or tries to fly, catch it in a hoop net 4½ feet in diameter attached to a line and thrown over the bird from the bow.

The bands we used at South Georgia were at first size 5 (13 × 90 mm) aluminum giant petrel (*Macronectes giganteus*) bands supplied by the Falkland Islands Dependencies Survey (now British Antarctic Survey), and later size 8 (15 × 115 mm) bands specially made for wandering albatrosses and supplied the USARP Bird Banding Program. Both types of bands are overlapped and bear a double inscription [*Sladen and Tickell*, 1958], similar to those originally designed for the Manx shearwater (*Puffinus puffinus*) [*Lockley*, 1942].

We also used colored plastic spiral bands in 11 colors, each color assigned a standard abbreviation for consistent recording as follows:

B, black	P, pink
G, green	R, red
L, light green	S, sky blue
M, mauve	W, white
N, navy blue	Y, yellow
O, orange	

The particular purposes for which these bands were used included:

1.   Identification of individual birds by a combination of three spirals on the leg opposite the numbered band and described by the color symbols read from top to bottom. Thus WBG wore a white top, black middle, and green bottom spirals.

2.   For distinguishing a particular class of birds. For instance, all birds outside the Wanderer Valley study area that had been banded or recovered in Australia carried one black spiral in addition to the numbered band.

3.   To distinguish birds of known age outside the

Wanderer Valley study area. When checked in the current year they were given a single year color spiral, which was changed when checked again in a subsequent season.

Individuals were also marked with cellulose paints in six colors. Combinations of three colored spots painted on the head in a line from the crown to the base of the culmen [*Tickell*, 1960] are described in that order with the same abbreviations used for the spiral colors. Both the colored spirals and the head spots were easily read with the naked eye at the distances the birds were usually read, even when they were flying low over the breeding grounds, and with 8-power binoculars when farther away.

A small observation hut was built on the inner study area the winter of 1963, and one of its principal uses was for weighing chicks. Elsewhere on the island we used empty packing cases to shield scales from the wind. Newly hatched chicks were weighed on a 1 kg × 5 g spring balance, older ones on a 30 lb × 1 oz pendant spring balance with a reduction beam for heavier birds. All weights were later converted to kilograms.

Specimens of *exulans* gonads were taken at various stages in the life cycle. Each sample was divided in two and fixed in Bouin's fluid before storing in alcohol. This material was later sectioned at 6 μ and stained with iron haematoxylin to determine the spermatogenic stages.

BREEDING GROUNDS

The climates of the islands on which the great albatrosses breed vary between extremes of the subantarctic and southern cold temperate zones. Characteristically there are persistent winds and few calm days during the year [*Westerskov*, 1963; *Richards and Tickell*, 1967]. Although the vegetation shows important differences from one island to another [*Wace*, 1960; *Greene*, 1964], all have a similar type of tussock-dominated grassland. Both the wandering and royal albatrosses require vegetation for nestbuilding and ample space for their social activities. The problems of landing and taking off also determine the nature of the breeding ground. Unless winds are near gale force, these birds need an appreciable run into the wind to become airborne, and they require appropriate "runways" for the purpose that are not too tussocky or obstructed by shrubs.

On the more northerly breeding grounds where shrub vegetation grows at lower altitudes, particularly in the valleys, both the great albatrosses nest at higher altitudes, as the wanderer does at Gough Island [*Swales*, 1965] and the royal at Campbell Island [*Westerskov*, 1963]. At South Georgia, the most southerly breeding station, absence of shrub vegetation permits *exulans* to nest at lower altitudes.

South Georgia lies almost 200 nautical miles south of the antarctic convergence of surface waters. Predominantly an island of high mountains and deep fjords, numerous glaciers flow into the sea, and between them are considerable expanses from sea level to about 500 meters that are free of permanent ice and from which the winter snows melt during the

Fig. 4. Banding an adult wandering albatross, January 1959. Photo by P. A. Cordall.

summer. The vegetation is dominated principally by the tussock grass *Poa flabellata* [*Greene*, 1964]. Many such areas apparently suitable for wanderer nesting are available around the coast, but the species breeds in comparatively few places, most of them in the northwest sector (Figure 2).

Bird Island (Figure 3) lying 500 meters off the northwest extremity of the South Georgia mainland has many more wandering albatrosses than all the rest of South Georgia together [*Wilkins*, 1923; *Rankin*, 1951; *Morris*, 1962]. Although the island is only 6½ kilometers long and 1½ kilometers wide, it is mountainous in the manner characteristic of South Georgia. The sea cliffs of the north coast form a forbidding wall along much of their length, rising to 365 meters below Roché Peak. The south side of the island is generally lower than the north, and the east is predominantly more mountainous than the west.

Most of the land below about 140 meters is covered by tussock grass, among which are scattered moss banks and small meadows of the antarctic grass *Deschampsia antarctica*. Surface water in quantity often makes the ground very wet and forms innumerable deep bog holes and many larger ponds. After heavy rains some of the streams become minor torrents, and the larger ones have cut deep gorges through the tussock-covered slopes.

The climate is generally cloudy and windy with frequent precipitation. Temperatures in summer vary from about −2° to 9°C and in winter between −10°

and 3°C [*Richards and Tickell*, 1967]. Snow cover builds up gradually as the winter progresses, and, when the ground freezes, stream water flows out over the surface to form substantial sheets of ice [*Tickell and Richards*, 1967]. Some snowbanks persist late into the summer above about 200 meters, but there is no permanent ice as on the mainland.

Immense numbers of birds and seals breed on Bird Island. The beaches teem with fur seals (*Arctocephalus tropicalis gazella*), while four species of albatross (*Diomedea exulans, D. melanophrys, D. chrysostoma*, and *Phoebetria palpebrata*) are conspicuous on the crags and slopes. Macaroni (*Eudyptes chrysolophus*) and gentoo (*Pygoscelis papua*) penguins are numerous, and the burrowing petrels (*Procellaria aequinoctialis, Pachyptila desolata, Peleconoides* spp., *Oceanites oceanicus*, and *Fregetta tropica*), inconspicuous during daylight, are heard everywhere at night among the tussocks and form the prey of a large number of brown skuas (*Catharacta skua lonnbergi*). The brown rat (*Rattus norvegicus*) was introduced long ago to parts of South Georgia, where it is reputed to prey on the smaller ground-nesting birds [*Murphy*, 1936]. Bird Island has no rats, and the dove prions (*Pachyptila desolata*), diving petrels (*Pelecanoides georgica*), South Georgia brown pintails (*Anas georgicus*), and and antarctic pipits (*Anthus antarcticus*) are particularly numerous.

The wandering albatrosses nest almost everywhere from level ground about Square Pond, no more than

Fig. 5. Bird Island, South Georgia, looking westward from the summit of Roché Peak. Wanderer Valley lies immediately below. January 1959. Photo by P. A. Cordall.

TABLE 1. Density of *D. exulans* Nests with Eggs in the 0.85 Hectare Study Area on Bird Island, South Georgia

	Nests with Eggs Jan. 31	+6.9% Egg Loss before Jan. 31	Nests per Hectare*
1958–1959	30	32	37.6
1960–1961	35	37	43.5
1961–1962	38	41	48.2
1962–1963	28	30	35.3
1963–1964	27	29	34.1

* Mean = 39.7 nests per hectare.

15 meters above sea level, to the rather limited flat land on the top of Molly Hill at 140 meters. The largest numbers are found between 30 and 100 meters. They prefer flat stretches and gentle slopes; especially popular are the crests of such broad ridges as Wanderer Ridge, Round How, and Gony Ridge, which are openly exposed to the prevailing winds. It is rare to find wanderers nesting on steep slopes, and those that do always have access to more level places such as the terraces or crests of ridges.

Smooth "greens" of the grass *Deschampsia antarctica*, the cushion plant *Colobanthus antarcticus*, and of mosses among the hummocky tussocks are characteristic of the Bird Island vegetation. Wanderer nests are numerous around the fringes and within easy walking distance of these greens, which the birds use as airfields. Some solitary nests, particularly those at very low or very high altitudes, may be hundreds of meters from their nearest neighbor, but in other sectors nests are sometimes less than 3 meters apart. The Wanderer Valley study area occupied approximately 14 hectares and included regions of low nest density at low and high altitudes as well as the more populous sites between. In effect it formed a transect from sea level to the upper limits of nesting. The inner study area of this valley comprised an 0.85-hectare plot near the center of the valley where the mean nest density, 40 nests per hectare (Table 1), was representative of the more favored nesting habitats on the island. The general impression is that nesting space is ample with only half the population breeding each season.

Yet in a population of biennial breeders the nesting grounds must be adequate to accommodate the whole breeding population, for a catastrophe that destroys all eggs in one season results in the whole population trying to breed the following season. To test this all eggs were removed as they were laid from a study plot in the most crowded nesting ground on Bird Island. In

this expanse of approximately 2.5 hectares, the birds laid 266 eggs in 1962–1963, a density of 106 nests per hectare. The following year this area supported 170 nests per hectare. All other parts of Bird Island have substantially fewer breeding pairs.

OCEANIC DISTRIBUTION

Determining the relative oceanic distributions of *exulans* and *epomophora* beyond feeding range of their breeding grounds is not a simple matter. Any assessment of their pelagic dispersal based solely on sight records suffers from the difficulty of distinguishing the older "chionoptera" individuals of *exulans* from *epomophora*. Although the darker-backed younger wanderers are easily identifiable, the chionoptera birds form too large a proportion (approximately 50% of breeding males) of the population to be ignored.

The most substantial work on the oceanic distribution of albatrosses is *Dixon*'s [1933] comprehensive analysis of 3500 days of observations (2002 days between latitudes 20° and 60°S) during 27 years (1892–1919) at sea in sailing vessels. Acknowledging his inability to distinguish between *exulans*, *chionoptera* (now *exulans*), and *regia* (now *epomophora*) at sea, he pooled all observations of these forms in his analysis. He is worth quoting at length:

My records indicate that these albatrosses are practically confined to the area between 30°S and 60°S. I have observed very few north of 30°S and none within the tropics. The farthest north I have seen one is 24°S, off the Brazilian coast in summer. This was unusual as most of those seen north of 30° were observed in winter or spring. There is a good deal of variation in the density of their numbers in different regions depending on season. Thus, in spring, 92% of the birds were seen between 30° and 50° and only 3% south of 50°, whereas in summer 73% were between 30° and 50°, and 26.5% south of 50°. In autumn there is a shift northwards again, and by winter 96.9% are between 30° and 50° and only 0.1% south of 50°. '[The table also indicates a] . . .' similar seasonal shift in longitude. In winter and spring the vast expanse of ocean from the 120th meridian of west longitude to the Horn is practically deserted, whereas in summer this region is the favored one. There seems to be a progressive shifting of the center of abundance from west to east each season as if the birds flew round the earth from west to east with the prevailing winds. For instance, in winter 82% are on the hemisphere 90°E–180°; in spring the center of abundance has shifted slightly towards the east so that 73.5% are in the hemisphere 90°E to 90°W; in summer 67% are in the hemisphere 180°–90°W–0, and in autumn 63% in 90°W to 90°E.

It is interesting to note that in the summer there is a considerable concentration of albatrosses in the region of the cold currents off the west coast of South America and up past the Falkland Islands in the Atlantic.

Fig. 6. Adult wandering albatross in flight. Photo by P. A. Cordall.

The greatest number ever seen together was estimated at fifty, close to Tristan da Cunha in summer. On one occasion, forty were seen near Crozet Islands. On twenty-five occasions in about as many years, flocks of ten were met with. Three or four is about the average number seen at once, and there are great numbers of instances when it was seen quite alone.

Holgersen [1957], *Bierman and Voous* [1950], and *Van Oordt and Kruijt* [1954] all present ocean transects of albatross observations made over one or two seasons (Table 2), and *Routh* [1949] assesses the relationship of the wandering albatross to the antarctic pack ice observed in one season as follows:

Wandering Albatrosses, in all probability, do not come close to the pack ice except as stragglers or in winter and early spring when the pack ice is far north. These birds were frequently recorded far out over open sea, but, excepting two birds which were seen in April at 63°14'S, a point at least 60 miles north of the ice edge at the time, there were no records south of 62°52'S. In the early spring, no doubt, they are nearer the edge of the pack ice, and are frequently to be seen near it, especially in areas of many icebergs. However, with one possible exception, none were recorded inside the ice edge. When the pack recedes, these birds do not appear to follow it down, and only about a dozen very scattered individuals were recorded during the course of the summer. Of these, only two were below 62°S.

Holgersen [1957] comments on the difficulty of distinguishing the two species at sea but accepts *Murphy*'s [1936] statement that *epomophora* keeps close to the continental shores and has a more northerly breeding distribution. *Bierman and Voous* [1950], also aware of the differences between the two species, note under the heading "Field Characteristics" that, "The shape of the nasal tubes which is a crucial distinguishing characteristic [of *exulans*] compared with *D. epomophora* was clearly observed on several occasions." Neither of these two papers reports seeing any royal albatrosses, and *Van Oordt and Kruijt* [1954] and *Routh* [1949] do not mention the species at all. Occasionally birds are reported farther north and south, and *exulans* has been identified among the most northerly and southerly records [*Holgersen*, 1957].

D. exulans occurs rarely in the northern hemisphere. The older records are all subject to doubt because seamen in sailing vessels were known occasionally to capture sea birds in the southern hemisphere and release them in the northern [*Bourne*, 1967]. Recently, however, E. P. Agate watched one for four hours at sea off the southwest coast of Portugal [*Bourne*, 1966], and another was killed off Sicily [*Orlando*, 1958].

TABLE 2. Some Extreme Northerly and Southerly Observations of Great Albatrosses

Authority	North		South	
Dixon [1933]	24°S (off Brazil)		60°S	
Holgarsen [1957]	21°S	40°W*	63°55'S	131°54'W
Bierman and Voous [1950]	18°00'S	04°00'E	62°00'S	25°30'W
Van Oordt and Kruijt [1954]	18°50'S	02°34'E	64°S	85°E
Siple and Lindsey [1937]			68°40'S	

* In immature plumage, therefore definitely *exulans*.

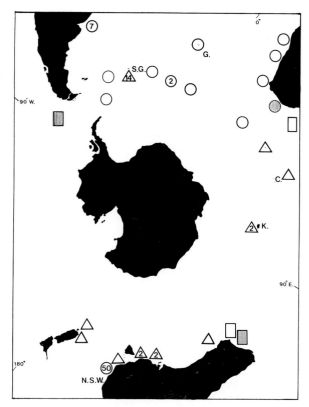

Fig. 7. Distant recoveries of *D. exulans* banded at: South Georgia (S.G.), open circle; New South Wales, Australia (N.S.W.), open triangle; Îles de Kerguélen (K), hatched rectangle; Îles Crozet (C), open rectangle; and Gough Island (G), hatched circle. Unless indicated by numerals, each symbol represents a single recovery.

On the basis of relative abundance, these reports probably refer more to *exulans* than to *epomophora*. Except for the more restricted breeding range of *epomophora*, no evidence suggests any differences in the two species' oceanic distribution.

Banding evidence of exulans *movements.* The first indication of how far a wandering albatross may fly from its nest was obtained when an albatross banded by J. Loronchet in 1913 on the Îles de Kerguélen was recaptured at sea near Cape Horn in 1916 [*Menegaux*, 1917].

In the early 1950's Australian ornithologists began to take an interest in the conspicuous numbers of albatrosses assembling off the coast of New South Wales in winter [*Gibson and Sefton*, 1955]. This interest led to the formation of the New South Wales Albatross Study Group, which began banding operations in 1956 and has been active in the field every winter since 1958 [*Gibson and Sefton*, 1959, 1960; *Gibson*, 1963, 1967; *Tickell and Gibson*, 1967].

Our banding studies at South Georgia were begun independently in 1958 and without knowledge of these activities. A wandering albatross banded at New South Wales in August 1958 was found at Bird Island in December the same year, and a wanderer banded at Bird Island that December was caught off Sydney the following July. These reciprocal recoveries brought the two groups into contact. The many recoveries accumulated since have revealed a distinct movement of birds between the South Georgia breeding grounds and the New South Wales winter feeding grounds.

More than 6000 wanderers have been banded at South Georgia and 1680 off New South Wales. From these, 94 birds have been recovered away from the place of banding, and a number have been reported more than once. All but 4 of the 61 recoveries away from Australia of birds banded there have been from South Georgia, and of the 32 recoveries of South Georgia birds to date, 14 have been in Australia and 18 elsewhere (Figure 7). Of 176 wanderers R. Tufft banded on nests at Île de la Possession in Îles Crozet [*Tilman*, 1961], two have been recovered elsewhere, and of 200 wanderers banded at Gough Island in 1955–1956, one was later recovered [*Swales*, 1965].

Gibson and Sefton [1960] note the wide range of bill and wing-span measurements among the albatrosses wintering off New South Wales, and the recoveries of their banded birds at South Georgia, Kerguelen, Auckland Island, and Marion Island indicate that several breeding populations are represented among them. This is further indicated by the recovery in western Australia of one wanderer banded in the Crozets. The fact that 50 Australian birds have been recovered at South Georgia compared to 1 at Marion Island and 2 at Île Kerguélen (Desolation Island) reflects the intensive field work at Bird Island and the lack of it elsewhere rather than the proportion of birds at the different islands.

The New South Wales banding has been done at two centers, Bellambi and Malabar, 22 kilometers apart, and it was first assumed that so much intermixing would occur that the birds could be considered as one population for statistical purposes. Analysis of the banding and recapture data, however, has shown that in general each place is visited regularly by different individuals. Some chance interchange does occur but not enough to obscure the distinctness. Furthermore, 4.9% of the 797 birds banded at Bellambi up to 1964 have been recovered at South Georgia against only 1.6% of the 680 banded at Malabar (chi^2 = 12.0, $p < 0.001$).

The albatrosses reach peak numbers at these two centers between mid-June and mid-September. The very small numbers of local recaptures suggest that individuals tend to move on after a short time, probably a few days. A few of the wanderers banded there have been recovered elsewhere around the Australian coast, but such coastal aggregations are apparently unusual, for no others have been reported. The importance of these winter feeding grounds is manifest from the number of birds that return there repeatedly, but it would be unwise as yet to postulate that this region is the South Georgia population's principal wintering ground.

Of the 65 passages recorded between Bird Island and New South Wales, 27 took place within 6 months or less, a wanderer present at Bird Island during the austral summer being reported off Australia the following winter or vice versa. Outstanding is the record of #140–02800, banded at Bellambi August 25, 1959, found at Bird Island February 26, 1962, and recaptured at Bellambi August 25, 1962. We have no record of it the summer of 1962–1963, but it was back in Australia in July 1963, and returned to South Georgia the following summer of 1963–1964.

Gibson [1963] tested the Australian retrapping data to see whether they might be correlated with the biennial breeding cycle of this species. As the wandering albatross appears to have an annual cycle of movement during its immature years, the degree of correlation will depend on what proportion of the birds visiting New South Wales are in fact established, successful breeders.

Most of the Australian birds at South Georgia are recovered at random away from nests during the daily tours of field workers about the island, and there is little opportunity to check whether or not they are actually breeding. In the annual census of the breeding population, however, when every occupied nest is examined shortly after the end of egg laying, we have a measure of the proportion of the Australian banded birds in the breeding population. In all we made 90 recoveries of 50 individuals, of which only 13, or 26%, were found incubating eggs. In the last two seasons of observations progressively fewer new birds were discovered, indicating that most of the Australian banded birds had been seen. As about 55% of the population breeds each year and only half the nesting birds are incubating at any one time, the number of Australian banded birds in the population should be approximately twice the sum of two season's counts. It is therefore reasonable to speculate that up to 1963–

1964 the total number of Australian banded birds at Bird Island probably numbered between 60 and 70, of which only about 25–30%, or 16 to 18, were actually breeding.

Although some of the South Georgia established breeders do appear to winter regularly off New South Wales, apparently a larger proportion of the South Georgia birds there are of the younger age groups not yet breeding. The retrapping evidence also shows the probability of an ordered migration pattern on an individual or group level, but where the birds spend the rest of the year still remains to be learned.

Of the 18 recoveries of South Georgia birds banded as fledglings, one reached southwest Africa about six weeks after leaving the island. Three others were reported in their first year, two in their second, and two in their fifth year. No distribution pattern is apparent for any of these age groups in the recoveries.

Young wanderers begin to return to the South Georgia breeding grounds three years after they fledge, much earlier than *Richdale* [1952] reports for the royals. From 1961 to 1964 more than 38% of the 656 fledglings banded in 1958–1959 had returned, a remarkably high survival rate. Only two birds banded as nestlings appear in the Australia–South Georgia records. One of these, 58311, banded at Bird Island in December 1958 just before it left, was captured off Bellambi in July 1959; on February 2, 1963, it was found at the age of four in the same valley where it was reared, where it was again seen the following summer.

Banding evidence of epomophora *movements.* The recoveries of banded royal albatrosses to date are comparatively disappointing. No distant recoveries have been reported from the small numbers banded at Taiaroa Head, New Zealand, since 1939, or from the 20 banded at the Enderby Island breeding ground in the Auckland Islands. One of the 45 royals banded since 1957 off the New Zealand coast at Cape Campbell, Castlepoint, Kaikoura, and Oamaru, none of which are near breeding grounds, was reported from Chile [*Kinsky*, 1959]. Table 3 shows the annual distribution of the 7216 royals banded at Campbell Island since 1943 and the 26 recoveries reported to date, which are mapped in Figure 8.

Most of these recoveries have been from South America, almost equal numbers from off the Pacific and Atlantic coasts. Apparently many of the Campbell Island yearlings go eastward on leaving the island,

TABLE 3. Bandings and Distant Recoveries of *D. epomophora* at Campbell Island, New Zealand [*Sorensen*, 1954; *Kinsky*, 1961, 1963; and Robertson, personal communication, 1964]

Season Banded	Banded			Recovered		
	Adults	Pul-lets	Total	Adults	Pul-lets	Total
pre–1950	112	262	374	0	1	1
1957–1958	79	0	79	0	0	0
1958–1959	0	10	10	0	0	0
1959–1960	51	91	142	1	0	1
1960–1961	37	293	330	0	3	3
1961–1962	628	150	778	2	0	2
1962–1963	250	629	879	0	12	12
1963–1964	180	174	354	1	2	3
1964–1965	1083	22	1105	0	0	0
1965–1966	2205	960	3165	0	4	4
Total	4625	2591	7216	4	22	26

and on reaching South America follow either the Peru or the Falkland current northward.

Because royal albatrosses have been collected in some numbers on waters of the South American continental shelf, *Murphy* [1936] suggests that the species has a greater predilection for coastal waters and shallow seas than the wanderer, but the recent New South Wales observations show *exulans* similarly attracted to inshore waters. *Murphy* [1936] also postulates a possible breeding colony at Tierra del Fuego on the evidence of P. Reynolds who reported "large white birds" nesting on "high ground" near Puerto Harberton in the Beagle Channel; this report has never been confirmed or investigated. The Campbell Island recoveries reduce the probability of this postulate.

Discussion. The numerous passages of individuals recorded between South Georgia and New South Wales demonstrate that wandering albatrosses favor particular winter feeding grounds far distant from the breeding grounds. Although the evidence is not so unequivocal for the royal albatross, the Campbell Island population seems to find similar habitual winter feeding grounds around South America. Although it stretches the evidence too far to suggest that any island breeding population has a regular or "traditional" wintering ground, obviously individual birds of both species are capable of remembering good feeding grounds and of navigating efficiently to and from them. Learning the marine environment is possibly an important function of the initial years before the birds return to the breeding grounds as subadults.

Whether the disparity in composition of the wanderer populations visiting Bellambi and Malabar, New South Wales, is predominantly due to age, birthplace,

or other factors is not known, but the ability of *exulans* to navigate precisely to remembered feeding places is well demonstrated, and the theory that they visit several such places annually (or biennially in the case of successful breeders) in a more or less ordered pattern is not an unlikely one. Young birds in their early years have ample opportunities to find feeding grounds that could become routine calling places in their later migrations.

The difference in the recovery areas for the two species is probably a consequence of the relative positions of their banding stations and a similar predominantly easterly movement from each. Although nothing indicates that the wanderer yearling fledglings tend to travel in any one particular direction when they leave South Georgia, the recoveries of the Campbell Island royals are consistent with this theory.

A minority of the recoveries of both species indicate northerly and northwesterly components in the migrations. Dead recoveries of New South Wales banded wanderers have all been from the southern and southeastern shores of Australia, except for one picked up

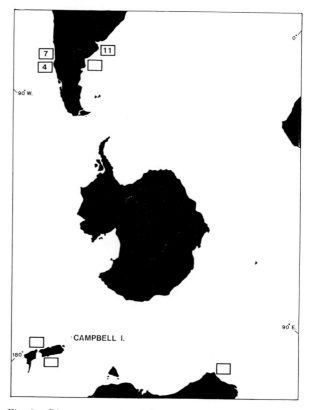

Fig. 8. Distant recoveries of *D. epomophora* banded at Campbell Island [*Sorensen*, 1954; *Robertson*, 1964; *Kinsky*, 1961, 1963, and personal communication]. Each unnumbered square represents a single recovery.

TABLE 4.   Measurements of *D. exulans* Adults at Bird Island, South Georgia

		N	x and SE	SD	Range	t	p^*
Weight, kg	males	20	9.768±0.196	0.875	8.193–11.907	9.282	<0.001
	females	22	7.686±0.119	0.559	6.719– 8.703		
Wingspan, m	males	17	3.096±0.012	0.051	2.997– 3.226	1.127	
	females	15	3.014±0.020	0.078	2.819– 3.124		
Culmen, mm	males	21	169.1±0.9	4.030	163–180	4.340	<0.001
	females	23	163.8±0.7	3.618	155–171		
Depth of bill, mm	males	21	40.1±0.4	1.656	37.2–44.6	7.850	<0.001
	females	22	35.8±0.4	1.918	32.0–40.0		
Tarsus, mm	males	21	118.0±1.0	4.614	110–127	3.811	<0.001
	females	23	112.9±0.9	4.270	106–123		
Mid-toe, mm	males	21	184.4±1.5	6.926	172–193	5.382	<0.001
	females	22	174.7±1.0	4.734	165–181		
Tail, cm	males	21	22.7±0.2	0.740	21.5–24.6	0.584	
	females	23	21.5±0.1	0.623	20.6–22.7		
Wing, cm	males	21	67.9±0.3	1.443	65.5–71.0	0.528	
	females	22	65.7±0.3	1.367	63.0–68.0		

* p = probabilities where difference between males and females is significant.

at the southern extremity of New Zealand. Regular patrols of approximately 800 miles of New Zealand beaches have yielded only four dead wanderers in 3 years [*Bull and Boeson*, 1963; *Boeson*, 1964, 1965]. The royal albatross is very rare in Australian waters [*Gibson*, 1963], and only one bird banded at Campbell Island has been recovered there in contrast to the many wanderers.

Considered comparatively, the available data demonstrate that both species travel great distances, but reveal no obvious differences in distribution beyond the breeding ground feeding ranges.

SEXUAL DIMORPHISM

Several authors have described size and plumage differences between males and females of both the great albatrosses [*Murphy*, 1936; *Matthews*, 1929; *Wilkins*, 1923; *Richdale*, 1950, 1952; *Sorensen*, 1950], but no qualitative assessment of these differences in one local population has been made.

Richdale was familiar with the small differences in plumage between the two sexes of *epomophora* at Taiaroa Head, but he says:

In the field I use the foregoing sex differences [measurements and plumage] only as an initial guide to the sex

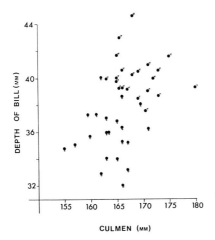

Fig. 9. Sexual dimorphism in *D. exulans* as indicated by length of culmen and depth of bill.

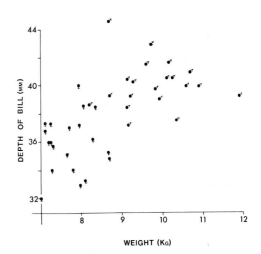

Fig. 10. Sexual dimorphism in *D. exulans* as indicated by body weight and bill depth.

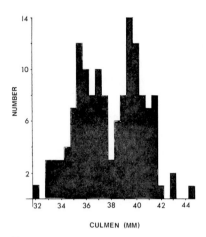

Fig. 11. Culmen lengths of 139 unsexed *D. exulans.*

of the bird. The noting of differences in behavior at the pre-egg stage is far more reliable. (1) A single bird at the nest more than two days before egg-laying is usually the male; if numerous observations indicate that the same bird is nearly always present, then there is no doubt that the bird is a male. (2) The female, unless ready to lay, appears only for coition and usually does not stay long. (3) A bird, often covered with mud, working feverishly at the nest a few hours before the egg is laid is the female . . . . Actually, if working continuously on the breeding area, the sexing of albatrosses is not difficult.

Exactly the same may be said about sexing the individual wandering albatrosses, but, in addition, sex differences in plumage are more pronounced than in the royals.

Table 4 gives measurements of wandering albatrosses at Bird Island made on live or freshly killed specimens in the field. Each bird was sexed either by dissection or palpation of an egg in the abdomen. Another 96 birds measured were unsexed, but their measurements show well-marked bimodal distributions consistent with those of the sexed individuals (Figures 9, 10, 11). Although some overlap is apparent between the largest females and the smallest males, the males are generally larger in all dimensions. This is particularly noticeable in the weights, the heaviest female being only 510 grams heavier than the lightest male.

The fledgling wandering albatross goes to sea with predominantly slate black dorsal plumage. The oldest breeding individuals are the completely white "chionoptera" stage with black wingtips. Many intermediate stages exist between these two extremes, and some authors [*Wilkins*, 1923] have assumed that the *exulans* plumage changes follow a simple 2- or 3-year sequence similar to those in such families as the Lari-

dae. Records of banded individuals followed over six seasons now show that the complete change in plumage may take many years, perhaps 20 to 30 or more.

During the 1962–1963 and 1963–1964 seasons at Bird Island the plumages of more than 400 marked breeding birds of known sex were recorded according to the plumage code devised by J. D. Gibson [*Murray,* 1962; *Gibson,* 1967] (Figure 12). The age of none of these birds was known, but in view of the Macquarie Island record of first breeding at 10 years of age (Ingham, personal communication), we may assume that the sample represents a cross section of the breeding population and includes all cohorts from 10 years upward in proportion to their frequency [*Leslie,* 1966].

Figures 13 through 17 show the frequency distribution of the numerical plumage classes among breeding males and females. More males are in advanced stages than females; 43% of breeding males are in full chionoptera plumage (represented by the code 6, 6, 5, 4), whereas 49% rate 5.5, 5.5, 4.5, 3.5 or higher. In contrast, none of the 208 females had reached the complete chionoptera stage, although 14 to 17% had completely white heads.

Of birds banded as chicks, 237 have been found back on the island 6 years after laying, 9 of which were killed, sexed by dissection, and coded (Table 5). These few records indicate that during the first five years away from the nest progressive whitening of the plumage goes ahead much more rapidly in males than in females. The 5- and 6-year-old females are immediately recognizable as young birds by their dark plumage, but the criteria are not precise enough to permit aging by plumage characteristics. Some breeding females and a number of breeding males, all presumably at least 9 or more years old, have been observed in plumages similar to those of the known 6-year-olds.

Thus although both males and females have predominantly black dorsal plumage when they fledge and leave the island, by the time they return at the ages of 4, 5, or 6 years, physiological processes are sufficiently developed to establish the secondary sexual characters. The end mechanism, loss or inactivation of melanophores, is probably similar in each sex, but presumably the speed at which this occurs is determined by a hormone more plentiful in males than in females, perhaps a male steroid hormone. It would be interesting to know whether the progressive changes in the females come to a stop at a given age, or whether the changes continue in both sexes but are so slow in

Fig. 12. Gibson code for describing *D. exulans* plumage phases. The circled numerals refer to arbitrary stages in the changing plumage, not to the birds' age.

TABLE 5. Gibson Codes for Plumage of *D. exulans*
Subadults of Known Age Observed on the Breeding Grounds

Age*	Sex	Back	Head	Wing	Tail	Total
5	♂	4.5	3.5	3.5	2	
5	♂	4.5	3.5	3.5	2	
5	♂	4	3	4	3	
Mean		4.3	3.3	3.7	2.3	13.6
5	♀	2.5	3	2	1.5	
5	♀	2.5	4.5	2.5	1	
6	♀	2.5	2.5	2.5	1	
6	♀	4	3	3	1	
6	♀	3	3	1	1	
6	♀	2	2.5	1.5	1	
Mean		2.7	3.1	2.1	1.1	9.0

* Years after laying.

females that most die before their plumage reaches
the chionoptera stage.

THE PRE-EGG PERIOD IN *D. exulans*

The pre-egg period includes events in the life cycle
of the mature adult bird that prepare for and culmi-
nate in the laying of an egg. Changes in the reproductive physiology begin at sea some time preceding the
return of birds to the island breeding grounds (see
Table 37), but just when is not known.

I watched the wandering albatrosses arrive at Bird

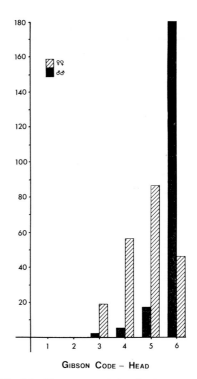

Fig. 14. Frequency of head plumage
patterns in breeding adult *D. exulans*.

Island in November 1963. None had appeared at Gony
Ridge on the 5th or 7th; one was present on the 9th,
and the numbers increased steadily thereafter (Figure 18). The breeding males return to the breeding
grounds first, and the females somewhat later. In the
inner study area, where the sexes of 169 marked individuals were known, the first male arrived November 12 and the first female not until the 24th. The
mean date of arrival of the males, November 29, is 11
days earlier than that of the females, December 10,
which return over a much longer period (SD ♂ ♂ =
2.27 days, SD ♀ ♀ = 9.44 days).

At the time of these observations, adults of the 1962–
1963 breeding season were still visiting the island to
feed their fledglings, and the albatross counts normally included these birds, which were not arriving
to start their breeding season but were at the very
end of it. At Gony Ridge, however, all 266 eggs laid
the previous season, 1962–1963, had been removed so
that no chicks hatched, and all the breeding birds left
early enough to return to breed the following season.
Thus the total breeding establishment of Gony Ridge
returned in 1963–1964, and the counts on which the
arrival curve (Figure 18) is based are not distorted
by adults in another demipopulation.

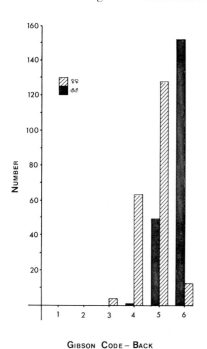

Fig. 13. Frequency of back plumage
patterns in breeding adult *D. exulans*.

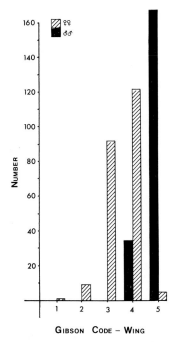

Fig. 15. Frequency of wing plumage patterns in breeding adult *D. exulans.*

gives a mean figure of approximately 34 days (range 29 to 45 days) for 44 pairs that laid eggs at Taiaroa Head between 1935 and 1945).

The male wandering albatross remains on or near its chosen nest during most of the pre-egg period (Figure 19, Table 7). The female appears only briefly on two or three occasions before she lays. In 2007 observations throughout the pre-egg period of 81 nests in which eggs were later laid, the males were present on 73.6% and females on only 7.4% of the occasions. In 1092 observations made after both individuals of the pairs had been seen at least once, female attendance was 13.1% compared with 76.2% of the days for males. *Richdale*'s [1950] observations of pairs arriving within an hour of each other led him to believe that royal albatross pairs associate at sea and return to land together, and "from that point either singly or together, when the internal urge drives them forward, proceed to the breeding area." I cannot draw similar conclusions from my observations of *exulans.* At 82 nests both members of a pair were seen together

Nest site tenacity. To determine changes in nest sites of marked pairs, all nests in the inner study area of Wanderer Valley were mapped to a scale of 1:305. The annual density in this area was given earlier in Table 1. Old nests litter the breeding grounds and, because of the excreta in their vicinity, quickly become overgrown with moss and tussock grass. Of a total of 65 nests observed, 20% of wandering albatross pairs used the same nest they had occupied previously; the nests of the remainder were up to 23 meters apart in consecutive breedings (Table 6).

When the female of a previously mated pair fails to return, the solitary male keeps territory in the same area until a new mate eventually joins him there. A solitary female that has lost her mate first returns to the former nest site of that mate and later looks for a new mate wherever males are alone on territory. Sometimes such remated females nest more than 150 meters from their previous nests.

Length of pre-egg period. At 82 nests where color-marked occupants were under daily observation, the time between the arrival of the first member of the mated pair (the male in all but two instances) and the laying of an egg varied from 8 to 45 days, with a mean of 26.6 days. (For *epomophora,* which nests about a month earlier than *exulans, Richdale* [1952]

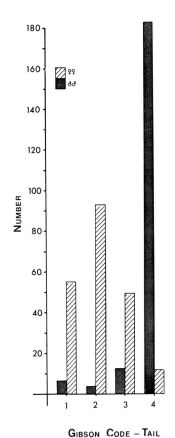

Fig. 16. Frequency of tail plumage patterns in breeding adult *D. exulans.*

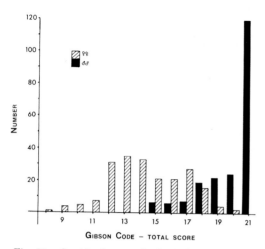

Fig. 17. Combined scores for back, head, wing, and tail patterns in breeding adult *D. exulans*. A yearling fledgling just out of the nest is scored as 4, a fully developed 'chionoptera' male as 21.

TABLE 6. Distances between Nests of Intact Pairs of *D. exulans* in Two Consecutive Breedings

Interval between Breeding	N	Distance between Nests		t
		$\bar{x}$, m	Range, m	
1 year	21	6.309	0–16.459	
2 years	40	7.376	0–23.470	0.833*
2+ years	4	6.949	0–15.850	—
Total	65	7.010	0–23.470	—

* Values of $t = 0.833$ occur by chance with a probability $p < 0.3$; therefore no significant difference is apparent in the distance if nests are moved after one or two years.

at the beginning of the pre-egg period only on one occasion.

Territory is confined to the area occupied by the nest itself (about 1 square meter) or the immediate vicinity of the bird when it is away from the nest. Adult males show little aggressiveness toward other wanderers near their selected nest site during the pre-egg period. They commonly indulge in bouts of nest building but never so vigorously as the females do in

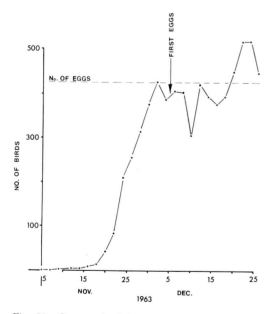

Fig. 18. Counts of adult *D. exulans* on the Gony Ridge nesting ground, Bird Island, during the pre-egg and laying periods of the 1963–1964 season.

the day or two preceding laying. The male's presence at the nest throughout the pre-egg period facilitates fertilization of the selected mate, and at the same time allows the female greatest freedom. If the male is certain to be present on the few occasions she comes ashore, copulation is possible without the female having to spend long ashore. This diversity of roles is seen in other species within the Procellariiformes [*Marshall and Serventy*, 1956; *Tickell*, 1962*b*; Pinder, 1966]. The urge to copulate is evidently very strong in the males during the pre-egg period, for they will often run from the nest and copulate with passing females other than their own mates.

Because most pairs shift their nest sites somewhat from year to year, it is essential that the male be conspicuous at the site when the female arrives. It is possible for a female to be unaware of the site the male has selected. One mated pair that had bred at least twice previously and hatched one chick, lost their egg early in the 1963–1964 season because the female (YBM) laid in a nest 36 meters from the one the male (PBG) had selected. This male was seen at the breeding grounds on only 5 days before the female laid; thus she may possibly have lost contact with her own mate before laying. YBM incubated for only 3 days before deserting, and 8 days later she was keeping company and copulating with PBG at the nest he had selected.

Copulation. Copulation was first seen on November 24, the first day females were recorded back at the breeding grounds. The last attempted copulation between a pair that subsequently produced an egg was seen December 27 (three instances at the end of January that did not lead to eggs are excluded). Between these dates copulation was observed frequently, and

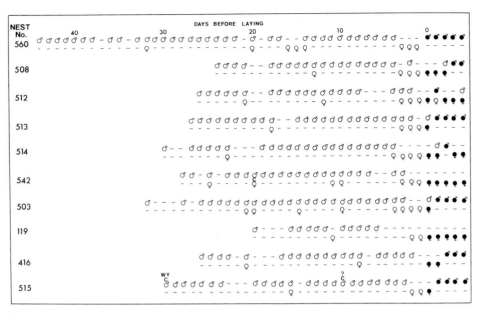

Fig. 19. Daily records of nest attendance by selected pairs during the pre-egg period: ♂ and ♀ = present at nest; the same symbols blackened = present at nest and incubating. C = copulating noted. — = bird absent.

47 instances were recorded in which both participants were identified (Figure 20). The mean date of copulation between 19 mated pairs was December 14, 10 days before the mean date of laying. In all these observations the roles played by the participants were subsequently proved by nest attendance to be correctly related to the sexes of the individuals. In no instance was the sex of a partner misidentified to the extent of apparent homosexual or inverted relationships.

Copulation is not restricted to mated pairs or pairs keeping company. Of the 47 instances noted above only 19 (40%) were between mated or prospectively mated pairs. The remainder were all between unmated birds, often from different mated pairs as well as birds without mates. In most cases the females were unwilling partners that fled from the approaching male, but were finally overpowered and compelled to submit. The presence of a lone female adult or subadult on the nesting grounds during the pre-egg period fre-

quently attracted the attention of many territory-holding males, and several might leave their nests to attempt copulation. A typical instance is quoted from my fieldbook for December 11, 1964:

NRB♀ [widow, bred 1961–1962] chased by RSM♂ [mate of SBY♀]. NRB makes a run to take off, but fails to get sufficient lift and crashes in tussock at end of meadow near nest 430. WSM♂ [mate of WOP♀] gets off nest 428 and goes after NRB, who runs away with RSM also in pursuit. As NRB passes nest 155, WS♂ [mate of YPN♀] leaps off and jumps on her. Successful copulation follows after which NRB goes to another nest nearby, but is again pursued by RSM and runs off.

Copulation occurs less frequently later in the season between pairs that have lost their eggs or between birds that have not yet bred. Complete coition was observed January 17 between two birds that were establishing a new pair bond (both had bred previously with different mates), and late on January 21 an unmated male tried to mount a widow but was rejected. The latest attempted (and incomplete) copulation was seen February 27 between a pair that returned and bred the following year. None of the 237 5- and 6-year-old birds that returned to the island was ever seen copulating.

THE EGG PERIOD

Clutch size. Characteristic of all Procellariiformes, the normal clutch of both the great albatrosses is a

TABLE 7. Attendance of Sexes at the Nest during the Pre-Egg Period by 81 Breeding Pairs of *D. exulans*

	No. of Observations	♂ ♂ Present	♀ ♀ Present
Entire pre-egg period	2007	1478, 73.6%	148, 7.4%
After both members of pairs seen once	1092	832, 76.2%	143, 13.1%

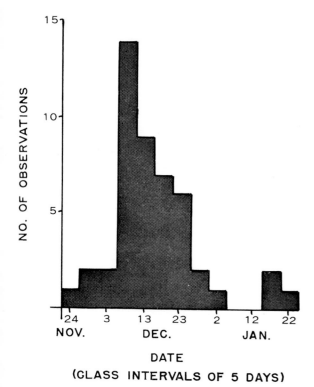

Fig. 20. Dates of copulations by *D. exulans* observed at Bird Island.

[*Verrill*, 1895] made the correct deductions in his observations of *D. exulans dabbenena* at Gough Island, although he erred in reporting multiple layings by the giant petrel (*Macronectes giganteus*).

Although *Richdale* [1939] reports two royal albatross nests with two eggs, the 64 eggs he records [1952] at Taiaroa Head were all laid singly. *Buller* [1905] reported a wanderer nest with two eggs at the Auckland Islands, but, in more than 5000 *exulans* examined shortly after laying at Bird Island, only once were two normal eggs found together. From what we know of similar occurrences in other albatrosses [*Richdale*, 1952; *Rice and Kenyon*, 1962; *Tickell and Pinder*, 1966] the reported two-egg clutches in the great albatrosses were probably laid by two different females in the same nest.

The one instance in which a wanderer seems to have laid a two-egg clutch at Bird Island was an abnormality. Female BON at nest 147 in the inner study area was found January 3, 1964, with an egg much smaller than normal and a peculiarly bubbly, soft, and crumbly shell; in the nest with it was a broken half shell of about the same size and texture and much albumen but no trace of yolk. The next day the runt egg had broken. As this nest was under daily observation and there was no indication of another female present, probably BON laid the two abnormal eggs a day apart. She and the same mate hatched a chick successfully during the 1961–1962 season, so that the

single egg, and, if that egg is removed, no other is laid the same season. *Wilkins*'s [1923] report of repeated laying by *exulans* at South Georgia is erroneous and can only be attributed to whaler's hearsay. Comer

Fig. 21. Wandering albatrosses displaying on the Bird Island breeding grounds. The birds may be adults, subadults, or both. January 1959. Photo by P. A. Cordall.

TABLE 8. Dimensions of Eggs of Great Albatrosses

In both the length and width, the difference between eggs of *D. exulans* from South Georgia and *D. epomophora* are significant at the 99.9% level when tested against *t* distribution.

		N	$\bar{x}$ and SE	SD	Range
D. exulans	Length	54	131.9 ± 0.8	5.572	114.0–142
(South Georgia)	Width	54	81.4 ± 0.2	1.846	78.5–86
D. exulans	Length	87	127.0	...	117.3–130.8
(Gough Island)	Width	87	77.0	...	71.6–82.3
[*Verrill*, 1895]					
D. epomophora	Length	54	123.6 ± 0.5	3.57	117.0–132
(Taiaroa Head)	Width	53	77.8 ± 0.3	2.44	73.0–84
[*Richdale*, 1952]					

phenomenon was not the result of physiological immaturity.

Egg dimensions and weights. The eggs of the two great albatrosses are the largest among the Procellariiformes. In color both are matte white with a more or less diffuse ring of peppery red-brown dots around the broad end, and the shape by *Palmer*'s [1956] scale is from long subelliptical to oval. Those of the Gough Island population are appreciably smaller than those of the South Georgia *exulans* (Table 8) and indistinguishable by size from those of *epomophora*.

In comparing egg weights, loss of weight during incubation must be taken into consideration, and statistical treatment of weights should include reference to the age of the eggs. These weight losses, as shown for individual eggs in Figure 22, are a linear function of time. The loss in *exulans* just before chipping amounts to 75 grams, or 15% of the new-laid egg weight, and approximates a linear regression. *Richdale* [1952] computes weight data from 41 *epomophora* eggs but without indicating age at time of weighing. The median values are 448.5 grams for *exulans* and 426.5 grams for *epomophora;* the difference of 22 grams between the medians is consistent with the differences in egg dimensions shown between the two species.

Laying date. Of 260 layings recorded at Wanderer Valley in 1958–1959, 1962–1963, and 1963–1964, 259 (99.6%) occurred between December 10 and January 8, with a mean of December 24 (Table 9). One pair laid January 17, and elsewhere on the island a few eggs were laid as early as December 5. The distribution of 640 known layings shows a normal distribution in which the mean date is also a satisfactory measure of the peak date (mode) of laying.

Comparison of the Wanderer Valley layings for the three seasons (Table 9) shows no significant difference between the means for 1958–1959 and 1962–1963. The

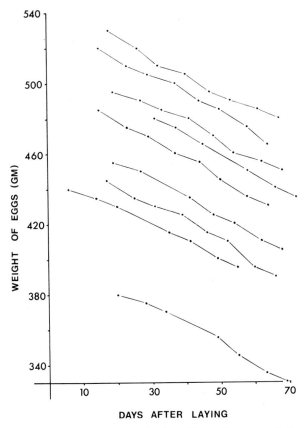

Fig. 22. Weight loss in individual *D. exulans* eggs during incubation.

TABLE 9. *D. exulans* Dates of Laying at Bird Island

Season	N	$\bar{x}$ and SE (Dec.)	SD, days	Range
1958–1959	79	22.6 ± 0.6	5.46	Dec. 13–Jan. 7
1962–1963	92	23.4 ± 0.5	5.24	Dec. 10–Jan. 3
1963–1964	89	25.4 ± 0.6	6.10	Dec. 12–Jan. 17
Combined	260	23.8 ± 0.3	5.73	Dec. 10–Jan. 17

TABLE 10. Differences Between Mean Laying Dates of
D. exulans at Bird Island

Season	Difference between Means, days	t	p
1958–1959 and 1962–1963	0.8	0.976	> 0.30
1962–1963 and 1963–1964	2.0	2.492	< 0.05
1958–1959 and 1963–1964	2.8	3.285	< 0.01

birds laying in these two odd-year seasons are mostly the same individuals. Those laying in 1963–1964 are predominantly different birds of the other demipopulation. The mean difference in laying dates between the demipopulations is manifestly greater than between individuals of the same demipopulation in different years. Statistically (Table 10) the difference between the mean laying dates for 1958–1959 and 1963–1964 is significant at the 99% level and thus sufficient to reject a null hypothesis, but the difference between 1962–1963 and 1963–1964 is significant only at the 95% level. As only about 55% of the breeding population lay eggs each year, it is not surprising that differences of this order can be demonstrated statistically, for at any one time 80% of the population is divided into two breeding isolates. The 20% that breed in consecutive years permit a continuous transfer of pairs from one demipopulation to the other, and proportionately larger numbers change when heavy mortality of eggs occurs.

Poor climate has no adverse effect; even after the most severe winter the nesting grounds thaw sufficiently by the time the wanderers start building their nests in mid-summer. The survival rate of exulans eggs is much higher than in those of melanophris and chrysostoma that are laid in October before winter conditions have ameliorated (Tickell, in preparation).

The Gough Island wanderers lay later than those at South Georgia. Comer [Verrill, 1895] reported first eggs there on December 26, and Swales [1965] on January 9 (Table 11). No data are available for statistical treatment, but evidently a difference of about 25–30 days exists between the two populations. The royals at Taiaroa Head lay from October 31 to November 23, with the laying peak November 12, but those at Campbell Island lay approximately three weeks later, from late November to early December (Table 13). On Campbell Island there are also a few wanderers, and from the few data available [Westerskov, 1963] these seem also to lay later, probably as late as the Gough Island birds. Thus, whereas the northern populations of exulans lay later than the southern populations, the reverse is true in epomophora.

Although we have one record of a female wanderer laying on the same day in different years, other individuals may show up to 21 days difference in laying dates in different seasons. The dates of 107 eggs laid by 50 female wanderers over 3 years indicate the range of individual layings (Table 12). Richdale's [1952] dates for 54 eggs laid by 10 royal albatrosses in 10 years emphasize how close together laying dates can be over the years. Although the data are not strictly comparable, the mean difference of 4.2 days between any two exulans layings does not differ significantly from that of epomophora.

Of the 50 female wanderers in this analysis, 36 had the same mate at each laying and 14 had different ones. No increase in laying date differences is evident with change of mates, and nothing indicates accommodation by the female to alteration in laying date. Richdale [1952] comes to the same conclusion for epomophora.

Both sets of data show that individual wanderer and

TABLE 11. Laying Dates of D. exulans at Different Breeding Stations

Breeding Station	First Eggs	Peak of Laying	Last Eggs	Authority
Gough Island	Dec. 24	...	...	Verrill [1895]
Gough Island	Jan. 9	...	...	Swales [1965]
Marion Island	Jan. 6	...	Jan. 31	Rand [1954]
Marion Island	...	Jan.	...	Crawford [1952]
Iles Crozet	Dec.	...	Jan.	Tuft [1960]
Îles de Kerguélen	Dec.	...	Jan.	Paulian [1953]
Campbell Island	Early Jan.	...	...	Sorensen [1950]
				Westerskov [1960]
Auckland Islands	...	Jan. 22	...	Fleming
Antipodes Islands	?	?	?	Turbott [1951]
Macquarie Island	Late Dec.	...	Early Jan.	Carrick [1955]
South Georgia	Dec. 5	Dec. 24	Jan. 17	This paper

TABLE 12. Difference between Laying Dates of the Same *D. exulans* and *D. epomophora* Individuals in Different Seasons
Pooled dates from 50 *D. exulans* and 10 *D. epomophora* females with the same and different mates at each laying.

Species.	N	$\bar{x}$ and SE, days	SD, days	Range, days	t
D. exulans	60	4.2 ± 1.2	8.998	0–20	
					0.411†
*D. epomophora**	167	3.8 ± 0.4	5.284	0–13	

* Computed from *Richdale* [1952].

† The probability of values of *t* of this order occurring by chance *p* > 0.06.

royal females lay over a shorter given span than the full extent of the laying period of the species. The period over which any individual lays is probably determined genetically and behaviorally. Though individual females may be subject to weather conditions such as storms at sea that might delay egg laying, no evidence as yet shows whether climatic conditions in any one season influence the laying date throughout the population.

Comparison of mean laying dates in two parts of Bird Island in 1962–1963 shows that time of laying in an area of high nest density (106 nests per hectare) does not differ from another of lower nest density (40 nests per hectare). Thus we may conclude that social facilitation does not affect the date of laying.

Data from 20 eggs laid in the 1958–1959 season suggest that it is more usual for eggs to be laid during the day than at night. When these eggs were laid the nests were inspected in the morning at approximately 0800 to 0900 and in the evening between 1700 and 2000. Fourteen eggs were laid between morning and evening inspections and only six between evening and the next morning, and some of the latter may have been laid between sunrise or sunset and the time of checking.

Incubation. At Bird Island the mean incubation period for 163 *exulans* eggs was 78 days with a range of 75–82 days (±24 hours) (Table 14). Two eggs that took 85 days to hatch were excluded from the calculation of this mean; in one the egg twice rolled

out of a badly built nest and remained unincubated from 1 to 48 hours until replaced by the observer. *G.V.T. Matthews* [1954] measures the time that Manx shearwater (*Puffinus puffinus*) eggs remain viable when not incubated, and *Tickell* [1962b] points out that incubation in the dove prion is prolonged in proportion to the number of days the egg is left unattended. Both these species are burrow nesters, and their neglected eggs are not destroyed. Eggs left unattended in the albatross's open nests are usually taken by skuas or sheathbills (*Chionis alba*). The wanderer egg mentioned survived only because it remained on the side of the nest within 30 centimeters of the sitting bird.

Richdale's [1952] records of 35 incubation periods for *epomophora* give a mean of 79 days with a range of 77 to 81 days (±1–24 hours). *Sorensen*'s [1950] five records for the Campbell Island royals are the same (Table 14). Although the differences between mean incubation periods of *exulans* and *epomophora* are less than one day, much less than *Richdale* [1952] predicted, they are statistically significant above the 99.9% level. The rate of an embryo's development is a precisely timed phenomenon, and *Richdale* [1952] in discussing the accuracy of his measurements concludes from selected samples in which the observed period was accurate to one hour that the true range of the royal's incubation period was from 77.5 to 80.25 days, a variation of not more than 2.75 days. A similar contraction of the *exulans* range could also result from more accurate measurements.

Incubation periods were obtained at Bird Island for eggs laid by nine females in two different seasons (Table 15). The variation of 0 to 3 days tested by variance ratio and *t* test shows that eggs a female lays in different years do not require the same length of incubation to hatch.

Share of the sexes in incubation. Both male and female albatrosses share incubation and, unless something unusual occurs, the incubating bird does not leave the nest and egg until relieved by its mate. After a wandering albatross egg is laid, change of incubating

TABLE 13. Laying Dates of *D. epomophora* at Different Breeding Grounds

Breeding Station	First Eggs	Peak of Laying	Last Eggs	Authority
New Zealand (Taiaroa Head)	Oct. 31	Nov. 12	Nov. 23	*Richdale* [1952]
Chatham Islands	About Nov. 8	About Nov. 15	?	*Dawson* [1955]
Campbell Island	Late Nov.	. . .	Early Dec.	*Westerskov* [1960]
Campbell Island	. . .	Early Dec.	. . .	*Sorensen* [1950]

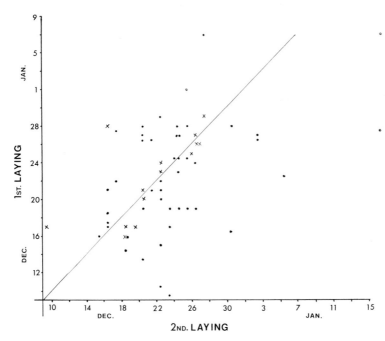

Fig. 23. Laying dates of individual *D. exulans* females in successive seasons
with the same (●) and different (×) mates. Distance from the diagonal shows
the degree of difference between the two laying dates for the same individual
female.

parent (nest relief) may occur from 5 to 14 times
before the chick hatches, and the average number of
changes is 9.7 (N = 166, SD = 1.8, range 5–14).
Richdale [1952] shows that 12 royal albatrosses
hatched their eggs in from 9 to 16 shifts.

Male wandering albatrosses more frequently take a
larger share of the incubation than females, which
accounts for the significant differences between the
mean lengths of incubation by males (41.8 days) and
females (38.5 days) shown in Table 16. In the royal

albatross [*Richdale*, 1952] the males' mean share in
12 completed incubation periods was 41.3 days compared with 40.5 days for the females.

The incubating regimes of individuals of the two
species in different years demonstrate the range of
variation in male and female dominance in incubation.
Table 17 gives two separate years' incubation records
for each of 11 pairs of *exulans* consisting of the same
birds each season. *Richdale*'s data [1952] for three
seasons' incubation records for two *epomophora* and
two years' records for another are similar.

In both the wandering and royal albatrosses it seems
that the males habitually assume the greater proportion in some pairs but the female is dominant in
others. In still other pairs either bird may take the
larger share of the incubating. Possibly these are less
experienced birds, which cannot be determined until
enough individuals of known age are available for
study.

The incubation shifts. Incubation is divided into alternating shifts by the male and female of each breeding pair. As it always begins with the female laying
the egg, the odd-numbered shifts are always by the
female and the even-numbered by the male. The
lengths of the shifts by both sexes vary throughout
the 75 to 82 days of incubation. From the variation

TABLE 14.   Comparison of Incubation Periods of
D. epomophora and *D. exulans*

	N	$\bar{x}$ and SE, days	SD	Range	t
D. epomophora, Campbell Island [*Sorensen*, 1950]	5	79.0 ± 0.3	0.707	78–80	
					0.643*
D. epomophora, Taiaroa Head [*Richdale*, 1952]	35	79.3 ± 0.2	0.990	77–81	
					4.708†
D. exulans	163	78.4 ± 0.1	1.166	75–82	

* Values of *t* of this order occur by chance with a probability
of *p* > 0.5.
† Values of *t* of this order occur by chance with a probability
of *p* < 0.001.

TABLE 15. Difference between Two Incubation Periods of Eggs Laid by the Same Pairs of *D. exulans*

Band no.		53298	58235	58972	58990	59016	59029	59030	10020	71179
Length of	1st	78	76	78	79	79	75	78	80	79
incubation, days	2nd	78	79	78	76	80	77	79	78	77
Difference		0	3	0	3	1	2	1	2	2

N	$\bar{x}$	s	t	p
9	1.56	1.064	4.135	> 0.01

in numbers of shifts needed to complete incubation as shown above, it follows that some pairs hatch their chicks in a few long shifts and others in more numerous shorter ones. In the three seasons that incubation records were kept at Bird Island, the lengths of 2009 separate incubation shifts were measured. Except for the first female shift that begins when she lays the egg, an incubation shift is defined as one that begins and ends with successful relief of a sitting bird. Shifts ended by egg breakage and desertion or lack of observation are excluded from the analysis.

In his analysis of *epomophora* incubation (Figure 24) Richdale [1952] points out that "the first span by the female is significantly different statistically from all subsequent groups of spans" and that "subsequent to the first span by the female there is no significant difference between the spans of the two sexes." The data for *exulans* (Figure 25) also show a notably short first female shift that likewise is significantly different statistically from all subsequent shifts, but the sample sizes of these data are large enough so that with three exceptions the mean lengths of all incubation shifts are significantly different from each other. Richdale remarks the inadequacy of his samples, and the *exulans* data show how great the variation in one species can be. The longest recorded shift in *epomophora* was 17 days, whereas one male *exulans* at Bird Island sat continuously for 38 days. Some shifts of either sex may be as short as two days.

Richdale concludes that in *epomophora* incubation

TABLE 16. Share of Sexes in Incubation of *D. exulans*
Total days incubated by males and females from laying to hatching.

Sex	N	$\bar{x}$ and SE, days	SD, days	Range and Per Cent Total Incubation	SE of Difference
Males	163	41.8 ± 0.4	5.38	22, 28%–57, 73%	0.586*
Females	164	38.5 ± 0.4	5.20	22, 28%–53, 68%	

* The difference between the means is more than 5 times the SE of Difference and therefore significant.

intensity warms up to a peak in the sixth or seventh shift, followed by a falling off to shorter shifts as hatching time approaches. Figure 25 shows that this is not so in *exulans*, where the means of the second through the sixth shifts are each about 10 days. However, the frequency curve for the lengths of the first through the fourth shifts are all positively skewed (Figures 26, 27), and the means are probably not a true indication of central tendency. The modes of the first four shifts (Figure 25) are a better approximation of Richdale's hypothesis with the peak around the sixth shift.

In the latter half of the *exulans* incubation the mean lengths of the shifts become progressively shorter, and it will be shown later that this shortening is carried

TABLE 17. Share of Incubation by the Same Pairs of *D. exulans* in Different Seasons

Pair	Total	Incubation, days				Excess	
		♂		♀		♂	♀
1	78	46	59%	32	41%	18%	
	77	35	45%	42	55%		10%
2	79	46	58%	33	42%	16%	
	76	44	58%	32	42%	16%	
3	75	42	56%	33	44%	12%	
	77	41	53%	36	47%	6%	
4	79	40	51%	39	49%	2%	
	80	40	50%	40	50%	Equal	
5	78	34	44%	44	56%		12%
	79	40	51%	39	49%	2%	
6	80	36	45%	44	55%		10%
	78	41	53%	37	47%	6%	
7	80	51	64%	29	36%	28%	
	79	46	58%	33	42%	16%	
8	78	41	53%	37	47%	6%	
	78	46	46%	42	54%		8%
9	79	30	38%	49	62%		24%
	77	36	47%	41	53%		6%
10	77	35	45%	42	55%		10%
	78	33	42%	45	58%		16%
11	80	42	53%	38	47%	6%	
	79	38	48%	41	52%		4%

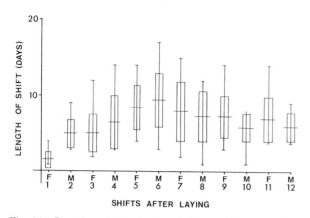

Fig. 24. Lengths of incubation shifts in *D. epomophora* throughout incubation. (Data from *Richdale* [1952], p. 27.)

into the brooding period. Richdale's figures for the eleventh and twelfth shifts do not show this so well for *epomophora*, but the small size of his samples may be responsible for misleading means.

It remains to examine the incubation shifts for any indication of individual birds habitually sitting for longer or shorter periods in any one shift. As the first shift is the shortest and laying gives a convenient starting point, the first shift is a useful one for comparison. The length of the first shift is known for 36 individual

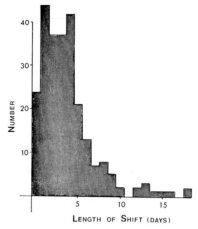

Fig. 26. Lengths of 250 first (female) incubation shifts in *D. exulans*.

wanderers for two seasons (Table 18). The differences between them in each female varied betwen 0 and 9 with a mean of 2.2 days. Nothing indicates that the lengths of the first shifts show any similarity for individuals in different years.

A second test was carried out to determine whether the observed differences between first incubation shifts of individual females in separate seasons differed from those that would appear if the lengths were determined strictly randomly. A random model was set up, and the numbers were drawn from a pool containing a total number of first shift lengths proportionate to the observed skewed distribution (Figure 26). Thirty-six pairs of numbers were drawn equivalent to the sample of observed shifts and differences. Three such samples were drawn and the results averaged and compared statistically by variance ratio and Student's *t* tests. This treatment (Table 18) showed no significant differences between the observed lengths of first female incubation shifts and those picked randomly. From this it may be concluded that the variables de-

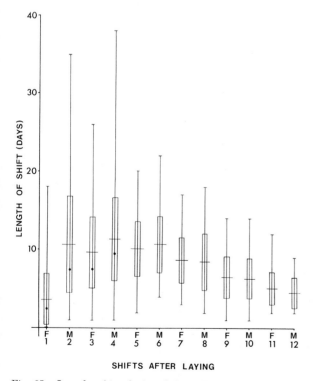

Fig. 25. Lengths of incubation shifts in *D. exulans* throughout incubation. The mode (●) is shown for the first four shifts.

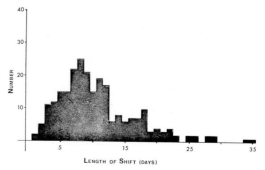

Fig. 27. Lengths of 249 second (male) incubation shifts in *D. exulans*.

TABLE 18. Mean difference between Lengths of
First Female Incubation Shifts in *D. exulans*

	N	$\bar{x}_1 - \bar{x}_2$	$\hat{\sigma}$	t	p
A*	36	2.167	2.036	6.386	< 0.001
B†	36	2.871	2.554	1.660	> 0.1

* 36 individual females incubating in two different seasons.
† 36 individual females picked randomly.

termining the length of time an individual female remains on the nest after laying are so numerous or so irregular or both that no consistency in the behavior can be identified from year to year.

Richdale [1952] points out that in all species of albatross for which information was available at the time, the female stayed ashore for only a brief period after laying. Although his 32 records for royal first shifts at Taiaroa Head were less than four days in length, two of *Sorensen*'s [1950] seven records for royals at Campbell Island were for 10 and 11 days. Of 250 first shifts of Bird Island wanderers, 12 were of 10 days or more, and 2 females sat 18 days after laying before the male took over.

Females of both species spend very little time ashore during the pre-egg period, while the males occupy nest territories during most of it. The question this raises is whether the very short first shift is determined more by the female's desire to return to sea or the male's persistent efforts to take over incubation, for in both species after the egg is laid the male repeatedly tries to urge the female off the nest, sometimes with considerable force. Moreover, in *exulans* (Figure 25) during the first half of incubation the means and upper ranges of male shifts are always higher than those of females, and the longest a female was known to incubate was 26 days compared to 38 days by a male.

Slight male predominance in incubation seems consistent with a dominant territorial role without need to postulate extra incubational urge. Albatrosses and petrels ashore during the breeding season must always be subject to a balance of two urges: (*a*) those such as incubation that keep the bird ashore, and (*b*) others such as hunger that stimulate it to go to sea. No adult albatross starves to death on the nest when its mate does not return; presumably, then, a point exists where the need to feed overrides all other urges, and the bird goes to sea.

Conceivably this threshold and other critical times are affected by the appearance of new stimuli associated with the breeding cycle. In this particular instance it is suggested that the first sight of the egg and later tactile stimuli on the brood patch might increase the threshold at which the urge to feed becomes dominant over the territorial urge. If the male is not present when the female lays, she is capable of remaining at the nest long enough for the mate to return (up to 18 days in *exulans* records). This suggests that the short first female shift is in large part due to the male's capacity to assume charge of the nest, rather than to any particular desire of the female to return to sea.

THE HATCHING PERIOD

Richdale's [1952] seven measurements of the hatching period in *epomophora* show that chicks take from 32.5 to 106 hours to hatch from the time the earliest chipping starts. The final stages of hatching seem to occur only during daylight hours. Observations made twice daily on 166 *exulans* eggs gave a range of 1 to 5 days (mean, 3.2 days) between first chipping and chick free of the shell. Seven of the recorded hatchings in less than 24 hours almost certainly resulted from the observer's failure to notice the early stages of chipping, but, with no means of testing the accuracy of the observations, I include all in the calculation. Although the data from the two species are not strictly comparable, wandering and royal chicks hatch out in approximately the same length of time.

Of 165 observed *exulans* hatchings, the chick appeared under the female parent in 82 instances and under the male in 83. *Richdale* [1952] presents similar data for *epomophora*. Thus in both species the sex of the incubating parent at hatching seems completely a matter of chance.

The mean peak of hatching of 336 *exulans* eggs in four seasons at Bird Island was March 11 with a range of 32 days from February 27 to March 30 (Table 19). This is approximately equivalent to the range of laying dates in the same area. The difference between the mean dates of laying and hatching, 77 days, corre-

TABLE 19. Dates of Hatching, Based on Daily Observations
of *D. exulans* Eggs at Bird Island

Breeding Season	N	Mean Date of Hatching (March) and SE	SD	Range
1960–1961	81	10.24±0.47	4.22	Mar. 2–Mar. 22
1961–1962	102	9.40±0.47	4.76	Feb. 28–Mar. 20
1962–1963	77	12.54±0.62	5.48	Feb. 27–Mar. 29
1963–1964	76	12.71±0.72	6.28	Feb. 27–Mar. 28
Total	336	11.07±0.29	5.36	Feb. 27–Mar. 29

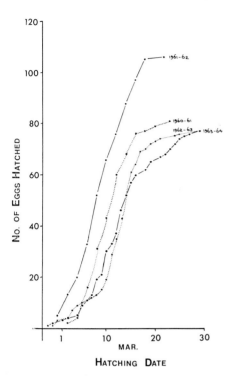

Fig. 28. *D. exulans* hatchings observed in the extended study area at Bird Island on daily visits during four seasons.

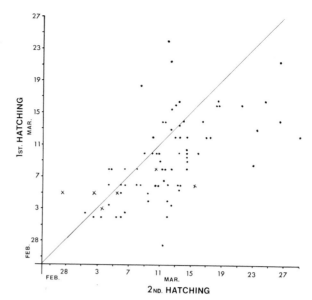

Fig. 29. Hatching dates of *D. exulans* chicks of females with the same (●) and different (×) mates in successive seasons.

sponds closely to the length of the incubation period calculated from individual records.

The hatching records for *exulans* in different years at Bird Island (Figure 28) show greater variation than is apparent in the laying records. The mean hatching date in 1962 (Table 19) was March 9.4, half a day earlier than in 1961 and more than two days earlier than in 1963 or 1964. Individual pairs tended to hatch chicks later in the second year (Figure 29), and, as no significant difference exists between the lengths of incubation periods in individual pairs from year to year, late hatching probably reflects late laying, but the corresponding laying dates for 1960–1961 and 1961–1962 pairs are not known.

Table 20 shows the differences in hatching dates of 50 pairs (each composed of the same mates) that hatched chicks after rearing their previous one successfully two seasons before. Of these pairs 66% hatched their chicks later on the second occasion. Of 13 wanderer pairs that lost chicks and returned to breed again the following year, 2 pairs hatched on the same day in each season, and the remaining 11 pairs hatched up to 14.5 days later (Tables 21, 22). This corroborates the evidence of laying dates discussed earlier and indi-

cates a tendency for individual *exulans* pairs to breed slightly later in subsequent seasons.

I proposed earlier that the individual laying period is shorter than that for the species. A female cannot continue to lay later each season indefinitely, but we have no evidence suggesting what determines whether a female will lay later or earlier than in previous seasons.

Brooding. The brooding period is defined here as the time between the chick's hatching and the day it is first left entirely alone by both parents. This definition differs slightly from *Richdale's* [1952] "guard stage." Toward the end of the brooding period parents on returning may stay at the nest for a day or so and sometimes cover the chick.

The brooding period lasts from 21 to 43 days, with

TABLE 20. Comparison of Dates of Hatching of 50 Pairs of *D. exulans* in Two Seasons*

Hatched in Second Season	N	Mean Difference between First and Second Hatchings, days	Range, days
Earlier	14	−3.9	−0.5 to −12.0
Same date	3	Nil	Nil
Later	33	+5.0	+0.5 to +16.0

* Every pair was composed of the same mates each season, and previous to the second hatching each pair had successfully reared a chick two seasons previously and had remained at sea during the intervening season.

TABLE 21. Comparison of Dates of Laying by 15 Pairs of *D. exulans* in Two Successive Seasons*

Laid in Second Season	N	Mean Difference between First and Second Layings, days	Range, days
Earlier	5	−2.9	−0.5 to −6.0
Same date	1	Nil	Nil
Later	9	+10.6	+1.0 to +14.5

* Every pair was composed of the same mates each season; after laying in the first season the egg was lost, and the pair returned to breed the following season.

TABLE 22. Comparison of the Dates of Hatching of 13 Pairs of *D. exulans* in Two Successive Seasons*

Hatched in Second Season	N	Mean Difference between First and Second Hatchings, days	Range, days
Earlier	0	0	0
Same date	2	Nil	Nil
Later	11	+6.2	+1.0 to +14.5

*Each pair was composed of the same mates in both seasons. In the first season the chick was lost during the first three months, and the pairs returned to breed again the following year.

a mean of 32 days (N = 74). This is divided by the parents into approximately 11 shifts (Table 23). The brooding period *Richdale* [1952] calculated for nine royal chicks at Taiaroa Head ranged from 28 to 43 days with a mean of 34.6 days. The difference between the mean brooding periods of the two species is thus only 2.5 days.

When the chick hatches, the brooding shifts become even shorter than the last incubation shifts. As in incubation, the *exulans* sexes share the brooding more or less equally. The mean totals for brooding and length of brooding shifts are both longer for males than for females. The differences are statistically significant, but trivial compared with the ranges of variability of individual behavior (Figure 30). Both, as mentioned earlier, probably reflect male dominance rather than a predominantly male role in brooding. *Richdale*'s [1952] brooding records for seven pairs of royals show a difference of 4.2 days between mean male and female participation, again emphasizing male dominance as in the wanderer. Also the mean length of the brooding shifts by both sexes in the royal of 1.9 days (N = 108, range 1–14 days) is significantly shorter than the wanderer's 2.6 days (N = 818, range 1–10 days).

Richdale's [1952] analysis shows the shifts toward the end of the brooding period to be significantly shorter than earlier ones. Unable to demonstrate any statistical difference between the male and female spans, he combines all shifts into three groups, the first four, the second four, and the remainder, each of which contain approximately the same proportion of male and female shifts (Table 24). Guardedly he notes a tendency to shorter shifts as brooding progresses. My similar treatment of the *exulans* data shows much less difference between the first and later shifts.

Table 25 reveals that 99.5% of wanderer shifts and 98.2% of the royal shifts last from 1 to 7 days. The very few shifts in excess of 7 days probably occur randomly from unusual causes. Excluding the shifts of more than 7 days from the analysis (Table 26) shows a difference of only 0.01 day between the first and second four spans in *exulans*, and of only 0.42 and 0.43 day, respectively, between the final and second and final and first four groups. Comparing these data with those for *epomophora* (Table 27) shows no significant difference between the lengths of brooding shifts in the first and second groups of four except that the mean of the final shifts in the royals is significantly shorter than that for the wanderers.

Analysis of the lengths of the individual brooding shifts in *exulans* (Table 23, Figure 31) numbered backward from the time the chick is first left alone, shows the shift in the middle of the period somewhat longer than those at the beginning or end. Each of the periods from 4 to 7 is significantly different from the adjacent one, with a maximum at period 6.

In conclusion, some increase in the frequency of

TABLE 23. Brooding Shifts of *D. exulans* Parents (Both Sexes) Calculated from the Time the Chick was Left Unattended for the First Time

	Brooding Shifts before Chick Left Alone									
	10	9	8	7	6	5	4	3	2	1
N	70	73	74	74	74	74	74	74	74	74
$\bar{x}$ and SE, days	2.7±0.1	2.8±0.1	2.7±0.1	2.5±0.1	3.1±0.2	2.8±0.1	2.5±0.1	2.5±0.1	2.5±0.1	2.4±0.2
Range	1–6	1–6	1–7	1–5	1–10	1–8	1–7	1–6	1–7	1–9

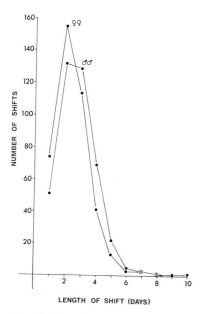

Fig. 30. Lengths of brooding shifts by
D. exulans males and females.

TABLE 25. Brooding Shifts in *D. exulans* and *D. epomophora*

Length of Shift, days	*D. exulans*			*D. epomophora**	
	No. and % of Observations		% Total Observations	No. and % of Observations		% Total Observations
1	125	15.3%		27	25.0%	
2	287	35.1		42	38.9	
3	243	29.7		22	20.4	98.2%
4	110	13.4	99.5%	12	11.1	
5	35	4.3		2	1.9	
6	8	1.0		1	0.9	
7	6	0.7				
8	2	0.3		1	0.9	
9	1	0.1	0.5%			
10	1	0.1				
11						1.8%
12						
13						
14				1	0.9	
	818			108		

* *Richdale* [1952], p. 72.

longer brooding shifts is apparent about the mid-point of the brooding period, and a decrease in their frequency in the last shift before its end. The differences are so slight, however, that any difference in mean lengths is difficult to demonstrate. Brooding by either parent for more than 8 days at a stretch is rare and is probably the result of unusual circumstances detaining the absent bird.

THE FLEDGLING PERIOD

Matthews [1929] estimated that the chicks at South Georgia remained ashore 9 to 10 months after hatch-

ing. Chicks were watched at Bird Island throughout the winter of 1963, and 35 measurements accurate to the nearest day show the mean length of the fledging period to be 278 days, with a range from 263 to 303 days (Table 28). Larger samples where the records of individual birds are unknown show the span between mean hatching and mean departure dates in 1963 to be a closely equivalent 272.7 ± 5.24 days. *Swales* [1965] indicates the fledging wanderers at Gough Island stay ashore approximately the same length of time, but, as mentioned previously, both laying and departure are correspondingly later there than at South Georgia.

From Matthews' and other data, *Richdale* [1952] predicted that the wanderer fledgling stays ashore an estimated 50 days longer than the royal fledgling, which comes very close to the observed difference of about 44 days. *Sorensen*'s [1950] figures for the Campbell Island royals, 216 to 252 days, correspond with the longest of Richdale's at Taiaroa Head, 242

TABLE 24. Comparisons of Pooled Brooding Shifts in
D. exulans and *D. epomophora*

Combined Brooding Shifts	N	x̄ and SE, days	SD	Range	SE of Difference
		D. exulans			
1–4	296	2.82±0.07	1.285	1–9	0.032
5–8	296	2.78±0.07	1.210	1–10	0.032† / 0.032†
9+	302	2.38±0.07	1.240	1–9	0.032†
		*D. epomophora**			
1–4	28	3.11±0.46	2.45	1–14†	
5–8	28	1.93±0.15	1.82	1–4†	
9+	53	1.33±0.13	0.95	1–5†	

* *Richdale*, 1952, p. 73.
† Difference between means more than 3 times SE of difference, therefore significant.

Table 26. Comparison of 818 Brooding Shifts of *D. exulans* from Which Four Observations (0.05%) of Shifts of Eight or More Days Have Been Excluded

Combined Brooding Shifts	N	x̄ and SE, days	SD	Range
1–4	293	2.77±0.07	1.163	1–7
5–8	295	2.76±0.07	1.160	1–7
9+	300	2.34±0.07	1.141	1–7

TABLE 27. Comparison of 107 Brooding Shifts of *D. epomophora** from Which Two Observations (1.8%) of Shifts of Eight Days or Longer Have Been Excluded

Combined Brooding Shifts	N	$\bar{x}$ and SE, days	SD	Range
1–4	26	2.50±0.48	2.45	1–6
5–8	28	1.93±0.15	1.82	1–4
9+	53	1.33±0.13	0.95	1–5

* Richdale [1952], pp. 72–73.

TABLE 28. Comparison of Fledgling Periods of *D. exulans* and *D. epomophora*

	N	$\bar{x}$ and SE, days	SD	Range
D. exulans	35	277.7±2.82	16.68	263–303
D. epomophora*	17	236.0±2.58	10.58	216–252

* Richdale [1952].

to 253 days. In view of the closeness of the means of their incubation periods (the wanderer's is 99% of the royal's), the fact that royal fledglings remain ashore only 84% of the time young wanderers do is striking. The difference in fledging times is difficult to account for as an accommodation of the species' breeding cycles to environmental conditions at different latitudes. Whereas the young royals have generally left the breeding grounds before the next season's breeding adults arrive, the young wanderers have not. Here we have the most clear-cut potential source of behavioral and morphological differences—adult wanderers are subject to encounters with begging chicks during the pre-egg period, but adult royals are not.

As the 34 wanderer fledging periods at Bird Island were all measured the same season, it is not possible to supplement *Richdale*'s [1952] claim that individual pairs tend to keep their chicks ashore about the same

length of time each season. However, as is shown later, some parents feed their chick substantially more than do others. Hence the time lapse before a chick departs may depend not on how long the parents "keep" it ashore but on how well it is fed, as the better-fed chicks develop more rapidly.

The first fledgling wandering albatrosses leave Bird Island the end of November, and the last definite recorded departures are mid-January. A few fledglings can still be found on the island up to the end of February (Figure 33), but most if not all of these die of starvation on the island. The mean date of departure of 63 fledglings from my study area was December 10, 1963 (range November 17–January 8). This is also the return peak of the season's breeding birds and the start of laying. Thus the breeding adults' arrival overlaps the departure of the previous seasons' young, the

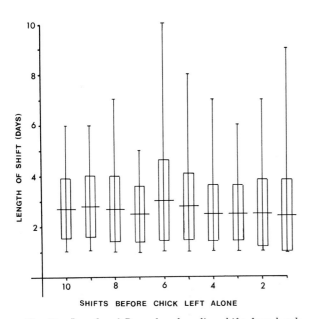

Fig. 31. Lengths of *D. exulans* brooding shifts from hatching until the chick is first left unattended.

Fig. 32. A fledgling wandering albatross about 260 days old soliciting food from an adult at Bird Island, December 1958. Photo by P. A. Cordall.

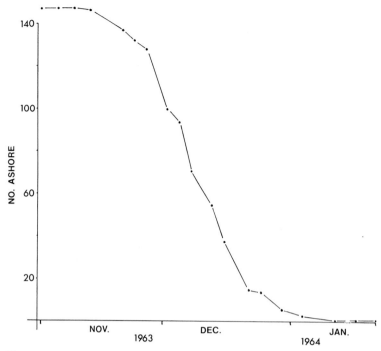

Fig. 33. Departure of yearling *D. exulans* fledglings from Bird Island,
1963–1964.

first breeding males appearing 8 days before the first
fledglings leave.

CHICK GROWTH

Weights of six wanderer chicks taken within six hours
of hatching were 303, 330, 340, 360, 368, and 410
grams. *Richdale* [1952] gives the weight of a newly
hatched royal as 305 grams. Weighings at 12-hour
intervals (Figure 34) show the wanderer chick is not
fed during the first 24 hours and is given only very
small amounts the next three days. Thereafter meals
get progressively larger, and within 15 days the chick
weighs 1000 grams. *Richdale*'s [1952] observations
on one royal chick show a very similar feeding frequency and growth rate. The wanderer chick is
brooded for an average of 32 days, and by the time it
is first left alone it weighs between 1.5 and 4.4 kilograms, with a mean of 2.9 kilograms for 70 chicks.

After brooding ceases, feeding visits are quite brief;
generally the adult remains ashore less than 30 minutes, though occasionally a bird may stay up to 24
hours, particularly if the visits of male and female
coincide. Comparison with Richdale's data shows no
apparent difference in the length of feeding visits between the two species.

To determine whether feeding visits occur more
often at one time of day than another, 20 chicks were
weighed three times daily, at 0600, 1200, and 1800
LT. Altogether 560 weighings were made between
April 22 and May 1, 1963, and 116 feeds were detected. A feed is defined here as a weight increase in
the chick after a period of 6 or 12 hours. It may have
been made by one or both parents, but as this procedure does not identify feeding parents other than those
present at the time of weighing, it is reasonable to
assume that double feeds occurred at random in all
periods and that they did not seriously affect the statistical validity of the conclusion. The chicks were
from 35 to 60 days old at the time, and they weighed
between 2 and 6 kilograms. As the weighings were
accurate to the nearest 15 grams and the recorded
overnight feeds after an unknown interval of 0–12
hours were between 137 and 990 grams, the chances
of a feed not being detected were negligible. The 0600
weighings began before sunrise, and those at 1800
finished at about sunset.

These weighings show that the chicks are fed almost
exclusively in the daytime. Of the 166 feeds detected,
only 10 were noted in the morning weighing, and in 8
of these the parents were seen to arrive after or during
the previous evening's weighing; 4 of these were not
present at dawn, and they probably fed the chick and
left the same evening. On the other two occasions a
parent was feeding its chick at the time of the dawn

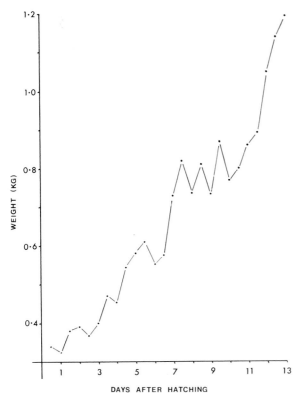

Fig. 34. Growth/weight curve for *D. exulans* chick 155 weighed every 12 hours during the period immediately after hatching.

Weights were read in pounds and ounces to the nearest half-ounce and later converted to kilograms. Each bird was carried a distance of from 5 to 50 meters into the shelter of an observation hut for weighing. As daily handling began shortly after hatching, the chicks were well conditioned to the procedure and never regurgitated as unconditioned chicks usually do.

The rate at which an albatross chick loses weight when not fed gives a characteristic curve (Figure 36) from which one can determine whether the differences between daily weighings are adequate to show that a bird has been fed during the previous 24 hours. In only 2% of the weighings was there any degree of uncertainty. This was due first to the fact that some chicks invariably were fed shortly after the daily weighings and digested and excreted most of the meal before the next afternoon. Second, their plumage sometimes held enough rain water or was encrusted with enough frozen snow to register slightly elevated readings, as may be seen in Figure 36.

This method provides no measure of the amount of individual feeds by each parent. A weight increment over 24 hours may have resulted from one or several feeds by either or both parents. During the weighings parents were often seen to arrive, feed the

weighing and could conceivably have arrived at first light before the observer. Thus, wandering albatrosses apparently do not return to the breeding grounds to feed their chicks during the night, though they may arrive at dawn and leave at sunset.

Of the daytime feeds, the weighings indicate that more chicks are fed in the morning (61%) than in the afternoon ($p < 0.01$). Richdale's intensive observations of the first royal chick reared at Taiaroa Head May 7–22, 1938, reveal the same tendency. He states that the parents arrived at the nest 8 times before noon and 5 times after, and 7 of the visits were before 10 a.m. The earliest was at 6:45 a.m. on May 20, only 15 minutes after daybreak; the latest was at 4:30 p.m., $1\frac{1}{2}$ hours before sunset.

To determine the growth rate of wandering albatross chicks a group of 23 selected chicks were weighed daily at Bird Island from March 25, 1963, until all had fledged and departed January 7, 1964. The chicks were weighed from about an hour before sunset in midwinter (June 22) to about four hours before sunset in midsummer (December 23). In all 4826 weighings were made.

Fig. 35. Adult wandering albatross brooding a young chick at Bird Island, January 1959. Photo by P. A. Cordall.

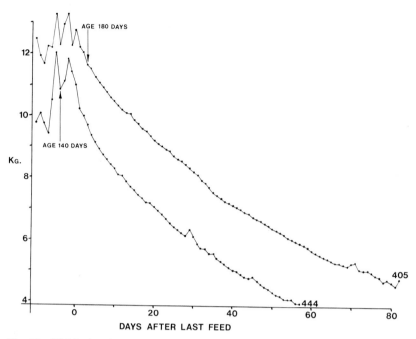

Fig. 36. Weight loss in two starving *D. exulans* chicks weighed daily from desertion to death.

chick, and leave. Whenever possible such chicks were weighed immediately before and after feeding; 53 measurements of individual feeds were thus obtained, the earliest on April 18, four days after chick 427 was first left alone, and the latest on November 27, 15 days before chick 431 flew away.

The smallest measured feed, 0.113 kilogram, was given November 10, and the largest, 2.112 kilograms, on October 7. Richdale's weights of 13 meals fed a royal chick from May 7 to May 22, 1939, are similar, from 0.227 to 1.928 kilograms.

Richdale [1952] shows that between May 5 and September 24 male royal parents visited the nest 59 (53%) times and females 52 (47%) times; both sexes were recorded during each month. My random records of marked adults seen feeding chicks in the study area between April 18 and December 31 show male wanderers fed 63 (53%) times and females 55 (47%) times. Thus both sexes share the chick-feeding duties in both species, the male probably averaging a slightly larger share.

The weight of any *exulans* chick changes greatly from day to day (Figure 37) depending on the amount of food it receives, which is equally true for the *epomophora* chicks [*Richdale*, 1952]. From one or two weight curves it is difficult to demonstrate any central tendency that may be characteristic for the species, and it is equally difficult to make such generalizations on

the basis of weighings made at longer intervals. Keith's curves of five wandering albatross chicks at Macquarie Island [*Carrick et al.*, 1960], drawn from weights taken at intervals of two or more weeks, show a marked drop in weight in October followed by a rise that is quite at variance with the Bird Island data. In view of the wide daily range of variation in the South Georgia weights and the few birds available to Keith, it is difficult to interpret his curves.

Complete or partial growth curves based on daily weighings of 24 wanderer chicks in 1963 allow the construction of a mean growth curve that effectively smooths out the individual fluctuations (Figure 38). The statistics on which the curve is based (Tables 44, 45) were calculated from the pooled daily weights for each 10-day interval. Because of the 40-day difference between the longest and shortest fledging periods, it was necessary to manipulate the data to bring equivalent parts of the individual curves together for analysis, as the ends of the growth curve are of particular interest. The figures of the first 200 days of the curve are reckoned from the date of hatching, those for the last 80 days from the day of departure. The total length of the combined curve, 275 days, approximates the mean length of the fledging period. The center portions contains some overlap, so that a few daily weights are used twice, but the resulting distortion is barely perceptible.

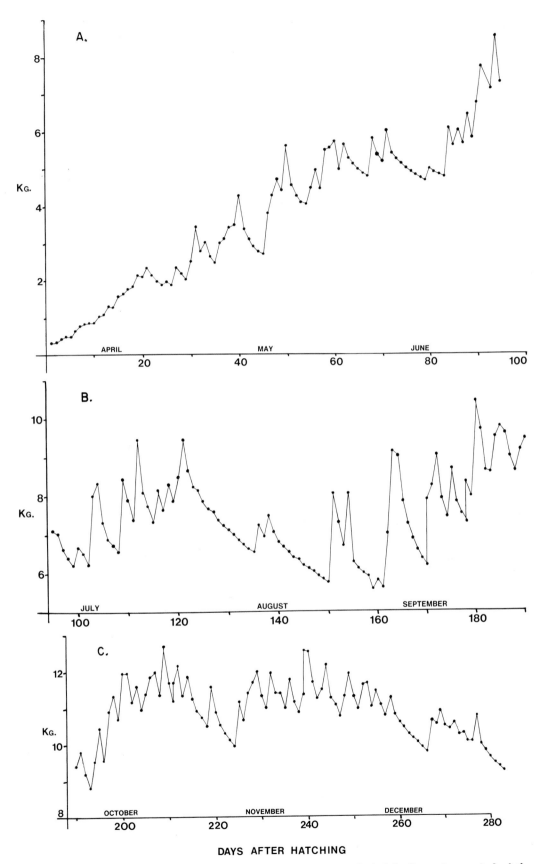

Fig. 37. A, B, and C. Growth/weight curve of chick 86 showing marked daily fluctuations typical of the entire chick period.

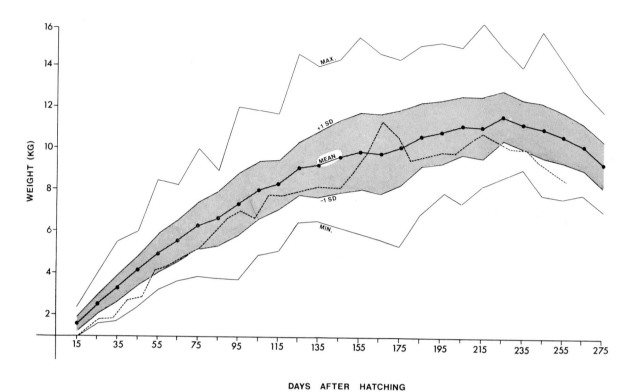

Fig. 38. Growth/weight curves of *D. exulans* (solid lines), means, standard deviations, and ranges computed from the pooled daily weights of 24 chicks over 10-day intervals, and of *D. epomophora* (dotted line) computed from weighings of 7 chicks at approximately weekly intervals [*Sorensen*, 1950].

Despite the pronounced fluctuations in the daily individual weights, the wanderer nestlings tend to increase in weight steadily through the winter. The mean peak weight, 11.7 kilograms, is reached 221–230 days after hatching (or 51–60 days before departure), and the mean weights decline steadily from then until departure. A more precise analysis of the weight loss during the fledglings' last 20 days on the island is shown in Figure 39 and Table 46. Only one of Keith's curves [*Carrick et al.*, 1960] gives any indication that the Macquarie Island wanderer chicks lose weight in the last days ashore.

Weight curves for seven royal chicks at Campbell Island [*Sorensen*, 1950] and one at Taiaroa Head [*Richdale*, 1952] show a general similarity. The means of Sorensen's seven royals plotted in Figure 38 remain with three exceptions consistently below those of the wanderers. Sorensen states his heaviest chick at Campbell Island weighed 32.5 pounds (14.8 kilograms), and Richdale's heaviest at Taiaroa was 27.5 pounds (12.5 kilograms). As these chicks were weighed only once weekly or less often, they possibly reached higher peak weights than those shown. The heaviest

wanderer chicks at Bird Island reached 16.13 kilograms.

The mean weight of the seven Campbell Island chicks the week before departure was 8.6 kilograms, and two at Taiaroa Head flew at weights of 7.3 and 10.2 kilograms, respectively. The wanderer fledglings average 8.93 kilograms the day before departure. In view of the uncertainty of the exact time of Sorensen's last weights, the mean weight of the wanderers during their last 10 days, 9.346 kilograms (N = 188, range 7.116–11.850) is probably better for comparative purposes. That the royal fledglings should on the average be lighter in weight than the wanderers when they leave the nesting grounds is consistent with their shorter fledgling period and with the evidence of the growth weight curves of the two.

Whereas wanderer chicks reach peak weight at 220–230 days after hatching, *Richdale* [1952] records that two royal chicks reached their peaks at 201 and 204 days, respectively, and Sorensen's curves for two individual chicks show peaks at 175 and 196 days. Growth-weight curves of individuals or of small samples can be misleading. Richdale writes of a plateau showing

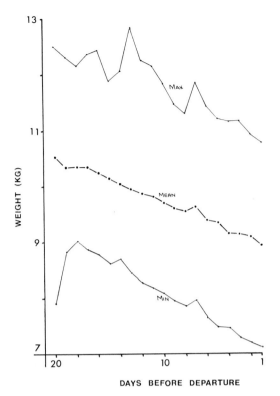

Fig. 39. Weights of 19 *D. exulans* fledglings during
the 20 days in the nest preceding departure.

much irregularity with a decline toward the end. I
came to the same conclusion when I first examined
the individual wanderer curves, but the curve based
on the larger samples shows no indication of such a
plateau. The mean chick weight rises fairly steadily
from hatching to its peak and then declines with similar consistency. My own interpretation of Sorensen's
royal data puts the peak of growth at about 210–220
days after hatching, which is more consistent with
what we know of growth in the wanderer. I believe
his peak in the 24th week (168 days) is probably due
to distorted means based on too small samples.

FEEDING AND FASTING

The following analysis of feeding is based on measurements of the total amount of food received by chicks
per day irrespective of whether it was delivered on
one or more occasions by one or both parents. Errors
inherent in this method were discussed earlier; they
resolve themselves into a few underestimates of the
amounts fed per day. The mean amounts of food
given to 25 nestling wanderers is surprisingly constant
from month to month (Figure 40, Table 47). When
tested they show significant differences only between
the amounts fed in April and May and in October
and November. No significant difference is evident
between the mean amounts fed monthly in the six
months May to October.

In his study of the king penguin at South Georgia,
Stonehouse [1960] suggests, "Slow growth rate is
probably an adaptational response to uncertain and
erratic food supplies in both species"; and "Food is
probably extremely abundant in December and again
in March and April, but may be short or difficult to
catch during the winter."

Food availability is an important aspect of the ecology of all species. During periods of shortage, the
amount of food delivered to the chicks decreases, as
does the frequency of feeding, because it takes parents longer to collect a given amount. Stonehouse's
remarks on the availability of food for the king penguin are based on the seasonal variation in catches
by plankton trawls from oceanographic vessels. This
assumes, of course, that the plankton is equally available to both nets and birds, which may not be so—
some organisms such as squid, which the birds pursue
and catch, may be able to swim away from plankton
nets.

The data here indicate that wandering albatrosses
had no difficulty in obtaining food within range of
South Georgia in any month between April and November. During the last weeks of fledging, adults feed
their chicks less frequently, but this is doubtless a
behavioral response to the approaching end of fledging
rather than a manifestation of food shortage in the
nearby waters. As the mean chick weight in mid-
April is barely 3.5 kilograms, it is understandable that
they do not consume as much then as in May, when
the mean weight rises above 5.5 kilograms. From
May to October, however, the young birds consume
equivalent amounts monthly irrespective of their size.

The total amount of food received by nestlings between the end of brooding and departure varies from
51 to 82 kilograms. At the rate of one feeding every
3 days, the average consumption in the nest is 0.264
kilogram per day. Except during the brooding period
and the final weeks at the nest, the mean frequency of
feeding remains fairly constant. During the 200 days
between the 41st and 240th days after hatching, the
mean frequency is 3.5 feeds every 10 days, with a
range from 0 to 7 feeds every 10 days, and a slight
decrease in feeding frequency is evident from days
175 to 225, before the decline prior to departure.

During the last 100 days in the nest (Table 29) the
amount of food and frequency of feeding both fall,

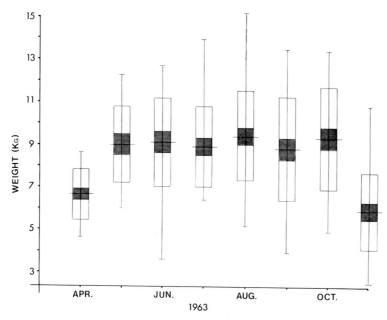

Fig. 40. Amounts of food received monthly by 24 *D. exulans* chicks at Bird Island.

but the precise point at which the decline begins cannot be determined. The change in mean weight of feeds, however, indicates that parental attention begins to wane about 70 days before the chick leaves.

No similar analysis of feeding has as yet been attempted for the royal albatross. *Richdale* [1952] weighed one chick twice daily for two periods in 1938, first from May 7 to 22, then from August 20 to September 21. His curves closely resemble those of individual wanderers over similar short periods.

The last feed occurs an average of 7 days before the chick departs, though some wait 3 weeks after the last meal before leaving, and others receive food on the day they fly. This last fasting spell is not exceptionally longer; in each of the 19 nests for which I have records, the chicks often went without food for

longer periods at other times. The mean lapse between the last feed and departure of the 19 fledglings was 6.8 ± 1.4 days (SD 6.1, range 0–21 days) compared with the previous longest period of fasting recorded for each chick of 14.0 ± 1.0 days (SD 4.2, range 10–24 days).

Although the mean feeding frequency was once every 3.5 days, wanderer chicks often waited much longer between feeds (Figure 41, Table 31). No demonstrable difference is apparent in the total amount per month fed to different chicks, but the frequency of longer than average fasts (Figure 42, Table 31) shows two distinct peaks. The data for December and January are omitted because chicks are then leaving. The figure for April, when brooding finally ends, is low, for the parents still visit the nest very often.

TABLE 29. Variation in the Amount of Food, Feeding Frequency and Size of Feeds of *D. exulans* Nestlings during the 100 Days before Departure, Measured in 10-Day Intervals

Day Intervals		91–100	81–90	71–80	61–70	51–60	41–50	31–40	21–30	11–20	1–10
No. of Observations		79	79	69	63	89	54	58	48	44	26
No. of feeds/10 days	Min.	1	2	1	0	2	0	0	1	0	0
	$\bar{x}$	4.2	4.2	3.6	3.3	4.7	2.8	3.0	2.5	2.3	1.4
	Max.	6	6	6	6	6	6	6	4	4	2
Wt of food/10 days, kg	Min.	1.049	0.666	1.134	0.964	1.928	0.425	0.482	0.283	0.539	0.085
	$\bar{x}$	2.963	3.255	3.175	2.824	3.751	1.857	2.293	1.905	1.202	0.578
	Max.	4.550	5.542	6.535	5.727	6.535	4.933	3.756	4.026	2.934	2.084
Mean wt of feeds/day, kg		0.712	0.782	0.873	0.850	0.799	0.652	0.751	0.754	0.519	0.422

TABLE 30. Number of Feeds and Amount of Food Delivered
to Nestling *D. exulans* at Bird Island between the End of
Brooding and Departure

Nest No.	No. of Feeds	Total Weight of Food, kg	Time between End of Brooding and Leaving, days	Food/ Day, kg	Feeding Frequency, 1 feed/ x days
411	80	57.054	242	0.236	3.0
425	74	66.821	235	0.284	3.2
243	81	51.158	252	0.203	3.1
143	77	62.824	262	0.240	3.4
428	85	55.764	246	0.227	2.9
91	72	69.727	237	0.294	3.3
431	74	66.892	243	0.275	3.3
86	94	73.540	255	0.288	2.7
87	80	71.073	231	0.308	2.9
401	67	62.412	233	0.268	3.5
430	96	81.733	249	0.328	2.6
92	85	68.579	259	0.265	3.0
433	79	61.307	234	0.262	3.0
434	75	61.293	247	0.248	3.3
154	90	64.269	270	0.238	3.0
Mean	80.6	64.963	246.3	0.264	3.08

The high number of fasting days in August might
indicate a reduced availability of food, though the
amount of food actually delivered to the chicks is not
reduced markedly. The November peak probably reflects a behavioral response of the adults to the end
of the fledging period.

Share of the sexes in feeding. As mentioned earlier,
both sexes were seen feeding young at the end of the
winter fledging period, but during the latter half of
the fledging period at least, some male parents apparently feed the chicks more frequently than the females
do. Table 32 gives the feeding data for three chicks
that lost one of their parents and were fed for several
months by only one adult. The two (N.V.902 and
N.V.904) fed by the male alone did not receive significantly less food than chicks being fed by both
parents, but the chick (N.V.903) fed solely by the
female did (Table 47).

Although these three incidents are highly significant
in themselves, they are not enough to demonstrate
overwhelming male dominance in feeding during the
fledging months. The combined evidence indicates
that, though both parents visit and feed the chick up
to the end of the fledging period, the male may take
a larger share from September onward.

Fasting. The circumstances of pelagic life are such
that even in the well ordered routine of the most

experienced breeding pairs that relieve each other
regularly during incubation, brooding, and feeding,
occasionally one or both parents are kept from returning, so that chicks are left for longer than average
periods. It is appropriate, then, to examine how well
the young bird is able to survive such periods of fasting at various stages of the fledging period.

Table 33 gives the weights of seven chicks that
underwent periods of fasting and starvation; three
occurred naturally in the course of feeding by both
parents; the other four were removed from their nests
experimentally. Two of the latter, 444 and 405, died
after going without food 58 and 81 days, respectively,
during which time they lost over 60% of their body
weight. Chick 216, after going without food 51 days
and losing 44% of its weight, was able to fly away
November 26—though it would be illuminating to
know how well it survived its initial experience at sea.

The curves for chicks 405 and 444 (Figure 36)
show how regularly a fasting chick loses weight. Excluding the five days following the last feed to allow
the stomach contents to be completely assimilated

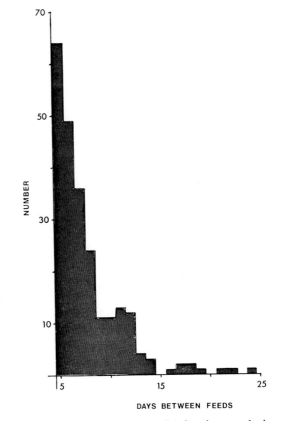

Fig. 41. Intervals in excess of 4 days between feeds
of 19 *D. exulans* chicks.

TABLE 31. Number of Fasts in Excess of 4 Days in 19 *D. exulans* Nestlings during Season 1963–1964

Length of Fast, days	No. of Nestlings in										
	April	May	June	July	Aug.	Sept.	Oct.	Nov.	Dec.	Jan.	Total
5	8	9	3	11	8	10	7	6	2		64
6	2	5	1	5	8	5	2	16	4	1	49
7	2	1	2	7	7	8	2	3	3	1	36
8		4	7	2	1	4	3	1	2		24
9		1	1	1	3	1	2	1	1		11
10			1	2	2	1	1	3	1		11
11	1	2	4		2			3	1		13
12		3		2	4		2	1			12
13			2	1					1		4
14				1	2						3
15											
16											
17									1		1
18								1		1	2
19			1					1	1		2
20									1		1
21											
22									1		1
23							1				1
24						1					1
Total fast days/month	77	181	180	230	290	211	170	274	158	30	

and/or excreted, the daily loss in body weight reflects the utilization and depletion of body reserves. The energy requirements of a comparatively sedentary chick can reasonably be assumed to be constant within certain limits. The relative amounts of time the chick spends sleeping, exercising by stretching its wings, moving about, or nest building is proportional to its stage of development but more or less constant in relation to the environmental conditions prevailing during any part of the fledging period.

The shape of the growth-weight curve during fasting shows a more rapid weight loss during the first part of any long fast. The mean daily weight loss of the two chicks that did not survive was 119 and 99 grams per day, respectively, but in each case the mean daily weight loss was greater during the first 25 days. The mean daily weight loss in chick 405 from 5 to 30 days after feeding is almost exactly twice the rate at 55 to 80 days.

Chick 444, which died after 58 days, was a month younger at the start of the fast than 405, which survived 81 days. The fact that 444's mean daily weight loss during the 5- to 50-day period was higher than 405's might have resulted from better insulation by its more advanced plumage and less demand on body reserves for basal metabolism. In this respect it is interesting that chicks 216 and 422 began their fasts later by one and two months, respectively, and each had a smaller mean daily weight loss (119 and 111 grams per day). Chick 426, whose fast started December 6 and who lost weight at 122 grams per day, could have used up additional energy in flight practice and therefore lost more weight during the days just before departure. Interesting though these speculations are, the examples from which they are drawn are too few to allow any generalizations.

Feeding range. A mean frequency of one feed in 3 days (Table 30) allows an average foraging adult to find enough food within one and one-half days flying distance from the nesting grounds.

The only information on the flight speed of great

TABLE 32. Monthly Amounts of Food Fed to 19 *exulans* Nestlings by Both Parents and to 3 Nestlings Receiving from Only a Single Parent

	Sept.	Oct.	Nov.
Mean amounts received by nestlings fed by both parents, kg (from Table 47)	8.746	9.157	5.840
NV902 fed by male only, kg	9.866	12.843	4.876
t	0.448	1.511	0.515
NV904 fed by male only, kg	8.392	8.845	6.577
t	0.141	0.128	0.398
NV903 fed by female only, kg		2.863	
t		2.580	

TABLE 33.   Weight Loss in *D. exulans* Chicks during Fasting

Chick No.	Last Feed		End of Fast		Length of Fast, days	Loss of Wt		Mean Daily Wt Loss, g/day			
								Days after Last Feed			
	Date	Wt, kg	Date	Wt, kg		kg	% at Last Feed	6–30	31–55	56–80	6–80
444*	July 30	12.757	26 Sept.†	4.394	58	8.363	65.6%	151	87		
405*	Sept. 1	13.764	21 Nov.†	4.975	81	8.789	63.8	137	96	68	99
216*	Oct. 6	12.942	26 Nov.‡	7.172	51	5.770	44.6	119	65		
422*	Nov. 1	13.154	25 Nov.‡	9.142	24	3.742	28.4	111			
411	Oct. 23	11.723	14 Nov.	7.711	22	4.012	34.2	123			
426	Dec. 6	12.474	27 Dec.‡	9.072	21	3.402	27.3	122			
143	Aug. 31	8.760	24 Sept.	5.429	24	3.331	38.0	102			

† Chick dead the following day.
‡ Chick flew the following day.
* Experimentally removed from nest.

albatrosses comes from the seafarer's habit of attaching messages to birds caught at sea, the classic instance being the wandering albatross that Captain Hiram Luther of the *Cachalot* shot at 43°24'S, 79°05'W on December 30, 1847, bearing in a vial hung around its neck the message, "Dec. 8th, 1847. Ship 'Euphrates,' Edwards, 16 months out, 2300 barrels of oil, 150 of it sperm. I have not seen a whale for 4 months. Lat. 43°S. long. 148°40'W. Thick fog, with rain." As *Snyder* [1958] points out, *Murphy* [1936] misquoted both the date and position of Captain Luther's capture and therefore estimated the bird flew 3150 nautical miles in 12 days. Actually, as computed by the Navigation Committee of the U.S. Naval Academy, the shortest great circle distance between the two corrected positions is 2950 miles, which the bird covered in 22 days, and the average is thus 134 nautical miles per day. *Dixon* [1933] notes individuals following

his ship 120 miles during the hours of darkness, but he does not distinguish between *exulans* and *epomophora*. The same reservation may be made of the identity of Captain Luther's albatross, but no other measurements of great albatross flight speeds are known to me. Dixon's record indicates a somewhat faster speed, but possibly a bird trying to keep up with a vessel might cover more distance than one alone. Using the estimate from Captain Luther's recovery, a reasonable one and one-half days' flying distance is approximately 200 nautical miles.

The abundance of whales in the vicinity of South Georgia has been known since the turn of the century, and the *Discovery* investigations showed dense aggre-

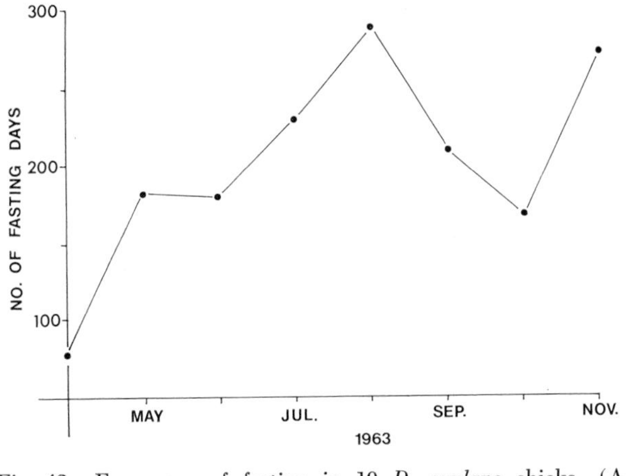

Fig. 42. Frequency of fasting in 19 *D. exulans* chicks. (A fast is defined as an interval of 5 or more days between two consecutive feeds.)

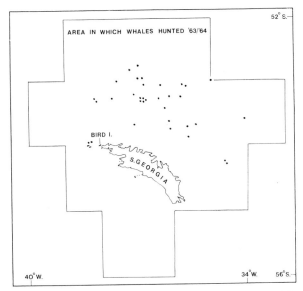

Fig. 43. Sightings by whale-catchers of *D. exulans* adults colored while incubating eggs or brooding chicks at Bird Island, 1963–1964.

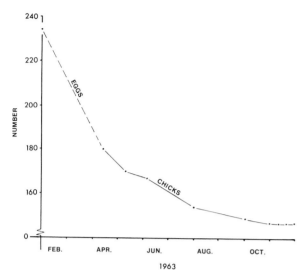

Fig. 44. Survival curve of *D. exulans* eggs and chicks at Bird Island.

gations of plankton in a belt about 100 sea miles north of South Georgia [*Hardy and Gunther*, 1935]. Whale catchers from the South Georgia shore factories operate regularly today up to 150 sea miles from the island, and expedition ships occasionally visit these waters in the austral summer. During the summers of 1962–1963 and 1963–1964 these vessels were asked to cooperate in reporting the positions of any color-marked wandering albatrosses encountered.

During these two seasons 305 wanderers, all of them incubating eggs or brooding chicks at Bird Island, were thoroughly sprayed with a dye made of rhodamine-B dissolved in isopropyl alcohol and water. Ships at sea reported a total of 52 sightings of these bright red birds, most of them from waters within 125 sea miles north of the island (Figure 43). This may have been because the catchers spent most of their time in the waters where whales were most plentiful, but it can be argued with equal conviction that the albatrosses frequent these waters in greater numbers

for the same reason the whales do—food [*Hardy and Gunther*, 1935].

Of the five colored birds reported more than 200 sea miles from Bird Island, three were within 933 sea miles or 14 days' flying distance to and from the breeding grounds. The two birds reported from South African waters may well have lost eggs or chicks and deserted the breeding grounds for the season. At any rate, the observations support each other well enough to indicate that in spite of their remarkable powers of flight, the wandering albatrosses at South Georgia are able to find adequate food within a few days' flight range.

MORTALITY

Mortality in albatross populations is considered in five categories:

1. Loss of eggs (viable and nonviable).
2. Mortality of chicks before they go to sea.
3. Mortality of young birds from the time of departure to their first return to the breeding grounds.
4. Mortality of subadults from the first return to the island until they first breed (age 4 to perhaps 10 years).
5. Mortality of breeding adults.

D. exulans egg and chick mortality at Bird Island (Figure 44) was determined by counts in two areas, Wanderer Ridge and Long Ridge (Figure 3), not subject to so much disturbance by observers as the main study area. The figures for each (Table 34) are very similar and show egg loss higher than chick mortality. The figures of eggs actually counted have been corrected to compensate for those lost before the count was made. This 7% correction is based on observed daily egg losses in the study area from the day of laying. As the figures are only for one season, we know nothing of annual variation in nesting ground mortality.

The climate causes marked annual differences in egg

TABLE 34. Relative Breeding Success in *D. exulans* and *D. epomophora*

	Location	No. of Eggs		No. of Chicks and % Eggs Producing Chicks		No. and % of Chicks That Survive		% Eggs Producing Flying Chicks
D. exulans	Wanderer Ridge	234	(250) *	180	72.0%	147	81.7%	58.8%
	Long Ridge	126	(135) *	98	72.6	79	80.6	58.5
D. epomophora†	Campbell Island	101				75		74.3

* Census of eggs made January 31. Figures corrected to compensate for 7% eggs lost before end of January.
† *Westerskov*, 1963.

survival among the black-browed and gray-headed albatrosses at Bird Island. Both these species return and lay in October when late winter weather can hamper nest construction, which is most important for egg survival in a crowded colony. One advantage of the wandering albatross' biennial cycle is that the eggs are laid in midsummer when the weather is ameliorated, and, as mentioned earlier, the breeding sites are never so crowded as the black-browed and gray-headed albatross colonies.

Westerskov's [1963] data on 101 royal albatross eggs at Campbell Island do not separate egg and chick survival but show that 75% of the eggs laid produced young that flew, a somewhat higher survival rate than the 59% rate in the South Georgia wanderers (Table 34). Included among the egg-loss figures are those that proved infertile. Out of 179 laid in one study area at Bird Island, 13 were lost by breaking and another 13 proved sterile. Thus, with a 14.5% egg loss, the infertility rate is about 7%.

Richdale's few royal albatrosses at Taiaroa Head were subject to so much animal and human interference from 1935 to 1952 that *Richdale* [1952] had difficulty in estimating "natural" mortality from his data. This disturbance plus the comparatively small number of birds probably account for the rather low infertility rate of 2.6% that he reports.

His figures for fledgling survival from the time of leaving the island to their return several years later are not quite so equivocal. As *Richdale* [1952] points out, the 6 returns from 11 chicks that flew from Taiaroa Head constitute a remarkably high 55% rate of return. Of the 656 chicks banded at Bird Island in the 1958–1959 season, approximately 38% were reported in their 4th, 5th, and 6th years by later expeditions to the island. I am sure we did not find all the banded birds of the cohort that returned, and others probably returned for the first time after our field work stopped. Hence it is reasonable to assume that probably about 50% of the wanderer fledglings leaving Bird Island survive to return to the breeding grounds.

Records of marked breeding pairs over the 6 years allow us to compute breeding adult mortality with more confidence. The principal criterion for considering a marked bird dead or alive is whether it is seen on the breeding grounds. As breeding birds return to the same vicinity to nest, the probability of an observer finding any marked breeding bird is higher than for subadults. A pair of birds that breed successfully in year x may be expected to nest again in year $x + 2$,

TABLE 35. Survival of Adult Wandering Albatrosses Breeding in the Inner Study Area at Bird Island during the Season 1960–1961*

	No.	Alive	Missing and Presumed Dead, %	Annual Mortality, %
Males	64	57	7, 10.9%	5.4%
Females	64	60	4, 6.2	3.1
Combined	128	117	11, 8.6	4.3

* Based upon daily records at the study area during 1962–1963 and 1963–1964.

and, if they do not, it may be surmised that one of the pair has died. This is further confirmed if breeding does not occur in year $x + 3$, and if the one remaining member of the pair remates.

The behavior of the remaining bird in years $x + 2$ and $x + 3$ indicates whether or not its former mate has been present. In most instances if the pair is not united on the breeding ground in year $x + 2$, the surviving bird starts to establish a new pair bond, so that it may breed successfully with a new mate in year $x + 3$. As no case of "divorce" and re-pairing has been observed in either *exulans* or *epomophora*, remating is substantial evidence for the death of the former mate.

Table 35 presents survival figures for 128 birds that bred in the inner study area in 1960–1961 (year x). Survival of birds that lost eggs or young in year x and bred successfully in year $x + 1$ are based on findings in year $x + 3$ (1963–1964). The combined mean annual mortality for both sexes of breeding adult *exulans* at Bird Island is 4.3%, a figure reasonably close to the 3% adult mortality observed in *epomophora* [*Richdale*, 1952; *Lack*, 1954].

THE BIENNIAL CYCLE

Of 22 pairs that bred in the Bird Island study area in 1958–1959 and remained intact and identifiable until 1963–1964, 14 (64%) bred three times (once every 2 years), 7 (32%) bred four or five times, and 1 pair produced an egg each season for 6 consecutive years and never reared a chick. Loss of the egg or chick ends the breeding cycle, the parents go off to sea, and many pairs have enough time to complete the physiological changes associated with the postnuptial season to return and breed the next season. If a pair is habitually unsuccessful this may be repeated for several seasons in succession so that the birds' gonads and endocrine glands cycle annually.

Among the marked breeding adults at Bird Island,

20% of the pairs that laid eggs in any one year were unsuccessful and bred again the following season. Thus in any one year the nesting population (x) is composed of a majority of pairs that bred successfully two seasons previously $(x-2)$ and a minority that bred unsuccessfully the previous season $(x-1)$, plus a very few that are breeding for the first time (disregarded in this analysis). Although *Richdale* [1952] noted this in the royal albatross, the proportion of pairs that breed again the following year has not previously been measured for either of the great albatrosses.

Most pairs that lose their eggs return and breed the following season. In the inner study area 15 pairs of marked birds either lost eggs or failed to hatch infertile ones in 1962–1963; 11 (73%) of these laid eggs the following season. Of the 4 pairs that did not lay again, the mate of one female failed to appear (and presumably died), and the other three returned but did not lay. The previous season these three pairs had incubated their eggs for 70, 81, and 88 days, respectively; all would thus have been brooding chicks when they left had their eggs been fertile.

The ability of pairs to return and breed again after one year was also tested by removing eggs experimentally. In 1962–1963 all 266 eggs were removed from Gony Ridge within two days of being laid, and 281 of the incubating and attendant birds were marked. The following season, 1963–1964, in three inspections of the area during the laying period we found 219 of these birds incubating eggs. This is 78% of the 281 originally marked, but as 4.3% should have died in the meantime, the expected return allowing for normal adult mortality is 259, of which the observed returns constitute 83%. It may reasonably be assumed that many of the missing 47 marked birds might have been found by further searching.

Thus, although loss of eggs at or soon after laying is probably always followed by laying the following season if both parents survive, a small proportion of those pairs that keep their eggs for more than about 50 days may return but fail to lay the next year. Clearly, by the time the chicks hatch, the activities of the adults on the breeding grounds are already affecting their breeding potential for the following season, and, as parental attention continues during chick growth, the adult's ability to breed the next year diminishes rapidly.

To determine how long adults retain the capacity to breed in the year $x+1$, we killed 20 chicks in Wanderer Valley between March and June and another 16 between July and October 1963. Of the first

TABLE 36. Number of *D. exulans* Pairs Breeding in the $x+1$ Year after Loss of Chicks

	No. of Chicks Lost, x Season	No. of Pairs Breeding, x + 1 Season, %	
March	3	2*	(100%)
April	10	7	(70%)
May	3	1	(33%)
June	4	2	(50%)
July	3	0	
August	6	0	
September	6	0	
October	1	0	
Total	36	12	

* One pair did not breed because of death of ♂.

20 (Table 36) 12 pairs bred again in 1963–1964. As the male of one pair disappeared and was presumed dead, 63% of the pairs that lost chicks before the end of June were able to breed again in December. None of the 16 parents whose chicks were removed in July or later bred again the following season.

Some of the pairs that lost eggs and did not lay again the year $x+1$ appeared on the breeding grounds for a short time during the early part of the $x+1$ season, and some of the males were seen copulating with other females. Similarly some pairs that lost chicks and did not breed the next year were seen near their old nests. In keeping with normal breeding behavior, the males of these pairs were present much more often than the females. Four parents of two chicks lost in September were not seen at all the next season, and at the nest from which the chick was removed in early October only the male appeared for the first four days of December.

Table 37 shows the results of histological examination of testes of five adult wanderers (Marshall, personal communication). Though we have no record of pairs breeding the next season after losing chicks as late or later than July, the testes of two adults (146 and 148) that lost chicks the previous July and August and were killed near their nests in December were enlarged and in potential breeding condition.

Breeding adults whose mates do not return because of presumed death at sea (very few adults die on the nesting grounds) generally establish a new pair bond toward the end of the same season their mate disappears. Surviving males build nests on or near the site of their previous one, but females wander about the breeding grounds at random visiting those males that happen to be advertising themselves. The new pair bond is usually established within one or two

TABLE 37. Histology of Testes from Male Wandering Albatrosses Breeding at Bird Island

Speci-men No.	Date	History	Testis Tubule vol cc, diam μ	Description (Marshall, personal communication)
131	28 Nov. 63	Adult newly arrived at nesting ground (mean date of arrival 29 Nov.)	12 292	Active interstitium and appearance of first sperma-tozoa. Sexual and (?) social stimulation would rapidly bring bunches of spermatozoa into being.
197	27 Jan. 64	Adult incubating an egg (mean date of laying 24 Dec.)	1.5 88	Classical picture of collapsed testis. The interstitium has big Leydig cells. These are either unexhausted cells of the current generation or rapidly developed ones of the new generation. Post-nuptial activity is already evident in some tubules.
150	21 Dec. 63	Adult bred previous seasons. Fledged young departed two days previously.	2 105	The vacuolated lumina were formerly filled with post-nuptial lipids, most of which had been dispersed before the animal was killed. The presence of an essentially new interstitium, and the relative lack of maturing Leydig cells, suggests that the big ones in specimen 197 are those of the current generation and *not* new ones quickly developed for the next. The random mitotic activity in specimen 150 is not uncommon in post-nuptial birds; the maturation process *perhaps* could go forward immediately (with the maturation of the new interstitium) if external conditions were propitious.
146	11 Dec. 63	Adult bred previous season but lost chick 21 July. Female mate not seen at nest.	8 210	Could breed almost immediately if provided with territory and mate. This is the condition African and s. European migrants are in when they reach England in spring.
148	13 Dec. 63	Adult bred previous seasons but lost chick between 1–22 Aug. Female mate not seen at nest.	12 252	Essentially the same as specimen 146 but slightly more advanced (= more exhausted interstitium and greater spermatogenic activity).

seasons. Of 28 known breeders whose mates failed to return, 86% remated and bred again within 4 years (Table 38). The two widows and two widowers that re-paired with another bird and bred again 2 years later could only have done so if they had lost their egg or young chicks while breeding with their previous mate, who did not return to the nesting grounds the next year. It is of interest to note that 64% of the males needed only one season to establish a new pair bond, and the females averaged only slightly slower (50%) at finding new mates.

POPULATION DYNAMICS

The total numbers of nests with eggs at Bird Island were counted in each of the four seasons from the end of January 1960, after all eggs had been laid, through 1964. Each nest was visited and counted only if it contained an egg or the remains of an egg. To compensate for the eggs that were lost before the census was taken a correction factor was computed from the daily observations in the Wanderer Valley study area. Of 262 eggs laid there during three seasons 18 (6.9%) were lost before January 31. Thus, to correct

the census figure, 7.4% of the egg count must be added (see below). As 20% of pairs that breed in year x also breed in year $x + 1$, we have the equation:

$$P = (C_x + fC_x) + (C_{x-1} + fC_{x-1}) - Y_x$$

where: P = total breeding population.

C_x = number of eggs counted at the end of January in *year x*.

C_{x-1} = number of eggs counted at the end of January in year $x - 1$.

f = egg-loss factor, $18/(262 - 18) = 0.074$.

Y_x = number of pairs breeding in year x that also bred in year $x - 1$.

Computed thus the mean of three estimates of the Bird Island breeding population is 3194 pairs, with a difference of only 48 (1.5%) between the minimum and maximum estimates (Table 39).

Estimates for the rest of South Georgia (Table 40) are much less reliable. *Wilkins* [1923] states that

TABLE 38.   Re-Pairing by Breeding Adult *D. exulans* Whose Previous Mates Failed To Return to the Breeding Grounds

	No. of Years Since Last Breeding with Previous Mate						Did Not Remate	Total
	2	3	4	5		Total		
♂ ♂	2 (14%)	7 (50%)	3 (21%)	0		12 (86%)	2	14
♀ ♀	2 (14%)	5 (36%)	5 (36%)	0		12 (86%)	2	14

four men collected 3500 wanderer eggs in three days at the northern end of the island, and *Matthews* [1929] writes of 2000 wanderer eggs collected in 3 days. *Rankin* [1951] assessed the wanderer nesting populations in his voyages around South Georgia, and, despite his underestimate of the Bird Island population (750–800 pairs), his figures are perhaps the best available for calculating the present South Georgia population. Possibly a few small unrecorded colonies exist, for instance on Saddle or Annenkov islands, but not in numbers that will appreciably alter the total estimate of 4500 pairs of wandering albatrosses on South Georgia, two-thirds of which are on Bird Island.

Until recent years whalers and sealers customarily took wanderer eggs for food [*Matthews*, 1951], which makes it impossible to determine whether the figures given by Wilkins, Matthews, and Rankin pertain to portions of a demipopulation or the total breeding population. The dynamics of the biennial cycle are such that a population exploited for eggs lays slightly less than twice (1.8) what it would without such interference (as 20% of the population breeds annually in any case). Wherever all eggs are removed annually at the time of laying, the whole population attempts to breed each year. Apparently landing difficulties prevented the South Georgia whalers from visiting Bird Island on their traditional Christmas egging trips as often as they did the other South Georgia breeding sites.

Matthews' [1951] entertaining account reveals that the *Sitka's* egging party of December 26, 1923, collected more than half its 2000 eggs in the Undine Harbor-Hope Valley area. In the same place *Rankin* [1951] found only 298 nests, and in October 1954 I found only one fledgling. Either the Hope Valley was egged extensively in December 1953, or the constant egging over the previous thirty or more years had the inevitable result of reducing the population substantially, if not to the vanishing point.

Estimates of nesting populations are available for all the breeding stations of the two great albatrosses except the wanderers at the Kerguelen and Amsterdam islands. Many of these estimates are of questionable accuracy, but they do serve to suggest the approximate level of the world breeding populations (Tables 41, 42). The Kerguelens are quite extensive, much larger than any other of the great albatross breeding islands, and with many small offshore islets eminently suitable for nesting. Hence the Kerguelen *exulans* population might well equal that of South Georgia (perhaps about 5000 pairs). If the waters around Kerguelen are richer in food, they might support an even larger population. The evidence suggests that the world breeding population of *exulans* is of the order of 20,000 pairs, approximately double that of *epomophora*.

Theoretical models of great albatross populations. In erecting his model for the world population of *epomophora*, *Westerskov* [1963] reasons as follows:

The basic biological facts used in these calculations are: (1) a breeding population in any one year of 2300 pairs [at Campbell Island], (2) each pair breeds every other year, (3) birds start breeding at 9 years of age, (4) equal number of males and females in population, (5) only one

TABLE 39.   Breeding Population of *D. exulans* at Bird Island, South Georgia

Breeding Season	Census	+6.9% Mortality Factor up to Jan. 31	Corrected Count	20% Bred Unsuccessfully the Previous Season	Population Estimates, Pairs	Mean Population Estimates, Pairs
1960–1961	1447	107	1554			
1961–1962	1790	132	1922	311	3165	
1962–1963	1551*	115	1666	384	3204	3194
1963–1964	1949	144	2093	546*	3213	

* 266 eggs were removed by the observer 1962–1963.

TABLE 40. Numbers of *D. exulans* Breeding at South Georgia*

Area	Nests Seen in One Season	Authority
Bird Sound, mainland	50	*Rankin* [1951] *Morris* and *Tickell* [1962]
Reef Point	10	*Rankin* [1951]
Hope Valley/Coal Harbor	298	*Rankin* [1951] *Morris* [1962]
Bay of Isles	421	*Rankin* [1951]
Jomfruene	1	*Morris* [1962]
W. Undine Harbor	5	*Morris* and *Tickell* [1962]
Total seen	785	

* Population, (785 + 785 − 157), 1413 pairs.
  Population of Bird Island, 3194 pairs.
  Total population of South Georgia, 4607 pairs.

egg is laid in each breeding cycle, and (6) combined egg and chick mortality amounts to about 25 per cent of the eggs laid. Furthermore it is assumed that: (1) after the first year of life, mortality is constant, and (2) the population maintains itself at a more or less constant size.

If a is the proportion of birds that survive to the stage when they can leave the nest, and b is the proportion of birds that survive a given year (adult mortality is assumed here to be independent of age once the bird has left the nest), then the proportion surviving to at least the jth year will be: ab^{j-1} since a bird j years old will have spent roughly $j - 1$ years away from the nest. The expected number of eggs from a particular female bird (this expectation being taken from the time when the bird itself is a newly laid egg) is:

TABLE 41. World Breeding Population of *D. epomophora*

	Nests Seen in One Season	Estimated Breeding Population, Prs.	Authority
New Zealand (Taiaroa Head)	8	15	*Westerskov* [1963]
Chatham Islands	2300	5940	*Dawson* [1955]
Campbell Island	2300	4140	*Westerskov* [1963]
Auckland Islands	9	22	*Westerskov* [1963]
Auckland Islands	10–15	7	*Kinsky*, personal communication*
Total		10,124	

* B. D. Bell, Auckland Islands Expedition, 1964–1965.

$$a\,(b^{8} + b^{10} + b^{12} + b^{14} + \cdots) = ab^{8}/(1 - b^{2})$$

If the total population size is not changing, then this figure must be 2 because on the average a pair will just replace itself.

As egg and chick mortality is 25 per cent, $a = 0.75$. Supposing, as above, that the expected number of offspring is 2, then we must have:

$$(0.75)\ b^{8}/(1 - b^{2}) = 2$$

This equation for b has a solution $b = 0.909$, indicating an annual mortality for birds that have left the nest of $100\ (1 - 0.909) = 9.1$ per cent.

Were this equation valid as it stands we would expect the substitution of the observed *exulans* parameters of egg-chick loss in first year $= 41\%$ and adult mortality $= 4.3\%$ per annum would be possible, but we obtain

$$(0.59 \times 0.703)/(1 - 0.916) = 4.94$$

Or solving for a, a 24% survival rate of eggs and

TABLE 42. World Breeding Population of *D. exulans*

Breeding Station	Nests Seen in One Season	Estimated Breeding Population, Pairs	Authority
Tristan de Cunha group	3	5	*Swales* [1965]
Gough Island	2000	3600	*Swales* [1965]
Marion Island	700	1360	*Rand* [1954]
Îles Crozet (I. Possession)	475	855	Tufft, personal communication
Îles de Kerguélen	?	?	
Amsterdam Island	?	?	
Campbell Island	5–7	11	*Westerskov* [1959]
Auckland Islands	500–1000	1360	Falla, personal communication Fleming, personal communication Kinsky, personal communication*
Antipodes Island	1000–1200	1980	Falla, personal communication Turbott, personal communication
Macquarie Island	25	45	Ingham, personal communication
South Georgia		4607	*Tickell* [1962a]
Total		13,823+	

* B. D. Bell, Auckland Islands Expedition, 1964–1965.

chicks (only one-third of the observed survival rate) is necessary to maintain a stable population with an adult mortality of 4.3% per annum.

Westerskov's first premise is not pertinent to the immediate discussion and may be ignored for the moment; his second does not allow for failed breeders nesting annually; his fourth and fifth are, from our present knowledge, applicable to any population of great albatrosses. His sixth premise is based on a field census, and it has been shown earlier (Table 34) that egg and chick mortality in the Bird Island wanderers is appreciably higher than that of the Campbell Island royals. Westerskov's assumption of a constant mortality after the fledglings leave the nest is the principal weakness in his argument. That mortality in birds is constant after the first year is well substantiated for annual breeders, particularly passerines [Lack, 1954]. For annual-breeding sea birds, however, Tuck [1960] gives evidence that juvenile mortality is higher in the first winter at sea than in the second, and higher in the second than in subsequent years. Leslie [1966] assumes constant mortality for common guillemots (Uria aalge) only after two years away from the nest.

As fledgling great albatrosses are in the nest most of their first year, one cannot equate their egg and chick loss in the nest with the first year mortality in annual breeders, for this disregards mortality during the period immediately following the young albatrosses' departure to sea. From what we know of mortality in other sea birds, albatross mortality in the first year at sea (and particularly in the first few months thereof) must be considerably higher than in later years.

Assuming great albatross mortality is constant after the third year, the initial mortality corresponding to the first year mortality in annual breeders is the sum of the mortality in the first year in the nest and the second and perhaps the third years at sea. It is simple to measure the egg and chick loss in the nest, but we have no way of measuring annual mortality at sea before the young birds return to the breeding ground.

Regarding Westerskov's last premise, that the breeding population of the Campbell Island royals is more or less stable, without some measure of the difference in annual mortality during the first 5 years it is impossible to compute an intrinsic rate of natural increase. From the annual differences in the wandering albatross population at Bird Island over a three-year period, which Table 39 shows to be an increase of 0.75% per year, if the population is changing size, it

is doing so very slowly and will take 93 years to double itself.

If it is assumed that mortality is constant after the second year away from the nest (at three years of age), then the second and third year survival may be derived by inserting a constant k in the Westerskov equation and by solving for k with observed values of a and b. Thus if $k =$ the juveniles surviving in the first and second years away from the nest, then

$$akb^6/(1 - b^2) = 2$$

and

$$k = 2(1 - b^2)/ab^6 = 0.389$$

In 1958–1959 we banded 656 fledgling exulans just before they left Bird Island. Their expected survival in year $j = 656 \, kb^{j-3}$.

Using the expression $P = Ea/1 - b$, where E is the number of royal eggs laid in any one season (2300), $a = 0.75$, and $b = 0.909$, Westerskov computed the world population (P) of epomophora as 18,960 birds. Modifying the expression for exulans to $P = Eka/(1 - b)$, where $E = 11,000$ (55% of the mean total number of breeding pairs), $k = 0.389$, $a = 0.59$, and $b = 0.957$, the world population of

TABLE 43. Life Table of D. exulans at Bird Island Based upon: Combined Egg and Chick Mortality 41%, Second Year Mortality (k), 61%, Adult Mortality, 4.3%, Total Population, 9,160

Age, yrs.	No. of Birds	Total Population, %	Surviving to Given Age, %
0	1714*		
1	1012	11.1	58.9
2	394	4.3	23.0
3	377	4.1	22.0
4	361	3.9	21.0
5	345	3.8	20.1
6	330	3.6	19.3
7	316	3.5	18.5
8	303	3.3	17.7
9	290	3.2	16.9
10	278	3.0	16.2
15	218	2.4	12.7
20	171	1.9	10.0
25	134	1.5	7.8
30	105	1.2	6.1
35	83	0.9	4.8
40	65	0.7	3.8
45	51	0.6	3.0
50	40	0.4	2.3
60	23	0.2	1.3
70	13	0.1	0.8
80	7	0.1	0.4

* Eggs (53.7% total breeding population).

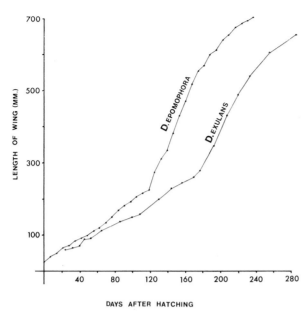

Fig. 45. Growth curves of wing-lengths in *D. exulans* and
D. epomophora [from *Richdale*, 1952].

exulans comes to 58,760. Taking into account the various probable errors, the figure is perhaps somewhere in the region of 50,000 to 60,000 birds.

Westerskov's [1963, p. 809] life table for the royal albatross shows that 0.1% of the population lives to 70 years of age, and an occasional bird may theoretically reach the age of 80. The life table in Table 43 for the wandering albatross is derived from expressions proposed by Westerskov for the royal and modified for observed values of the parameters. It indicates that, although fewer birds survive the earlier years, a larger proportion of older birds are present in the population.

DISCUSSION

The data presented on the breeding biology of the two great albatrosses show them to be remarkably similar in most respects. Most of the slight differences demonstrable between the two amount to no more than those between the subspecies of either species. In fact, similarities so far outnumber differences that one cannot escape the obvious conclusion that *exulans* and *epomophora* are closely related and most certainly evolved from a common ancestor in the not too distant past.

The two notable exceptions to these similarities are the striking differences in the juvenile and subadult plumages of each and the variance in the lengths of their respective fledging periods. The 44-day excess in the mean length of the wanderer's fledging period

could be due to (1) more severe weather conditions on land during the fledging period, forcing the chick to use more food for thermal regulation, (2) a shortage of available food during the winter, or (3) the need to delay the fledglings' departure until a period of appreciably more favorable weather or feeding conditions.

The food supply is adequate in the waters about South Georgia throughout the winter, and weather conditions at sea in South Georgia latitudes are not appreciably different in December than in November [*Richards and Tickell*, 1967]. Thus it seems more plausible that the longer fledging period of *exulans* is associated with the more severe climate through which the chick must survive, or, conversely, that *epomophora's* shorter period is associated with less severe winters. This theory is supported by the growth curves of the nestlings (Figure 38, Tables 44, 45), which show the young royal albatross to be consistently lighter than the wanderer chick. Comparison of the wing growth in the two species (Figure 45) shows that the royal's primaries first appear about 120 days after hatching and that those of the wanderer do not begin to show until about 180 days, i.e. 60 days later. Thus in both species the fledgling leaves the island approximately 100 to 120 days after the primaries start to grow, and the wanderer's longer fledging time results mainly from a prolongation of the earlier part of the chick period while the weather is still comparatively inclement.

The other fundamental difference between the two species is in coloration: the young royals fledge into a mainly white plumage almost indistinguishable from that of the adult, whereas the young wanderer is largely black and unlike the adult. When the royals arrive at the nesting grounds to start breeding, most if not all the previous season's fledglings have left, but the arrival of the wanderer breeding adults just precedes the departure of the fledglings. At this time of year the fledglings are very active in a number of ways that could interfere with and disturb the early nesting procedures of the newly arrived adults. For instance, all fledglings beg for food from any adult in their vicinity, and the fledglings' continual wing-stretching and exercising is accompanied by aggressive displays similar to those of the adult courtship rituals. Thus having fledglings that are immediately recognizable at a distance may be of selective value. With the earlier departure of the royal young this ready distinction is not so important.

The marked differences in populations and the geo-

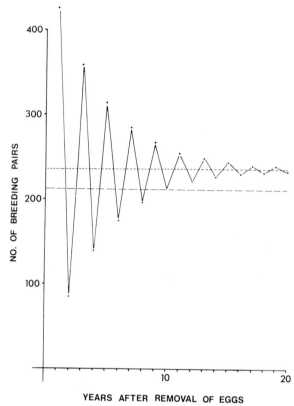

YEARS AFTER REMOVAL OF EGGS

Fig. 46. Predicted fluctuations in the numbers of *D. exulans* pairs breeding each year on Gony Ridge following the experimental removal of all eggs shortly after laying in 1962–1963. Approximately 20% of the pairs breed in both the x and $x + 1$ years. Long-dashed line, mid-population level. Short-dashed line, stabilized demipopulation level.

graphical differences of their breeding grounds must be a significant reflection of the two species' evolution. In that they share a common ocean habitat, at least over part of their ranges, the two species are sympatric at sea. But as most of the breeding islands of the southern oceans are far distant from each other, the likelihood of gene flow between local populations (demes) may be quite low and possibly approaches the probability of mutation within the deme. Hence the demonstrable but slight morphological differences on which the named races of the two species have been established.

The polar and cold-temperate fauna of the southern hemisphere is characterized by comparatively few species represented by very large numbers of individuals breeding in widely dispersed populations. This is generally regarded [*Mayr*, 1963] to be indicative of a rigorous environment that has changed very slightly over long periods and hence rarely allowed any new ecological niches to develop.

TABLE 44. Weights of *D. exulans* Chicks during 200 Days after Hatching

Age, days	N	$\bar{x}$ and SE, kg	SD, kg	Range, kg
11–20	80	1.556±0.040	0.354	0.936– 2.339
21–30	179	2.500±0.034	0.452	1.602– 3.941
31–40	200	3.256±0.042	0.595	1.758– 5.415
41–50	170	4.132±0.052	0.683	2.410– 6.095
51–60	149	4.966±0.074	0.909	3.189– 8.491
61–70	200	5.547±0.068	0.970	3.657– 8.250
71–80	200	6.323±0.077	1.098	4.139–10.008
81–90	210	6.620±0.087	1.261	3.884– 8.944
91–100	210	7.364±0.106	1.539	3.785–12.077
101–110	210	8.029±0.097	1.408	4.890–11.850
111–120	210	8.331±0.085	1.238	5.160–11.666
121–130	210	9.077±0.088	1.279	6.478–14.600
131–140	210	9.312±0.114	1.649	6.606–14.118
141–150	200	9.685±0.117	1.662	6.294–14.430
151–160	210	9.972±0.123	1.785	6.039–15.422
161–170	209	8.834±0.131	1.890	5.741–14.629
171–180	190	10.182±0.134	1.847	5.429–14.373
181–190	199	10.748±0.105	1.474	6.932–15.054
191–200	190	10.837±0.099	1.373	7.825–15.196

The alternating glacial and interglacial periods of the Pleistocene have been shown to be responsible for dramatic avifaunal changes in the Palearctic [*Moreau*, 1955]. In the southern hemisphere the antarctic ice sheet extended farther north, and some new niches must have appeared and disappeared in the lands and waters adjoining it during the Pleistocene. The environment that favored the ancestral species of the great albatrosses was probably altered or fragmented sufficiently to allow the two separate lines to evolve in isolation. This could well have been caused by the extension of the antarctic ice sheet eliminating the more southerly islands as breeding grounds by increasing their ice cover, reducing vegetation, and changing the character or abundance of the plankton fauna within easy flying range of the land. Today no wandering

TABLE 45. Weights of *D. exulans* Chicks during the 100 Days Preceding Departure

Days before Departure	N	$\bar{x}$ and SE, kg	SD, kg	Range, kg
1–10	188	9.346±0.080	1.098	7.116–11.850
11–20	190	10.167±0.082	1.131	7.910–12.843
21–30	190	10.641±0.091	1.252	7.711–14.345
31–40	190	11.024±0.096	1.319	7.881–15.763
41–50	190	11.261±0.085	1.165	9.114–14.062
51–60	189	11.703±0.089	1.218	8.661–14.940
61–70	190	11.118±0.109	1.498	8.278–16.131
71–80	190	11.197±0.102	1.410	7.456–14.940
81–90	190	10.883±0.109	1.509	7.853–15.196
91–100	189	10.578±0.100	1.381	7.768–15.054

TABLE 46. Weights of *D. exulans* during Last 20 Days before Departure

Days before Departure	N	$\bar{x}$ and SE, kg	SD, kg	Range kg
1	19	8.930±0.241	1.053	7.116–10.773
2	18	9.103±0.246	1.045	7.201–10.915
3	19	9.154±0.258	1.125	7.286–11.156
4	19	9.154±0.237	1.035	7.456–11.145
5	19	9.348±0.243	1.058	7.484–11.198
6	18	9.403±0.248	1.052	7.626–11.411
7	19	9.632±0.267	1.164	7.966–11.850
8	19	9.557±0.233	1.014	7.853–11.297
9	19	9.603±0.230	1.004	7.952–11.468
10	19	9.706±0.249	1.088	7.080–11.822
11	19	9.810±0.263	1.145	7.179–12.134
12	19	9.885±0.274	1.196	7.278–12.247
13	19	9.960±0.259	1.127	7.448–12.843
14	19	10.049±0.243	1.058	7.703–12.049
15	19	10.139±0.242	1.056	8.618–11.879
16	19	10.243±0.278	1.211	8.788–12.417
17	19	10.363±0.255	1.112	8.874–12.361
18	19	10.363±0.240	1.045	9.029–12.162
19	19	10.348±0.282	0.947	8.817–12.304
20	19	10.527±0.270	1.178	7.910–12.502

albatrosses breed on islands colder than South Georgia, nor royal albatrosses on islands colder than Campbell Island.

At first sight it would seem that the size of the great albatrosses alone might account for the long fledging period and that this in turn might be sufficient to account for biennial breeding. However, recent evidence [*Tickell and Pinder*, 1967] indicates that one of the smaller albatrosses, *chrysostoma*, is also a biennial breeder, although the closely related *melanophrys* is not.

Lack [1947, 1948] has argued that clutch size, which is related to the number of chicks that a pair can rear successfully, is controlled by the availability of food. *Wynne-Edwards* [1962] thinks, however, that such environmental pressures as food shortages result in special (epidietic) behavioral responses by the species, which in turn regulate fertility. *Ashmole* [1963] from his work on marine species at Ascension Island supports Lack in thinking that the amount of food within range of islands could operate directly to reduce fertility.

Either way, it is generally accepted that availability of food within range of the breeding grounds is important in determining the reproductive rate of a species, and biennial breeding reduces the reproductive output below that of a bird laying one egg annually. At the same time biennial breeding allows much more flexibility in response to the environment than does the annual cycle. In this respect it is necessary to emphasize that biennial breeding is related directly to and controlled by the probability of producing offspring successfully.

Physiologically the great albatrosses have reproductive systems that cycle annually and are presumably linked to the same extrinsic features of the physical environment such as climate and day length, as in other Procellariiformes such as *Puffinus tenuirostris* [*Marshall and Serventy*, 1956]. This cycle is interrupted in the great albatrosses by the demands of prolonged parental care. Presumably the stimulus of frequent chick-feeding gives rise to neurological or hormonal feedback that either inhibits the release of FSH (follicle-stimulating hormone) from the pituitary or blocks its action at the gonads, with the net result that the reproductive system fails to respond to the extrinsic environmental stimuli.

In an annual breeder the theoretical maximum number of offspring is equal to the maximum number of eggs produced (ignoring egg sterility and nestling mortality). During any two-year period the great albatrosses have an egg potential almost twice that of the maximum theoretical number of offspring, or, in other words, though both types have the same potential fertility, the biennial breeders' productiveness is approximately half that of the annual breeders.

A biennial breeding cycle superimposed on an annual reproductive cycle permits breeding birds of one demipopulation to move to the other. This balances the numbers of birds breeding each year, and the

TABLE 47. Amounts of Food Received by *D. exulans* Chicks Each Month at Bird Island

Month	N	$\bar{x}$ and SE, kg	SD	Range, kg	t
April	19	6.529±0.273	1.192	4.536– 8.477	
					4.603*
May	16	8.908±0.461	1.843	5.982–12.162	
					0.077
June	21	8.959±0.457	2.093	3.530–12.573	
					0.188
July	21	8.842±0.424	1.942	6.308–13.863	
					0.737
Aug.	24	9.287±0.425	2.084	5.145–15.096	
					0.796
Sept.	20	8.746±0.546	2.441	3.785–13.395	
					0.539
Oct.	20	9.157±0.532	2.381	4.791–13.268	
					4.780*
Nov.	18	5.840±0.429	1.821	2.381–10.674	

* Difference significant, $p < 0.001$.

TABLE 48. Predicted Fluctuation in Numbers of Eggs Laid Each Year on 'Gony Ridge', Bird Island, after Removal of All Eggs in 1962–1963 and Assuming That 20% of Breeding Pairs Are Unsuccessful in Year x and Breed Again in Year $x + 1$

Year x	Number Which Bred Successfully in Year $x - 2$		Number Which Bred Unsuccessfully in Year $x - 1$		Number of Eggs in Year x
1962–1963					266*
1963–1964	(160)	+	266	=	426
1964–1965	Nil	+	85	=	85
1965–1966	341	+	17	=	358
1966–1967	68	+	72	=	140
1967–1968	286	+	28	=	314
1968–1969	112	+	63	=	175
1969–1970	251	+	35	=	286
1970–1971	140	+	57	=	197
1971–1972	229	+	39	=	268
1972–1973	158	+	54	=	212
1973–1974	214	+	42	=	256
1974–1975	170	+	51	=	221
1975–1976	205	+	44	=	249
1976–1977	177	+	50	=	227
1977–1978	199	+	45	=	244
1978–1979	182	+	49	=	231
1979–1980	195	+	46	=	241
1980–1981	185	+	48	=	233
1981–1982	193	+	47	=	240
1982–1983	186	+	48	=	234
1983–1984	192	+	47	=	239

* All eggs removed within two days of laying.

mechanism that maintains it depends on the fact that egg and chick loss always occurs. The magnitude of such mortality determines the size of the demipopulation relative to the total population and the rate at which demipopulations are restored to equilibrium.

Theoretically, if all eggs developed into advanced fledglings, exactly half the total breeding population would breed in each year and, as described above, no exchange between demipopulations would occur. At the other extreme, if all eggs are destroyed in one season, then the entire population breeds together the next year. In undisturbed circumstances the number of birds breeding each season will stabilize somewhere between these two extremes in direct relationship to the magnitude of the mean annual loss of eggs and young chicks. The higher this mortality is, the larger the demipopulation in relation to the total population. This characteristic is also important in that after catastrophic egg loss it will, by a series of decreasing oscillations, bring the demipopulations back to their previous levels (Figure 46, Table 48).

Acknowledgments. I am grateful to W. J. L. Sladen for his generous encouragement throughout the study and to F. B. Bang for providing facilities for writing in the Department of Pathobiology at the Johns Hopkins University.

The paper was read by Helen Abbey, O. L. Austin, Jr., E. Gould, C. Robbins, W. J. L. Sladen, and C. Southwick, to all of whom I am indebted for helpful criticism and discussion.

I am grateful to the late Professor A. J. Marshall for examining histological material and to J. D. Gibson of the New South Wales Albatross Study Group who allowed me to use his unpublished method of recording plumages. Both the British Antarctic Survey and the USARP Bird Banding Program generously supplied the bands essential to the study.

At Bird Island I enjoyed the friendship and willing cooperation of W. N. Bonner, P. A. Cordall, H. B. Clagg, H. Dollman, C. F. leFeuvre, R. Pinder, and R. W. Vaughan, all of whom assisted in the field work.

Except for my first expedition to South Georgia in 1958–1959, the work described here was financed by grants from the National Science Foundation (G19590 and G23943). The Falkland Islands government and British Antarctic Survey provided logistic support for four expeditions, and I wish to thank Sir Edwin Arrowsmith, K.C.M.G., and Sir Vivian Fuchs for this cooperation.

REFERENCES

Ashmole, N. P., Regulation of numbers of tropical oceanic birds, *Ibis*, 103b, 458–473, 1963.

Bell, L. C., Notes on the birds of the Chatham Islands, *Notornis*, 6, 65–68, 1955.

Bierman, W. H., and K. H. Voous, *Birds Observed and Collected during the Whaling Expedition on the* Willem Barendz *in the Antarctic 1946–1947 and 1947–1948*, Brill, Leiden, 123 pp., 1950.

Boeson, B. W., Sea birds found dead in N. Z. in 1962, *Notornis*, 10, 404–411, 1964.

Boeson, B. W., Sea birds found dead in N. Z. in 1963, *Notornis*, 12, 169–175, 1965.

Bourne, W. R. P., Observations of sea birds, *Sea Swallows, 18*, 9–36, 1966.

Bourne, W. R. P., Vagrancy in petrels, *Ibis, 109*, 141–167, 1967.

Bull, P. C., and B. W. Boeson, Sea birds found dead in N. Z. in 1961, *Notornis, 10*, 265–277, 1963.

Buller, W., On the wandering albatross; with an exhibition of specimens, and the determination of a new species, (*Diomedea regia*), *Trans. N. Z. Inst., 23*, 230–235, 1891.

Buller, W., Supplement to the "Birds of New Zealand," London, 1905.

Carrick, R., K. Keith, and A. M. Gwynn, Fact and fiction on the breeding of the wandering albatross, *Nature, 188*, 112–114, 1960.

Dabbene, R., Una nueva especie de albatros y dos especies nuevos para la Argentina, *Physis, 8*, 563, 1927.

Dawson, E. W., The birds of the Chatham Islands 1954 expedition, *Notornis, 6*, 78–82, 1955.

Dixon, C. C., Some observations on the albatrosses and other birds of the southern oceans, *Trans. Roy. Canad. Inst., 19*, 117–139, 1933.

Falla, R. A., Distribution of the royal albatrosses in the New Zealand region, *Canterbury Mus. Rec., 4*, 213–217, 1938.

Gibson, J. D., Third report of the New South Wales Albatross Study Group, *Emu, 63*, 215–223, 1963.

Gibson, J. D., The wandering albatross, *Diomedea exulans*, results of banding and observations in New South Wales coastal waters and the Tasman Sea, *Notornis*, in press, 1967.

Gibson, J. D., and A. R. Sefton, Notes on some albatrosses of coastal new South Wales, *Emu*, 55, 44–48, 1955.

Gibson, J. D., and A. R. Sefton, First report of the New South Wales Albatross Study Group, *Emu*, 59, 72–82, 1959.

Gibson, J. D., and A. R. Sefton, Second report of the New South Wales Albatross Study Group, *Emu*, 60, 125–130, 1960.

Greene, S. W., The vascular flora of South Georgia, *Brit. Antarc. Surv., Sci. Rept. 45*, 58 pp., 1964.

Hardy, A. C., and E. R. Gunther, The plankton of the South Georgia whaling grounds and adjacent waters, 1926–27, *"Discovery" Rept.*, 11, 1–456, 1935.

Holgersen, H., Antarctic and subantarctic birds, *Norweg. Antarctic Expeds. 1927–28 et seq., Sci. Res.*, 23, Oslo, 99 pp., 1945.

Holgersen, H., Ornithology of the "Brategg" Expedition 1947–48, *Sci. Res.*, 4, Grieg., Bergen, 1957.

Idrac, P., Experimental study of the "soaring" of albatrosses, *Nature, 115*, 1924.

Jameson, W., *The Wandering Albatross*, London, 99 pp., 1958.

Kinsky, F. C., *9th Annual Report of the Banding Committee for the Year Ending 31st March 1959*, Ornithological Soc. of N. Z. Inc., Wellington, N. Z., 1959.

Kinsky, F. C., *11th Annual Report of the Banding Committee for the Year Ending 31st March 1961*, Ornithological Soc. of N. Z. Inc., Wellington, N. Z., 1961.

Kinsky, F. C., *13th Annual Report of the New Zealand Bird Banding Scheme for the Year Ending 31st March 1963*, Dominion Museum, Wellington, N. Z., 1963.

Lack, D., The significance of clutch size, *Ibis*, 89, 302–352, 1947.

Lack., D., The significance of clutch size, *Ibis*, 90, 25–45, 1948.

Lack, D., *The Natural Regulation of Animal Numbers*, Clarendon Press, Oxford, 1954.

Larsen, C. A., Nogle optegnelser af sael-og hvalganger "Jasons" reise i Sydishavet 1893 Øg 94, *Norsk. G. S. Aarbog*, 5, Kristiania, 1894.

Leslie, P. H., The intrinsic rate of increase and the overlap of successive generations in a population of guillemots (*Uria aalge* Pont), *J. Anim. Ecol.*, 35, 291–301, 1966.

Lockley, R. M., *Shearwaters*, London, 1942.

Lockley, R. M., and R. Russell, *Bird Ringing*, London, 1953.

Marshall, A. J., Reproduction, in *Biology and Comparative Physiology of Birds*, Vol. 2, pp. 169–213, edited by A. J. Marshall, Academic Press, London, 1961.

Marshall, A. J., and D. L. Serventy, The breeding cycle of the short-tailed shearwater, *Puffinus tenuirostris* (Temminck), in relation to trans-equatorial migration and its environment, *Proc. Zool. Soc. London*, 127, 489–510, 1956.

Matthews, G. V. T., Some aspects of incubation in the Manx shearwater, *Procellaria puffinus*, with particular reference to chilling resistance in the embryo, *Ibis*, 96, 432–440, 1954.

Matthews, L. H., The birds of South Georgia, *"Discovery" Rept.*, 1, 561–592, 1929.

Matthews, L. H., *Wandering Albatross*, Macgibbon and Key, London, 1951.

Mayr, E., *Animal Species and Evolution*, Cambridge University Press, London, 1963.

Menegaux, A., Reprise d'un albatross bague, *Rev. Franc. d'Ornith.*, 5, 64, 1917.

Moreau, R. E., Ecological changes in the Palearctic region since the Pliocene, *Proc. Zool. Soc. London*, 125, 253–295, 1955.

Morris, R. O., Bird life in N.W. South Georgia, *Sea Swallow*, 15, 43–49, 1962.

Murphy, R. C., *Oceanic Birds of South America*, Am. Mus. Natl. Hist. Vols. 1 & 2, New York, 1254 pp. 1936.

Murray, M. D., The wandering albatross, a brief review of studies in progress, *Austr. Natl. Hist.*, 14, 75–78, 1962.

Oliver, W. R. B., *New Zealand Birds*, 661 pp., Reed, Wellington, N. Z., 1955.

Van Oordt, G. J., and J. P. Kruijt, Birds observed on a voyage in the South Atlantic and southern oceans in 1951–52, *Ardea*, 42, 245–280, 1954.

Orlando, C., Cattura di un Albatros urlatore (*Diomedea exulans exulans*, Linnaeus) in Sicilia, *Rev. Ital. di Ornith.*, 28, 101–113, 1958.

Palmer, R. S., *Egg Profiles*, Am. Ornith. Union, Pullman, Washington, 1956.

Paulian, P., Pinnipedes, cetaces, oiseaux des Îles Kerguelen et Amsterdam, *Mem. de l'Inst. sci. de Madagascar*, Series A 8, 111–234, 1953.

Pinder, R., The cape pigeon, *Daption capensis* Linnaeus, at Signy Island, South Orkney Islands, *Bull. Brit. Antarct. Surv.*, 8, 19–47, 1966.

Rand, R. W., Notes on the birds of Marion Island, *Ibis*, 96, 173–206, 1954.

Rankin, N., *Antarctic Isle*, Collins, London, 1951.

Rice, D. W., and K. W. Kenyon, Breeding cycles and behavior of Laysan and black-footed albatrosses, *Auk*, 79, 517–567, 1962.

Richards, P. A., and W. L. N. Tickell, A comparison of weather at King Edward Point and Bird Island, South Georgia, *Bull. Brit. Antarct. Surv.*, in press, 1967.

Richdale, L. E., A royal albatross nesting on the Otago Peninsula, New Zealand, *Emu*, 38, 467–488, 1939.

Richdale, L. E., The pre-egg stage in Buller's Mollymauk, *Biol. Monographs*, 2, Dunedin, N. Z., 1949.

Richdale, L. E., The pre-egg stage in albatrosses, *Biol. Monographs*, 3, Dunedin, N. Z., 1950.

Richdale, L. E., 1951.

Richdale, L. E., The post-egg stage in albatrosses, *Biol. Monographs*, 4, Dunedin, N. Z., 1952.

Robertson, C. J. R., *14th Annual Report of the New Zealand Bird Banding Scheme for the Year Ending 31st March 1964*, Dominion Museum, Wellington, N. Z., 1964.

Routh, M., Ornithological observations in the antarctic seas, 1946–47, *Ibis*, 91, 577–606, 1949.

Siple, P. A., and A. A. Lindsey, Ornithology of the second Byrd antarctic expedition, *Auk*, 54, 147–159, 1937.

Sladen, W. J. L., and W. L. N. Tickell, Antarctic bird-banding by the Falkland Islands Dependencies Survey 1945–57, *Bird Banding*, 29, 1–26, 1958.

Snyder, D. E., Correcting an old albatross error, *Auk*, 75, 478–479, 1958.

Sorensen, J. H., *The Royal Albatross*, Cape Expdn. Sci. Res. of the N. Z. sub-antarctic expdn., 1941–45, Bull. 2, 1950.

Sorensen, J. H., Royal albatross A-99 (*Diomedea epomophord epomophora*), *Notornis*, 6, 25–27, 1954.

Stonehouse, B., The king penguin *Aptenodytes Patagonica* of South Georgia, *Falkland Is. Dependencies Surv., Sci. Rept.,* 23, London, 1960.

Swales, M. K., Sea birds of Gough Island, *Ibis, 107,* 215–299, 1965.

Tickell, W. L. N., A new method of colour marking petrels and albatrosses, *The Ring, 22,* 201–203, 1960.

Tickell, W. L. N., Ornithological investigations at South Georgia 1960–62, *Polar Record, 11,* 282–283, 1962*a*.

Tickell, W. L. N., The Dove Prion, *Pachyptila desolata* Gmelin, *Falkland Is. Dependencies Surv. Sci. Rept.,* 33, London, 1962*b*.

Tickell, W. L. N., and P. A. Cordall, The South Georgia biological expedition, 1958–59, *Polar Record, 10,* 145–146, 1960.

Tickell, W. L. N., and J. D. Gibson, Movements of wandering albatrosses *Diomedea exulans, Emu,* in press, 1967.

Tickell, W. L. N., and R. Pinder, Breeding frequencies in the albatrosses *Diomedea melanophris* and *D. Chrysostoma, Nature, 213,* 315–316, 1967.

Tickell, W. L. N., R. Pinder, and H. B. Clagg, Biological studies on Bird Island, South Georgia, 1962–64, *Polar Record, 12,* 601–602, 1965.

Tickell, W. L. N., and P. A. Richards, Earth temperatures at Bird Island, South Georgia, *Bull. Brit. Antarct. Surv.,* in press, 1967.

Tilman, H. W., *Geographical Journal, 127,* 310–316, 1961.

Tuck, L. M., The Murres, *Canad. Wildlife Serv., Bull.* 1, 1960.

Turbott, E. G., Antipodes penguins were not amused, *The Weekly News,* Auckland, Jan. 24, 1951.

Verrill, G. E., On some birds and eggs collected by Mr. Geo. Comer at Gough Island Kerguelen Island and the island of South Georgia, *Conn. Acad. of Arts and Sci., Trans., 9,* 430–478, 1895.

Wace, N. M., The botany of the southern oceanic islands, *in* A discussion on the biology of the southern cold temperate zone, *Proc. Roy. Soc. (London),* Ser. B, *152,* 475–490, 1960.

Westerskov, K., The nesting habitat of the royal albatross on Campbell Island, *Proc. N. Z. Ecol. Soc., 6,* 16–20, 1959.

Westerskov, K., Field identification and sex determination of the royal albatross, *Notornis, 9,* 1–6, 1960*a*.

Westerskov, K., Birds of Campbell Island, *Wildlife Publication 61,* N. Z. Dept. of Int. Affairs, Wellington, N. Z., 1960*b*.

Westerskov, K., History of discovery and taxonomic status of the royal albatross, *Emu, 61,* 153–170, 1961.

Westerskov, K., Ecological factors affecting the distribution of a nesting royal albatross population, *Proc. XIIIth. Int. Ornith. Congr.,* 795–811, 1963.

Wilkins, G. H., Report on the birds collected during the voyage of the "Quest" (Shackleton-Rowett Expedition) to the southern Atlantic, *Ibis, 5,* 474–511, 1923.

Wynne-Edwards, V. C., *Animal Dispersion in Relation to Social Behavior,* Oliver & Boyd, Edinburgh, 1962.

THE AVIFAUNA OF HASWELL ISLAND, ANTARCTICA

MADISON E. PRYOR

Institute of Polar Studies, Ohio State University, Columbus[1]

Abstract. Seven species of birds nest on Haswell Island. During the breeding season of 1962 population numbers varied from 63 for the south polar skua to some 35,000 for the Adélie penguin. Approximate numbers of other species in the avifauna were: silver-gray fulmar 8,000, antarctic petrel 1,000, pintado petrel 750, snow petrel 700, and Wilson's storm petrel 500. Except for the south polar skua, population figures for all species breeding in the area were considerably larger in 1962 than estimates made in previous years.

Adélie penguins, south polar skuas, silver-gray fulmars, and Wilson's petrels also nested on other islands in the Haswell group. A colony of 18,000 emperor penguins bred on the sea ice just east of Haswell Island. Nesting sites for all species were confined to areas on or near the offshore islets. Ice-free nunataks on the continent in the immediate vicinity of Mirnyy station were not inhabited.

The summer breeding season of 1962 began October 7 with the arrival of the silver-gray fulmar, antarctic petrel, and the snow petrel. The arrival of Wilson's petrel on November 10 marked the end of the annual migration. Most summer residents left the area before the beginning of May, and large numbers of birds departing the area in early April met the southward-moving emperor penguins, then returning to start their breeding cycle.

Population numbers for some species have apparently varied somewhat in recent years. There is evidence that the numbers of Adélie penguins, silver-gray fulmars, and antarctic petrels have increased, but the 1961 and 1962 populations of south polar skuas were considerably below those of previous years. Adult mortality was negligible for all species. Brood mortality rates varied considerably among the several species and were generally higher among those that nested at lower elevations where they were exposed to the movement and accumulation of meltwater. Major factors contributing to brood mortality in the emperor penguin were inclement weather and under-nutrition. The primary and secondary effects of local climate fluctuations were of great importance to brood mortality in all species, but skua predation was of little moment.

Arrival dates for each species may vary considerably from year to year, and also between areas of equal latitude in a given year. The timing of the start of egg-laying seems far less variable between both seasons and places.

INTRODUCTION

Haswell Island lies in the Davis Sea at 66°32'S, 93°02'E (Figure 1). It is the largest and northern-most of a closely assembled group of islands just off the Queen Mary Coast (the Pravda Coast in Queen Mary Land by Russian nomenclature). The Russians include 16 islands in the group, but the number of islands appears as 14 in most other literature.

Haswell Island, which is approximately 1 km² in area and 93 meters high at the highest point, lies about 2.5 km north of the Russian coastal research facility, Mirnyy station. The other islands in the group are smaller, and together they comprise less than one-fifth the total area of Haswell Island. The summits of most of the smaller islands vary from 6 to 15 meters above sea level; Fulmar Island rises to 38 meters. The exposed cliffs of the larger islets are steep, frequently almost vertical, and the predominantly flat-topped summits bear evidence of glacial erosion. On the lee side of some islets are snow-ice fields that offer easy access from the sea ice.

The group of islets was discovered in 1912 by members of Mawson's Australasian Antarctic Expedition. During an overland journey from the western edge of the Shackleton Ice Shelf to Gaussberg, the western coastal party detoured from the heavily crevassed mainland and spent 5 days exploring, collecting, and making general observations on Haswell Island. Although work was frequently interrupted by unfavor-

---

[1] Present address: Division of Biological Sciences, Morehead State University, Morehead, Kentucky 40351.

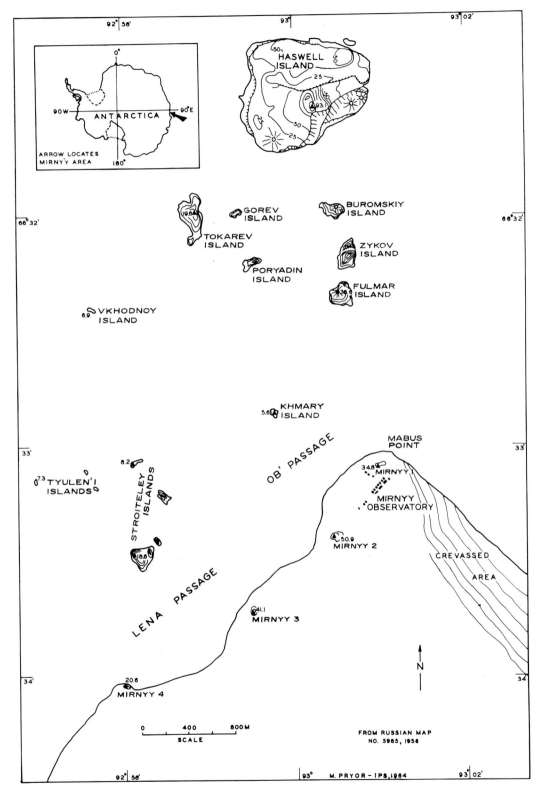

Fig. 1.   Map of Haswell Island and environs.

able weather, biological and geological data gathered by the 3-man party and published by Mawson in 1915 stimulated interest in a variety of scientific investigations applicable to the area.

Haswell Island was next visited by the United States Navy Operation Windmill in 1947. In December two ships forced their way into the ice of the Davis Sea at 92° E to obtain ground control data for the aerial photography of the previously completed Operation Highjump. A landing was made on Haswell Island, but no specific biological observations were recorded.

The Russians selected Mabus Point, an ice-free point on the continent south of the island, as the site of a research facility for the International Geophysical Year. They began building Mirnyy station in January 1956 and have maintained a full program of scientific research in the area ever since.

Its Australian discoverers assigned the name Rookery Islands to the offshore group from the large colony of emperor penguins they found near Haswell Island. In 1955 the Committee of Australia on Antarctic Names proposed that the name Haswell be extended to the entire group because of its wider familiarity. Since the beginning of the International Geophysical Year, Soviet expeditions have assigned individual names to each of the smaller islands.

The first ornithological notes on Haswell Island were made by members of Mawson's Australasian Antarctic Expedition in November 1912. A field party from the expedition's West Base on the Shackleton Ice Shelf discovered the second known emperor penguin colony near Haswell Island. The same party was also the first to find there one of the nesting grounds of the antarctic petrel, and, although it was too late in the breeding season to observe courtship and behavior, they obtained interesting and valuable information on the nest sites and breeding habitat. These data, with observations made on other birds nesting on the island, were first published by Mawson in 1915 and later reviewed by *Falla* [1937]. After these exploratory observations, research specifically associated with the avifauna of the islands was not resumed until the establishment of Mirnyy station. *Korotkevich* [1958] published the results of ornithological investigations made during the first wintering of the Soviet Antarctic Expedition. Since then Korotkevich, Makushok, and others have continued to publish data gathered on succeeding expeditions. Their observations have been rather general; published reports have been concerned primarily with enumerating species and estimating population sizes for some of the more abundant

groups; however, more detailed investigations were made in the emperor and Adélie penguin colonies.

Eight species of birds nest in the immediate vicinity of Mirnyy station. Seven species are summer residents on Haswell Island, and the emperor penguin breeds in winter on the ice east of the island. Resident and visiting species observed in the vicinity of the island during the summer of 1962–1963 are listed below. Scientific names are those in general use and trinomials have been avoided.

STATUS OF HASWELL ISLAND BIRDS

Emperor penguin, *Aptenodytes forsteri*
   Gray .... resident 10 months
Adélie penguin, *Pygoscelis adeliae*
   (Hombron & Jacquinot) .... summer resident
Chinstrap penguin, *Pygoscelis antarctica*
   (Forster) .... rare visitor
Giant petrel, *Macronectes giganteus*
   (Gmelin) .... rare visitor
Pintado petrel, *Daption capensis*
   (Linn.) .... summer resident
Antarctic petrel, *Thalassoica antarctica*
   (Gmelin) .... summer resident
Silver-gray fulmar, *Fulmarus glacialoides*
   (Smith) .... summer resident
Snow petrel, *Pagodroma nivea*
   (Forster) .... summer resident
Wilson's storm petrel, *Oceanites oceanicus*
   (Kuhl) .... summer resident
South polar skua, *Catharacta maccormicki*
   (Saunders) .... summer resident

The opportunity to study so many species within so compact an area is rare at high latitudes. Haswell Island is far enough from the mainland where man has been active to leave the avian populations relatively undisturbed, yet close enough for easy access. The area has not been affected appreciably by the encroachment of man, and the bird populations are large enough to support specialized investigations. Each species breeds in places easily accessible to the worker. With certain modifications in observational procedures and facilities for reaching the island in late summer, Haswell would afford ideal conditions for ornithological research.

The data presented in this paper were collected between November 6, 1961, and January 6, 1963, while the author was engaged in research on arthropod ecology as a U.S. exchange scientist at Mirnyy station. The proximity of the nesting grounds to study areas visited daily during the summer made it both possible and practical to maintain a program of observations on various phases of breeding cycles and population fluctuations.

The sea ice surrounding Haswell Island begins to deteriorate in mid-January, and the observer without transportation facilities for such conditions must end his research and leave the island. Consequently, data on the final stages of breeding cycles are unobtainable or, at best, incomplete. As in past years, the study reported in this paper was terminated early in January, and accurate information on the subsequent stages for some species are not available for the breeding seasons of 1961 and 1962.

This paper is presented as a more thorough and expanded analysis of data the author previously submitted in five short reports for publication in Russian. It aims (1) to supply accurate dates for some phases of the breeding cycles of the local species, (2) to present a brief résumé of population dynamics, (3) to consider certain aspects of breeding behavior, and (4) to make available to future workers the results of a small-scale bird-banding program. It also includes a fairly complete analysis of brood mortality in the emperor penguin colony.

EMPEROR PENGUIN

Two estimates for the world population of emperor penguins have been presented recently. *Budd* [1962] concluded that more than 240,000 birds bred in at least 21 "rookeries." *Korotkevich* [1963] suggested that the total population may be as high as 350,000, and as many as 29 colonies may exist in Antarctica. In his summary report Korotkevich lists a total population of 100,000 for Coulman Island (73°16′S, 169°40E). If this estimate is accurate, the Coulman Island colony is by far the largest known. The smallest colony found to date is in the Dion Islands (67°52′S, 68°43′W), where approximately 300 birds occupy a low, snow-covered shingle on one corner of the largest islet [*Stonehouse*, 1953]. These figures represent the extremes in colony size; the numbers of birds in most colonies discovered to date vary from 2,000 to 12,000. Population figures for 12 of the 19 colonies Korotkevich reviews fall within this range.

When the colony east of Haswell Island was discovered in November 1912, the only other colony known was the much smaller one at Cape Crozier (77°27′S, 169°34′E). The discoverers of the Haswell colony estimated the population at 7,500, most of which were chicks that ranged in height from 6 to 15 inches [*Mawson*, 1915]. Harrisson, the biologist

at Mawson's West Base, from verbal reports by members of the discovering party estimated the number of chicks to be 7,000 and the total number of emperors to be approximately 20,000 [*Falla*, 1937]. Although the chick/adult ratios observed in November of 1961 and 1962 are close to those shown by the 1912 estimates, the validity of calculating past population figures on the basis of later data is questionable. One must assume that factors affecting mortality then were similar to those today. Current mortality figures show that such assumptions cannot be accepted without reservations.

From data collected in two other colonies, *Budd* [1962] suggests that the combined November adult/chick population may be approximately equal to the number of breeding pairs of birds in a colony. Applying this ratio to the 1912 Haswell Island observations, Budd suggests that approximately 7,500 pairs bred there that year. Although the breeding population may well have been 15,000 in 1912, recent Haswell Island figures show that the total number of birds counted in a November census does not always reflect the number of breeding pairs in the colony. In November of 1961 and 1962 the total population varied from approximately 7,000 to 15,000, and extreme values were frequently observed on the same day. The lower value is slightly less than the number of breeding pairs noted in 1962; the higher value is much greater than the total number of pairs in the colony. As factors affecting population numbers vary from day to day and from colony to colony, a November adult/chick count is of limited value in computing breeding populations.

A population count made in the Haswell Island colony May 9, 1962, showed that the total number of birds in the colony was 18,000 ± 500. This total is slightly lower than the estimate of 20,000 *Korotkevich* [1958] reported for 1956–1957, and much higher than the 12,000–14,000 *Makushok* [1959a] reported for April 1958. Because it is quite possible that migration had not ended by April 23rd, the 1958 population may have been larger than 14,000. Recent population data summarized by *Korotkevich* [1963] for all known emperor penguin colonies show the Haswell Island colony to be the fourth largest in total numbers of birds.

March 1962. The first emperor penguins seen in the Mirnyy station area were a small group of 16 birds sighted on the sea ice approximately 5 km north-northeast of the observatory on March 26. On March 31 larger groups of 50–60 birds were wandering nearer

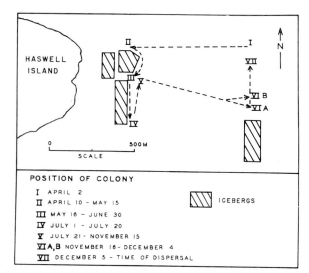

HASWELL ISLAND

500M

SCALE

POSITION OF COLONY

I   APRIL 2
II APRIL 10 - MAY 15
III MAY 16 - JUNE 30   ICEBERGS
IV JULY 1 - JULY 20
V   JULY 21 - NOVEMBER 15
VI A,B NOVEMBER 16 - DECEMBER 4
VII DECEMBER 5 - TIME OF DISPERSAL

Fig. 2. Areas occupied by the emperor penguin colony and approximate dates of local movements during the 1962 breeding season.

the shore. Many of the early arrivals were rather inactive and spent much time resting on the sea ice.

Although this study makes no attempt to consider the intricacies of navigation or homing, it is of interest to note that landmarks are abundant in the immediate vicinity of the colony site. In addition to the Haswell Islands and ice-free areas on the nearby continent, the region has a large number of grounded icebergs. The birds occupied sites near four large bergs at various stages of the breeding cycle, and their directional movements to and from the colony altered at each site. The dark summit of Haswell Island was clearly visible from a point about 8 km north-northeast of Haswell Island, although large bergs blocked the smaller islets and ice-free sections of the continent from view. At a number of other points along the migration route closer to the colony site, Haswell Island was always visible on the southwest horizon. Although the numbers and positions of icebergs change constantly in this part of the Davis Sea, several large grounded bergs and the exposed peak of Haswell Island made stable landmarks for orientation in 1962.

In addition to the occasional local shifts during the breeding season (see Figure 2), the exact location of the colony may vary considerably from year to year. Its position seems to be influenced greatly by the number and position of the icebergs near the offshore islets. *Korotkevich*'s [1959] sketch map of the 1956 colony shows that the birds remained within an irregular area roughly 1500 meters square and about 800 meters south of the 1962 sites. The icebergs that year

were smaller, more numerous, and much closer to the mainland. They completely hemmed in and shielded the 1956 breeding ground, within which the birds shifted to at least seven different sites during the season.

In 1956 Korotkevich reported that the emperor penguins first appeared off Mirnyy on March 25. In 1958 *Makushok* [1959a] first observed them there near the end of March.

April 1962. Scattered groups of birds began to consolidate on April 2 and formed a rather compact unit about 4 km north-northeast of the observatory approximately where the colony had been the preceding December (Figure 2). Most of the colony remained in that particular area for one week, but on April 6 more than 2,000 birds moved as a unit about 800 meters west of the main body and regrouped in the north side of a large iceberg near Haswell Island. The colony remained divided approximately 4 days, the smaller group occupying the site that would be used during much of the egg-laying period. On April 10 at the peak of the migration the main body of birds moved westward and joined the smaller group. What caused this local dividing and regrouping was not apparent.

After the colony had regrouped near Haswell Island, localized changes in position were noted between April 13 and 26. During this period the configuration of the colony varied from a roughly circular compact "huddle," to use *Stonehouse*'s [1953] term, to an elongate, contracting and expanding mass stretching some 200 meters west to east. These configuration changes generally coincided with local weather conditions; during periods of high winds the colony huddled compactly; in calm spells the birds spread out more loosely.

The colony seemed to become stabilized around a large iceberg east of Haswell Island on April 30. Data from other years show the colony settled down about mid-April in 1956 and on April 13, 1958.

Air temperatures recorded at Mirnyy station varied from $-29.2°$ to $-3.3°C$ during April. Average wind speed was 14.2 m/sec, and prevailing winds were from the south-southeast. Wind velocities were frequently 20% lower at the colony site than at the observatory, so that the effects of katabatic winds there were greatly reduced and sometimes negligible. Icebergs near the colony site broke the full force of prevailing winds. One large berg grounded approximately 1 km southeast of the site was a major windbreak,

Fig. 3.   Breeding site occupied by emperor penguin colony near large icebergs east of Haswell Island during most of the 1962 breeding season.

and, although a considerable amount of snow accumulated in its lee in 1962, the total accumulation was much less than Makushok recorded for 1958. Air temperatures were even more variable, being lower on the continent. Local differences in the climates of both areas were noted at all seasons, but especially in the austral spring and summer.

May 1962.   Migration reached a peak between April 5 and 15, but birds continued to arrive until May 15. The first egg was found May 6, when 5 birds from a total of 300 pairs observed were incubating. Approximately 50% were incubating on May 15, and on May 28 perhaps 80%. In the period immediately following the onset of egg-laying, accurately estimating the numbers of birds in various stages of the breeding cycle was impossible, as new arrivals were still coming in while mated females were leaving. The percentages presented above are based on random counts of incubating birds made from the periphery of the colony. Even careful movement through the massed birds would have disturbed them and contributed to egg mortality.

In the second half of May many females left the colony for feeding grounds to the northward, decreasing the population markedly. The colony then moved as a body 150–200 meters southward and regrouped between and near the walls of large icebergs, where it remained until July 1st.

Climatic data recorded at Mirnyy station for May were rather similar to those recorded in April, except for increased wind velocity. Air temperatures varied from −28.5° to −9.4°C, and the mean temperature was −15.2°C. Wind velocities greater than 40 m/sec were frequently recorded, and the average wind speed of 16.2 m/sec was the highest for the year.

In 1956 *Korotkevich* [1959] noted mass migration from April 5 through May 5; birds continued to arrive in the area until May 15. Egg-laying began May 1–2, and large numbers were laid between May 10 and 20. Females began to leave the colony May 5, but no mass departure was noted until May 15.

In 1958 *Makushok* [1959a] found the first egg May 16 and egg-laying continued through the rest of the month. By the end of May the departure of females after laying had reduced the colony population to 7,000–8,000.

June 1962.   June was the month of maximum incubation, and during the first half of the month movement and displays in the colony were at their lowest. Incubating males were extremely passive and little movement was noted in the undisturbed colony, except for infrequent stretching and inspection of eggs. Incubating birds usually huddled into a single compact group, but on calm days as many as seven individual groups were discernible.

Small groups of unattached birds frequently moved

into the close-packed masses of incubating males. They often interrupted the tranquility of the colony by battling to gain possession of temporarily misplaced eggs.

Great fluctuations in the population occurred during the last week of the month, as females began to return after June 25 and relieve incubating males. On June 30 the migration route north of the colony was crowded with birds moving both ways.

Air temperatures in June varied from −27.4° to −6.2°C and averaged −17.3°C. Prevailing winds from the south-southeast averaged 14.4 m/sec, and a maximum wind speed of 38 m/sec was recorded.

July 1962. During the first week of the month and just before hatching began, a large part of the colony moved from the seemingly protected area among the icebergs to a more exposed site 150–200 meters southward, but still near the group of bergs east of Haswell Island. There the colony regrouped into a long narrow mass, oriented north to south along the east side of a large berg. Reasons for the move were not clear, but the numbers of frozen eggs and dead chicks showed the move involved a large segment of the population. Possibly the move to a more exposed site was made to avoid being inundated by snow deposited by the varying wind patterns around the grounded bergs. That the move was not part of an established pattern was suggested by the short occupancy of the exposed site.

Hatching began July 9. Between 40 and 50% of the birds in the colony were brooding chicks on July 18. Scattered observations showed the average incubation period to be about 65 days, confirming those reported in other colonies by *Stonehouse* [1953] and *Prevost* [1961].

A considerable number of chicks (approximately 35% of total chick mortality) died during the first 3 weeks after hatching. Specific causes of death were difficult to determine, but undernutrition and harsh weather were major factors. Some were crushed and killed by adults fighting to take possession of chicks that had fallen accidentally from their parents' incubation pouches.

On July 20 the southern half of the colony moved 150–200 meters northward and rejoined the major group. After a short period of readjustment, movement within the colony slowed to the minimum expansion and contraction observed during the occupation period. The east-west oriented colony remained in this semienclosed area until mid-November.

July was the coldest month in 1962. Air temperatures varied from −35.1° to −7.8°C and the mean

temperature was −20.8°C. Winds from the south-southeast averaged 13.2 m/sec.

In 1956 the first chick was seen July 14 [*Korotkevich*, 1959]; large numbers of chicks were hatched between July 17 and August 10, and hatching continued until August 15.

In 1958 [*Makushok*, 1959a] hatching began July 10 accompanied by the return of the female birds; almost all chicks had hatched by the end of July.

August 1962. Activity increased during the month as adults moved to and from the feeding grounds to meet the increasing food needs of the growing young. The chicks remained in the adults incubation pouches approximately 45 days. The first young completely free of the pouches were seen August 23. By August 30 many young birds were noted standing in front or beside an adult. *Korotkevich* [1959] reports chicks first seen free of the pouch August 20, 1956.

Temperatures varied from −36.8° to −3.0°C during August and the mean air temperature was −17.8°C. Maximum wind speeds of 48 m/sec were recorded, but the prevailing east-southeast winds were much less violent and averaged 14.2 m/sec.

September 1962. The major change in colony structure during the month was the formation of creches. The first stages of creche formation were noted September 7, when many small groups of chicks were spread throughout the colony. Large, compact, and mobile creches were observed September 26. During the latter part of the month a great increase in both activity and noise level was noticeable. The trumpeting of the adults was almost inaudible above the almost constant "whistling" of the chicks.

A mean air temperature of −16°C was recorded for the month and winds from the south-southeast averaged 11.6 m/sec.

Korotkevich [1959] reports that in 1956 creches were formed during the first 10 days of September.

October 1962. The creches increased in size throughout the month, and the adult population fluctuated greatly. On some days almost 80% of the adult population seemed present; on other days only about 20%. A count made October 26 showed a live chick population of 6000 ± 200. At that time the young birds were in nine large creches, which simplified the counting and made a higher degree of accuracy possible. As the creches were easily set in motion, the observers had to move cautiously through occupied areas.

Air temperatures varied from −24.5° to −3.5°C during the month, but the mean temperature of

Fig. 4. Molting emperor penguin chick in area occupied by the colony until November 15, 1962. Note condition of ice surface in foreground.

−12.1°C showed a warming trend. Winds from the east-southeast averaged 12.4 m/sec, and a maximum of 34 m/sec was recorded.

November 1962. The first stages of molt were noted among the larger chicks on November 8. Signs of molt first appeared on the undersides and tips of the flippers and at the base of the tail. As the molt progressed the characteristic "whistle" of the young chicks was replaced by an immature trumpeting similar to but somewhat less melodious than that of the adults.

The colony site, which had been occupied since July 21, was abandoned November 15. Probably the move was precipitated partly by the poor state of the ice surface. Snow began to melt wherever it was stained with excreta or marked with accumulated down. Close clusters of pot holes up to 50 cm deep made it almost impossible for birds or man to move through the older colony site, shown in Figure 3. This long-distance view from the east shows the general positions of the three large bergs that gave the colony

some protection during the egg-laying and the early growth periods. A molting chick is shown in Figure 4. The picture was made December 8 approximately 3 weeks after the older site was abandoned. Pot holes can be seen in the foreground, and one part of the colony in its new site appears in the background.

For a short time the birds remained in two large groups some 700–800 meters east of the older site. According to *Falla* [1937], in late November 1912 the colony was divided into three large groups, each of which occupied an area of 3 to 4 acres. Pictures accompanying Falla's text suggest that the birds may have occupied the same general sections of sea ice in November 1912 as in 1962.

The first signs of molt in adults were noted November 28 in a very few individuals.

The population fluctuated greatly during the month. On some days fewer than 2,000 adults were present, but at other times at least half the breeding adult birds could be seen. Maximum and minimum populations were often observed during a single 24-hour period. The migration route north of the colony was often crowded at midday, and activity usually decreased in the late evening hours.

Air temperatures increased greatly during November, varying from −14.1° to +3.6°C, with a mean of −6.4°C. On days when temperatures were near 0°C, rapid melting occurred in many parts of the colony. Winds from the east-southeast averaged 9.6 m/sec.

In 1956 [*Korotkevich*, 1959] chick molt began in late November or early December. In 1958 *Makushok* [1959b] first noted chick molt November 10 and adult molt about November 23.

December 1962. The adult population fluctuated widely in December as some birds began to use tide cracks north of Haswell Island to reach open water. Observations from the summit of the island showed that birds traveling over the sea ice to open water probably found places to feed within 25 km of the colony, which reports from ships substantiated.

On December 4 the two groups of birds rejoined on the sea ice approximately where the colony had been the previous December and where it now remained until it dispersed.

Immature birds began to leave the area December 7, when a few of the largest birds in most advanced molt were seen moving northward. On December 8 seven immature birds from the preceding season, easily recognized by their brownish backs and white throats, were seen in the vicinity of the 1962 colony.

Large numbers of adults and immatures began to leave the area December 25. Most of the young were then in full juvenile plumage, but a few were still molting. The adults were in various stages of molt, and some appeared quite ragged as they left. The departure route was almost due north as far as it could be traced, about 12 km. In 1956 chicks began to leave the colony December 20 [*Korotkevich*, 1959].

Air temperatures varied during the month from $-11.7°$ to $+4.3°C$ and the mean temperature was $-3.4°C$. Winds from the east-southeast averaged 10.7 m/sec.

January 1963. On January 3 a group of 23 downy chicks was seen on the sea ice southwest of Haswell Island. There were no adult birds in the group, and the chicks were last seen moving toward the northwest. All other adults and immatures went due north as they left the colony.

Remnants of the colony were still in the vicinity on January 6, when I left Mirnyy station. Observations in January 1962 showed small numbers of adult and immature birds present until the last week of the month. These dates parallel those Korotkevich recorded in 1956.

In 1958 *Makushok* [1959a] took an albino chick from the colony to the Mirnyy station for detailed study. The chick was covered with silky down and the beak, claws, and palate were white; the skin of the feet was rose-colored.

In 1962 two chicks were seen with a recurved maxilla that moved in horizontal apposition to the mandible. One molting chick had the head completely bare, and another chick approximately 4 months old had neither down nor feathers on the lower part of the abdomen. Both these chicks were smaller than normal, and neither survived.

Chronological Résumé, 1962

Arrival of first emperor penguins in general area	March 26
First indication of colony formation	April 2
First egg observed	May 6
First chick hatched	July 9
First chick free from incubation pouch	August 23
First creche formed	September 7
First indication of chick molt	November 8
First indication of adult molt	November 28
First return of juvenile birds	December 8
Beginning of colony dispersal	December 25
Breeding grounds abandoned	*January 6, 1963

* Major part of colony had left by this date, though a few scattered groups still remained when observations ended January 6, 1963.

Brood Mortality

Data collected in the emperor penguin colony from early May 1962 to the end of the first week of January 1963 showed a total brood mortality of approximately 20%. This figure was obtained by collecting eggs and dead chicks (cumulative census) and expressing the total loss for the year as a percentage of the number of eggs produced. Total mortality calculated from direct census figures (number of live chicks in December subtracted from the total number of eggs in May) was approximately 21%. Neither of the two methods used to determine mortality is free from error, and the validity of each depends on the accuracy with which the observer calculates the number of eggs produced. In a most thorough program on the ecology of the emperor penguin, *Prevost* [1961] showed mortality figures computed by cumulative census to be much more reliable than those collected by direct census. For this paper I have drawn heavily upon the collection methods and statistical techniques Prevost developed. Although the Haswell Island colony was not visited daily, the frequency of visits and the prevailing climatic conditions allowed enough data to be collected for formulating statistically significant conclusions.

No specific studies of brood mortality were made previously in the Haswell Island colony, but *Makushok's* [1959] estimates of egg and chick mortality in 1958 and *Korotkevich's* [1958] estimates for 1956 furnish approximate mortality figures for two other years. Makushok's estimates of observed egg and chick loss indicate that brood mortality may have been as low as 20% in 1958. This value does not account for an unknown quantity of chicks presumably covered by snow during storm periods, but, even if half the dead chicks were covered and not counted, brood mortality did not exceed 30%. Approximations for egg and chick losses during the 1956 breeding season [*Korotkevich*, 1959] show total brood mortality that year may have been about 48%.

In 1962 the colony was visited almost every second day, and mortality data were gathered on each visit. The two values calculated, 20.1% by cumulative census and 21.3% by direct census, show that brood mortality for the 1962 season was extremely low. The significance of this low value becomes more apparent when it is compared with mortality rates computed or estimated for other colonies.

Budd [1962] analyzes mortality data gathered from four colonies in East Antarctica between 57° and 65°

east longitude. Mortality rates (extrapolated from direct census results) varied from a low of 23% at Kloa Point (66°38'S, 57°19'E) to a high of 73–90% at Fold Island (67°20'S, 59°23'E). Data collected at the Auster colony (63°23'S, 64°02'E) showed a mortality of 43% for chicks 16 weeks old in 1959 and of 39% for chicks 15 weeks old in 1960. In the colony at Taylor Glacier (67°28'S, 60°53'E) in 1957, 1959, and 1960, mortality may have varied from 29 to 74%. The high value, 74%, was recorded for chicks 11 weeks of age, and Budd suggests that this figure may be an underestimate resulting from an early egg count.

Thus mortality rates in the Auster colony appear to be significantly lower than those at Fold Island or Taylor Glacier. Except at Kloa Point, mortality was relatively high in other colonies near Mawson station. The available data suggest that colony size may influence mortality, the rates being somewhat lower in large colonies.

In his re-evaluation of data collected at Cape Crozier (77°29'S, 169°34'E) in 1902, 1903, 1957, and 1958, *Budd* [1962] shows that mortality there has varied from 15 to 100%. These rates, calculated from censuses of broods 3 to 13 weeks old, are based on the assumption that 90% of the females in the colony produced eggs. The exact numbers of birds in the colony in those years are not known, and Budd calculated values for probable minimum and maximum populations of 560 and 1,000 pairs. Percentages based on the minimum population vary from 15 to 100%; those based on the maximum population vary from 53 to 100%.

At Cape Crozier, *Caughley* [1960] found brood mortality may have been as low as 16% for chicks 13 weeks old in November 1958. He notes that a change in the colony's location provided a more favorable environment, which may have accounted for the apparent decrease in brood mortality.

More recent data suggest that population fluctuations and mortality rates for the Cape Crozier colony should be re-examined. An individual count made October 17, 1962, showed a live chick population of 1,120 (Penney, personal communication). As this total is almost 3 times greater than previous estimates of a surviving brood population, it raises a number of questions: (1) Has the Cape Crozier population been consistently underestimated? (2) Has mortality been overestimated? (3) Has a more recent change in the location of the colony contributed to increased survival? (4) Is the 1962 live chick population a one-season phenomenon? Estimates of brood mortality in this colony have

been highly variable; further studies are needed to confirm them.

In 1948 and 1949 in the Dion Islands (67°52'S, 68°43'W), *Stonehouse* [1953] found that egg and chick mortality was considerably less than the 77% recorded at Cape Crozier. Up to the time his observations terminated August 15, 1949, mortality was no higher than 10%. The oldest chicks were then 2 weeks old, and we may assume that mortality increased in the following months. Assuming the same number of eggs were laid in 1948 as in 1949, a live chick count made in the Dion Islands colony in October 1948 suggested a mortality of about 50% for a brood 10 weeks of age, but approximations based on data from overlapping years (i.e. 1948 chick count and 1949 egg count) are highly unreliable, and the 1948 brood mortality may have been greatly overestimated. *Stonehouse* [1953, p. 20] comments on the 1948 mortality figure: "it is difficult to imagine the reasons for so high a death rate between August and October in any year." The Dion colony is very small, and it may be more difficult for so few birds to protect eggs and young against high winds and low temperatures. If this is true, brood mortality in the Dion Islands may be consistently higher than that in larger colonies.

The most complete study of the ecology of the emperor penguin was made by Prevost on the Géologie Archipelago (66°40'S, 140°01'E) in the Adélie Coast in 1952 and 1956. His results, published in 1961, include an excellent analysis of egg and chick mortality rates and causes. Daily visits to the colony allowed Prevost to tabulate mortality monthly throughout the breeding cycle. Figured cumulatively, total brood mortality at Géologie Archipelago was approximately 26.6% in 1952 and 24.9% in 1956. Prevost (personal communication, 1964) believes the 1952 figure may have been closer to 30%. On the basis of direct census results, *Budd* [1962] indicates that these same data from the Géologie Archipelago show mortality rates of 31% for 1952 and 33% for 1956. Although values computed each year from direct and cumulative census data differ slightly, brood mortality on the Géologie Archipelago has manifestly been considerably lower than in most other colonies.

From a live-chick count of 750 made by Sapin-Jaloustre in October 1950 and other correlative data collected by Cendron in June 1951, *Budd* [1962] calculated a mortality rate of 83% for the 1950 breeding season on the Géologie Archipelago. As this high mortality rate is for chicks 13 weeks of age, he suggests it as a possible minimum value. The figure seems much too high for the area and, if valid, must have resulted from some catastrophe earlier in the breeding cycle. The validity of the estimate is largely dependent on the accuracy of the 1950 chick count. Prevost (personal communication, 1964) believes the presence of such a small number of chicks can be explained in two ways: (1) part of the colony may have been carried to sea when the shore ice broke up, or (2) some members of the colony may have been hidden in a nearby glacial recess, where Prevost found them in 1952. He added that Sapin-Jaloustre agrees to the likelihood of either of the two possibilities.

Data collected through October of 1962 suggested total brood mortality for that year would probably equal or be slightly higher than that Prevost computed for 1952 and 1956 (Arnaud, personal communication, November 1962). A final analysis of Arnaud's data showed a total brood mortality of 24% for 1962.

Egg loss in the Haswell Island colony was relatively low in 1962. From the beginning of May to the end of August, 673 eggs (approximately 9% of the total produced) were frozen, broken, or otherwise destroyed. Although no studies were made to determine specific causes of egg mortality, at least 25% of total egg losses occurred when incubating birds were in, or moving into, close huddle formations. Eggs accidentally dropped in the crowded huddle are not easily retrieved and are soon broken or frozen.

Data collected in August 1962 show that approximately 22% of total egg losses resulted from failure to hatch; whether these addled eggs were infertile or frozen earlier could not be determined. The two causes account for almost half of the total eggs lost in 1962. Causes of the remaining losses were not determined.

Previous observations at Haswell show that egg mortality was not overly high in past seasons. *Makushok* [1959] recorded an egg loss (abandoned, frozen, addled) of about 525 in 1958; *Prevost* [1961] reported that at least 500 eggs were lost during the incubation period in 1956. Approximate figures presented by *Korotkevich* [1959] show that total egg loss in 1956 may have been as high as 1,000. Calculations based on the 1962 ratio of adult birds to eggs produced suggest these approximations account for 12% of the eggs laid in 1956 and 10% in 1958, assuming an adult population of 20,000 in 1956 and 13,000 in 1958. The 1958 estimate is considerably lower than that recorded in other years. It was made on April 23, possibly before migration was completed, and may have been greater than 13,000 later.

At other colonies where cumulative records have been maintained, egg mortalities have varied greatly from year to year. On the Géologie Archipelago [*Prevost*, 1961] 4.35% of the eggs laid were lost in 1952, 11.3% in 1956, and about 7% in 1958. These losses accounted for 16.4% of the total mortality in 1952, 45.4% in 1956, and about 20% in 1958.

Egg loss accounted for almost 45% of total brood mortality in the Haswell colony in 1962, remarkably close to Makushok's figure of 48% in 1958. However, *Korotkevich*'s [1959] data show that egg loss comprised only 25% of total mortality in 1956.

In 1962 approximately 43% of total egg loss at Haswell occurred during May, the first month of incubation; later losses were 16% in June, 19% in July, and 22% in August. It is noteworthy that the colony was situated in a semi-open area and exposed to the effects of high winds in May, when average wind speeds in the Mirnyy station area were highest (16.2 m/sec). Quite possibly movements made by incubating birds to avoid wind chill contributed to the high mortality.

Monthly rates of egg loss vary so greatly from year to year and from colony to colony that they are applicable only to a particular colony in a particular year. Prevost found on the Géologie Archipelago that maximum losses may occur at either end of the incubation period; 44.9% of his total egg loss occurred in August in 1952 and 50.1% in May 1956.

With respect to chick mortality, from the onset of hatching in July to the start of colony dispersal in late December 1962, 833 emperor chicks (12.2% of the total eggs hatched) perished at Haswell. *Makushok*'s [1959] figures indicate that approximately 12% of the chicks hatched here perished in 1958; this figure was calculated from total population estimates and observed egg and chick losses; as it does not account for an unknown number of chicks covered by snow, chick mortality at Haswell may have been nearer 25% in 1958. *Korotkevich*'s [1959] averages for the 1958 breeding season suggest that about 40% of the chicks hatched died from a variety of causes.

On the Géologie Archipelago chick mortality has been less variable. *Prevost* [1961] reports that chick mortality expressed as a percentage of eggs hatched was 23.2% in 1952, 15.3% in 1956, and approximately 27% in 1958. At both Géologie and Haswell, 60 to 85% of the chicks hatched in recent years have survived to the first stages of colony dispersal, considerably more than in most other colonies.

Chick mortality at Haswell in 1962 was highest during July and September and almost negligible in December. Approximate monthly mortality rates were: July 35%, August 15%, September 30%, October 10%, November 9%, December 1%. No effort was made to determine specific causes of monthly fluctuations, but the high rate for July resulted directly from activity accompanying a local change in colony orientation. On July 20 approximately 10 days after hatching started, birds in a southern arm of the colony moved 150–200 meters northward and rejoined the major part of the colony. The area vacated by the northward-moving adult birds was littered with approximately 200 dead chicks. Monthly mortality rates were not compiled at Haswell previously, but a graph prepared by *Korotkevich* [1959] gives some information on general mortality trends in 1956. A maximum live chick population of approximately 7,000 the third week of August decreased to about 5,000 by the end of the first week in September. Just prior to colony dispersal, which began during the third week of December, the area contained about 4,000 chicks. Thus in approximately three weeks (mid-August through the first week in September) almost 2,000 chicks perished in the colony, or about 65% of total chick mortality. The gradual decrease in live chick population from mid-September to mid-December suggests that the remaining 35% mortality was spread evenly throughout this 3-month period.

Chick loss at Haswell was 55.3% of the total mortality in 1962. Korotkevich's figures show chick mortality was about 75% of the total mortality in 1956. Makushok's averages show that chick loss probably was 50–70% of the total in 1958. Chick loss in terms of total brood mortality has been higher at Géologie Archipelago: 83.6% in 1952, 54.6% in 1956, and approximately 80% in 1958 [*Prevost*, 1961]. Arnaud's figures show chick loss was approximately 50% of the total in 1962, which is close to those recorded at Géologie Archipelago in 1956 and at Haswell in 1962. Prevost (personal communication, 1964) suggests that these percentages may approximate rates for "more normal" years; the higher rates at Géologie Archipelago in 1952 and 1958 are attributable to unfavorable weather conditions. If bad weather also contributed to the higher percentages Korotkevich noted for 1956, he makes no mention of it.

Of many factors that affect brood mortality directly or indirectly, inclement weather and undernutrition are the major causes of chick loss in the Haswell colony. In 1962 the combined effects of these two factors accounted for at least 85% of total chick losses. An additional 10% were killed by falls in tide cracks or were crushed by adult birds battling for temporarily misplaced chicks.

Avian predation is not a factor affecting brood mortality in the Haswell colony. The south polar skua returns to the area in mid-October and feeds in the emperor penguin colony until other food sources are available. In this case the skua is not a predator, but a scavenger feeding on spoiled eggs and the carcasses of chicks and adults dead from other causes. The compact creches of large, healthy chicks are apparently immune to skua attack and may even offer refuge to weaker and smaller chicks. Chicks near starvation that cannot keep pace with the local creche or group movements may fall prey to the skua, but the number so dispatched appears small. Only two such incidents were observed during the whole breeding season.

Avian predation is a more important but variable factor in other colonies. *Prevost* [1961] found on the Géologie Archipelago that chick mortality from predation by the giant petrel varied from 5% in 1952 to 34.2% in 1956.

It is of interest that approximately 85% of total chick loss on the Géologie Archipelago in both 1952 and 1956 was attributed to the combined effects of blizzards, starvation, and predation. In 1962 weather and its side effects accounted for about 85% of chick loss at Haswell. Thus in areas free of predation, weather alone may directly or indirectly cause a chick mortality as high as that in colonies where both weather and predation take their toll. This suggests those chicks that do fall prey to avian predation are weak, undernourished birds that probably would have died from other causes anyhow. Although this was apparent at Haswell, *Prevost* [1961] states that the giant petrel sometimes attacks well-fed chicks on Géologie Archipelago.

The primary effects of weather on chick mortality may be exaggerated in some areas. Although many chicks perish from freezing, undernutrition when storm periods prevent adults from foraging may be equally important. This was most obvious at Haswell during the 1962 breeding season. Three rather severe storms on September 9, 15, and 18 produced no immediate increases in mortality, but for three days after two storms mortality increased daily. Autopsies of 24 dead chicks collected at random on September 21 showed the stomachs of 17 completely empty. Thus secondary effects may be as important or more important than the primary effects of inclement weather on chick mortality.

In the following summary of mortality data covering three seasons at Haswell Island the figures for 1956 and 1958 are approximations; those for 1962 represent actual counts made in the colony. Values suggested for 1956 and 1958 are based on the assumption that the ratios of adult birds to eggs produced were similar to that observed in 1962. Values for live chick population and chick mortality are from a graph prepared by *Korotkevich* [1959]; additional data on egg mortality show that these ratios were almost identical in 1956 and 1962. Computations for 1958 are based on numbers observed by *Makushok* [1959], who notes that some dead chicks were covered by snow and could not be counted. Thus two rates are computed for 1958, one on Makushok's actual count, the other assuming that only half the dead were counted.

	1956	1958		1962
Adult population	20,000	13,000	13,000	18,000
Number of eggs produced	8,350	5,400	5,400	7,485
Egg loss	1,000	525	525	673
Chick loss	3,000	525	1,150	833
Per cent total brood mortality	48	20	31	20
Chick loss as per cent mortality	75	52	69	55
Egg loss as per cent mortality	25	48	31	45
Per cent brood surviving to time of dispersal	52	80	69	80

ADÉLIE PENGUIN

The Adélie penguin is the most abundant summer resident on Haswell Island. A nest count made November 24, 1962, showed a total breeding population of 35,600 ± 500. This figure is much greater than the total adult population of 15,000 that Korotkevich reported for the 1956–1957 breeding season. Unless something greatly decreased the population immediately preceding his observations, his estimate seems extremely low. Although nesting colonies overlap on parts of the island, estimating the number of nests accurately is facilitated by topographic features that limit the breeding birds to five or six colonies. Most of these sites appeared old and long occupied, and no clear evidence indicated any recent expansion of the breeding grounds.

The biologists who visited Haswell in November 1912 estimated the Adélie penguin population at 10,000 [*Falla*, 1937], divided into rookeries of about 50 nests each. In 1962 most of the colonies contained 300–400 birds, and some large colonies on the north side of the island supported 500 to more than 1,000 birds. A few small colonies occupied by 100 birds or less were located in more isolated parts of the island. Comparison of the figures for 1912, 1956–1957, and 1962 suggests that both the total number of birds and the

sizes of occupied areas may have increased considerably in recent years. However, a map *Korotkevich* [1959] prepared shows that areas occupied in 1956 were approximately equal in size and number to those in 1962.

The first Adélie returned October 14, 1962. This bird remained at its nest site until October 17, when the population suddenly increased to approximately 100. These first arrivals went directly to nest sites at various elevations on the island, where they stood or lay down in their territories waiting for their mates to arrive. During this phase of recolonization the rookery was usually very quiet. The occupation period [*Sladen*, 1958] reached a peak on October 23, when birds were arriving at the rate of approximately 300 per hour. On October 28 a sharp decline in the number of arrivals was evident, but small groups continued to come until November 14. A steady flow of arrivals was observed only on October 23; at other times the birds came in irregularly in groups of from 2 to perhaps 150; rarely during the peak period did a bird arrive alone.

From as far as they could be seen with binoculars, the birds came across the sea ice from the north-northwest. Groups of from 2 to approximately 100 moved southward in single file and congregated near an iceberg 2 km north of Haswell Island. From that point smaller groups broke away from the massed formation and went to particular spots on the north and west sides of the island. When still about 0.5 km away, the groups often divided again as single birds or very small groups moved toward access points on the north side of the island. Birds heading for nest sites on the south and west sides of Haswell Island or on other islands nearer the mainland passed along the northwest corner of Haswell. Cliffs on the east and northeast faces of Haswell are too steep for the birds to climb, and Adélies were rarely seen on the sea ice east of the island. During the early stages of occupation most birds moved more rapidly and called more frequently as they came nearer to the island's access points.

The Adélies gained access to Haswell Island at four main points (Figure 5), each a snow-ice field of varying steepness that afforded a fairly permanent and easily negotiable route. Most of the returning birds came onto the island through a narrow ravine on the west side and then crossed a broad ice field to the central and eastern sections. Birds occupying heavily populated sites on the northern edge of the island went directly to them from the sea ice on the north

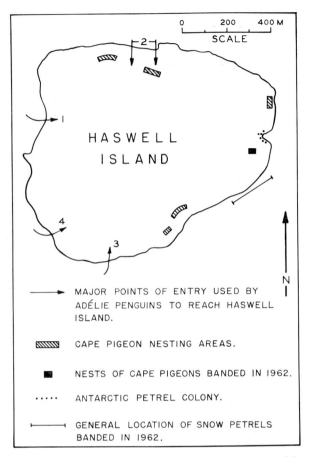

Fig. 5. Haswell Island, showing major points of entry used by the Adélie penguins and the general locations of the nesting grounds of the pintado petrels, antarctic petrels, and snow petrels in 1962.

side. Smaller numbers used two snow-ice fields on the southwestern corner of the island to reach the compact, smaller nesting grounds on the south side. Aerial observations when the island was surrounded by water showed that these four sites also provided the best facilities for landings from the sea. Birds nesting at higher elevations on the east side entered from the west and moved eastward along the frozen floor of a long saddle. When thawing filled the floor of the saddle with fresh-water drainage ponds, birds altered their routes to avoid the water covering their old trails. Figure 6 shows the saddle as it appeared in January; the Adélie nesting grounds are the light-colored guano-stained expanses at lower levels. Although paths and trails existed across the summit, many seemingly easily negotiable routes were never used. Late in the breeding season the Adélies wandered into many areas they had previously avoided, but they were never seen at high elevations on the southeast corner.

Copulation was first observed October 22, but the pair involved did not produce the first eggs. The first egg was found November 12 when one nest of a total of 182 examined had a single egg. Nests with two eggs were first found November 15. By November 24 the population was essentially one of incubating males, and a degree of quietness once again characterized the rookery. In places where stones were plentiful enough for nest building, the birds incubated in the normal prone position. In one particularly high and open area on the south side where no stones were available, the birds laid their eggs in shallow depressions in the bedrock. Four birds here were frequently seen standing and holding single eggs on their feet the way the emperor penguin does.

Females began to return to the rookery November 16 from the open water, which was then about 10–15 km away to the northward. The second relief, or return of the males, began December 5. The first eggs hatched December 15, and in this instance the incubation period was approximately 34 days. At 100 nests timed, the incubation period ranged from 33 to 36 days, which is consistent with periods previously reported by *Sladen* [1958] and others. As previously noted, no data were obtained in late January 1963, but in 1962 loose creches were first formed January 15, when small numbers of chicks began grouping together. Not until the last week in January were large formations in evidence. Observations made from the mainland indicated that much of the population left the rookery the first week of March. Fourteen adults passed the later stages of their molting on rocky sites on the mainland. The last of these left March 28, 1962.

Mortality on the island during breeding season was almost entirely confined to eggs and young birds. In none of the nesting areas were adults threatened by overhanging ice or snow, or by rockfalls or slides. Just prior to creche formation the combined egg-chick mortality was 50–55%, approximately 70% of which occurred in the egg stage. Skuas are well established in the area, but egg and chick losses from other causes provided them with so much food to scavenge that they seldom had to prey aggressively on the Adélies. The major climatic factor contributing to egg loss in the 1962–1963 season was thawing that flooded nests in badly drained areas, most evident on the north side. Egg and chick losses in exposed parts of the south side were primarily by breakage and crushing to which the general topography and the lack of nesting materials made them vulnerable.

Adélie penguins also nest on five other islands be-

Fig. 6. Adélie penguins nesting in Haswell Island saddle, January 1962.

tween Haswell and the mainland (Figure 1). Nest counts made during the third week of November 1962 showed their respective breeding populations as follows: Tokarev 5280 ± 200, Fulmar 1860 ± 50, Zykov 1720 ± 50, Buromskiy 298 ± 10, and Stroiteley Islands 228 ± 10. Tokarev, the most heavily populated of the smaller islands, is the most interesting. Approximately 425 meters long, 175 meters wide, and 19.5 meters high, nesting Adélies are concentrated on its southern end and at lower elevations on its north and west sides. Most areas the nesting birds occupy have guano deposits more than 25 cm deep. On Fulmar Island the Adélies nest mainly in the easily accessible areas on the south side. Buromskiy and Zykov are small islands easy of access, but they lack suitable territory for more than a few birds. Some 200 birds occupied the small group known as the Stroiteley Islands and built elaborate nests on the islands nearest the mainland. Molting birds took refuge on all the smaller islands late in the summer.

A few plumage abnormalities were noted during the 1962–1963 breeding season. One isabelline male (?)

occupied a nest site on the north side of Haswell. One bird with the head, neck, and undersides of the flippers glossy blue-black was seen in several places during the breeding season; between black and white areas on the breast was a horizontal band of mottled feathers about 6 cm wide. One breeding male (Figure 7) had a horizontal band of white feathers above each eye; white feathers were approximately 3 cm long and stood erect on the otherwise normally colored head. The chicks of those birds with plumage variations were all normal in color. *Korotkevich* [1959] reported a "smoke-colored" Adélie on Haswell Island in which all the usually dark colored areas were silver-gray.

In 1956 *Korotkevich* [1959] saw the first Adélie in the area October 19, and the colony was formed by October 25; the period of mating continued from October 30 until the last of November; egg-laying began November 8 and the first chick was observed December 15; molt period for chicks began near the end of January 1957 and was completed at the end of February; after the molt the young birds left. Adult birds completed their molt during the first 10 days of March

Fig. 7. Plumage abnormality in a breeding male Adélie
penguin.

and the last Adélie departed about March 20, 1957.
In 1958 [*Makushok*, 1959*b*] the Adélies returned
about October 20; the first egg was observed November 12 and toward the end of the month (approximately November 23) Adélies began to leave for feeding grounds north of the island. Observations made
by Makushok at the end of the previous breeding season showed that adults began to molt about the first
of February. The last Adélie left on March 29, 1958.

Chronological Résumé

	1962, Pryor	1956, Korotkevich
First Adélie arrived	Oct. 14	Oct. 19
Peak of occupation period	Oct. 23	Oct. 25
First copulation	Oct. 22	
First egg	Nov. 12	
First females return from feeding	Nov. 26	
First males return from feeding	Dec. 5	
First chick hatched	Dec. 15	Dec. 15
First creches formed	Jan. 15*	
First departure	Mar.1–7*	
Departure of last bird	Mar. 28*	

* Dates recorded at the end of the 1961–1962 breeding
season.

CHINSTRAP PENGUIN

One chinstrap penguin approached Haswell Island
with a group of 9 Adélies November 21, 1962. When
four of the Adélies left the group and came to the
snow ramp on the north side of Haswell, the chinstrap
came as far as the base of the ramp with them. When
the Adélies climbed onto the island, the chinstrap
moved back and forth on the sea ice north of the
island for about an hour. It then joined a large group
of female Adélies leaving for open water to the northward and remained with them until they disappeared
from view.

Korotkevich [1959] states that one specimen was
collected on Haswell Island but gives no date. Arnaud
(personal communication) saw three chinstraps in
the Adélie Coast the summer of 1961–1962. *Sladen*
[1958] summarized records of chinstraps around the
antarctic continent and suggested that their distribution is becoming circumpolar, which occasional reports
such as this seem to confirm.

GIANT PETREL

Twice, November 8 and 10, I saw a single bird flying over the north side of Haswell Island. Brownbodied, with a dull gray-white head and neck, it was
probably the same bird both times. It landed momentarily on top of the northwest cone but took flight
as I approached. *Korotkevich* [1959] states the species has been seen frequently in the Haswell area.

PINTADO PETREL (CAPE PIGEON)

The pintado petrel nests in four general areas on Haswell Island (Figure 5) from sea level to the summit,
but it seems to prefer large flat spaces and wide ledges
to more precipitous slopes. One such site on the northeast corner was approximately 4 meters above sea
level, one on the southeast corner was nearer the summit at about 60 meters. All nests were in exposed
positions that gave the birds maximum access. They
did not use inland cliffs and ledges and rarely flew
into or over the center of the island.

A single pintado petrel returned to Haswell October 22, flew directly to the north side of the island,
where it seemed to search for a particular nest site,
and then left October 26. On the morning of October 27 larger numbers of birds arrived and occupied
nest sites on the north and east sides. More birds continued to appear over the next 6 days. The nesting
territories on the south side were the last to be occupied. Birds coming in from the north flew around the
west side of the island to the south side in preference
to flying directly across the island. They seemed reluctant to fly inland and always flew around the edge of
the island. A number of nest sites occupied in 1961
were not used in 1962, when a number of new sites
were established.

Fig. 8. Breeding habitat of antarctic petrels on southeast corner of Haswell Island.

Nests were typically shallow scoops in wind-blown gravel with varying amounts of small stones and, in most instances, were near large rock outcrops. At the start of the season the birds often had to remove snow that completely covered the sites. Taking snow in the beak, they tossed it away to one side. Sometimes they seemed to be excavating nests in snow drifts, but later thawing revealed the old established nests. To reach sites under overhanging ledges the birds often tunneled through deep snow drifts. When thawing started, nests inundated by winter snow were often wet and flooded. The nests were not elaborate, but birds frequently added gravel to them while incubating. The sitting birds stretched their necks to pick up nearby pieces of gravel or small stones and quickly dropped them close to their bodies. Before the species arrived, nest sites were easily identified by the scoops surrounded by stones.

From the first day of establishment, most birds defended their territories and ejected stomach oils at the slightest provocation. They could eject the oily amber-yellow fluid accurately as far as 2 meters. Some individuals displayed shyness and timidity and left their nests when startled by minor disturbances. The departure of one bird frequently led neighboring birds to leave. Usually one member of the pair seemed more aggressive than the other. After the eggs were laid both parents became more protective, facing the intruder with wings slightly spread and ready to eject oil if need be. The defense of territory against other species was seldom necessary. On rare occasions when the more clumsy silver-gray fulmar flew accidentally

into pintado petrel territory, it was quickly driven away by the threats of the incubating birds. Skuas breeding near the pintado petrels were apparently not aggressive predators; egg shells or remains of the smaller species were never found in the skua territories.

Copulation was first observed October 28, and mating reached a peak on November 5. The courtship rituals were similar to those of the antarctic petrel, but somewhat more reserved and less intense. Head wagging was accompanied by a pleasant trilling call similar to but more melodious than that of the antarctic petrel.

After establishing the nesting site and courting for perhaps a week, each pair left the island for a period of about 2 weeks before laying. Their departure was not the mass movement made by the antarctic petrel. Some birds were still in various stages of courtship when others returned to lay. Birds began to leave November 10, and their population fluctuated greatly from day to day until November 27, when the population was at its lowest. The birds began to reoccupy their nesting sites November 28, and a marked increase was evident November 29. Six marked pairs were each absent approximately 2 weeks. Their whereabouts and activities during this time are unknown. Departing birds flew north-northeast; the return flights came from the same direction. Probably the birds spent much of their time feeding. The color of the stomach oil ejected changed from amber-yellow before departure to red-pink on return. Flights to and from the north-northeast continued throughout the incubation period.

The first egg appeared November 28, one month after the major part of the population first arrived. Most birds were incubating single white eggs by December 3. Measurements of 30 eggs showed an average egg size of 64.2 × 44.2 mm. Eggs were removed from three nests two days after they were laid, and, though each pair continued to occupy and defend its nest, none produced a second egg and each abandoned its nest approximately 6 days later. On December 25 some pairs were still courting, but more than 80% of all nests had eggs. By December 31 almost 30% of these eggs were destroyed, mostly by flooding.

A nest count made December 11 showed the 1962 breeding population to be 740 ± 50. Approximately 83% of the birds bred successfully. A section of the northeast side of Haswell was the most heavily populated. No pintado petrels were seen on or flying over the other islands or the mainland.

No young hatched before the study terminated January 6, 41 days after the first eggs were laid. *Pinder* [1966] gives the incubating period at Signy Island as 43–48 days with a mean of 45 days, which corroborates *Prevost*'s [1964] findings of 43–46 days on the Géologie Archipelago.

Observations from shore early in 1962 showed that the pintado petrels left the Mirnyy region near the end of the second week in April.

Twenty-five incubating birds were banded (as part of the USARP Bird Banding Program) from 25 different nests on the south corner of Haswell Island.

Observations on this species by previous expeditions have been very general. *Korotkevich* [1959] reports that the pintado petrel arrives during the first 10 days of October, egg-laying begins about November 25, and the first chicks appear about mid-January. His population estimate for Haswell Island is less than 25% of that noted in 1962. *Falla* [1937] states egg-laying began about November 27 in 1912, when S. E. Jones reported, "These birds were very numerous."

ANTARCTIC PETREL

The breeding of the antarctic petrel was first discovered at Haswell Island by the Mawson party November 28, 1912. According to *Falla* [1937] "several hundred birds" were then present, apparently in much the same places they occupy today. They now nest in two semi-isolated compact breeding colonies about 50 meters apart near the top of almost vertical cliffs on the southeast corner of Haswell (Figure 5). One colony in a deep wedge-shaped cleft is rather overcrowded (Figure 8); the other on more level terrain has enough room for wider spacing of nests. Both colonies have large, flat rock ledges for landing platforms. The accumulation of guano showed both sites to be old and long occupied. Some evidence indicates that the more level area is expanded periodically to receive the excess of a slightly fluctuating population. In December 1962 the breeding population of the more sparsely settled area increased 13% over the December 1961 population, but the number of nests in the cleft site decreased slightly.

During periods of high winds, incoming birds soared high and invariably overflew the nest sites, then turned into the strong updrafts to make their approach. The antarctic petrels were notably adept in taking advantage of local wind patterns, and their graceful approaches to the nesting sites were in marked contrast to the frequently awkward crash landings of the neighboring silver-gray fulmars. Exposure to high winds

could well be a factor in nest site selection by the species.

Because of the openness of the sites, the antarctic petrel rarely had to contend with snow accumulation. Nests were generally crude scoops in the wind-blown gravel or, in more exposed areas, simple berths behind rock outcrops. Before the birds arrived, sites were easily identified by the scoops that were neither leveled by winter winds nor covered by snow. Immediately on arrival birds occupied the scoops and began making repairs. Wherever stones or gravel were available, the nests were lined with them. The process of nest construction was not elaborate; the bird simply stretched its neck, took a small stone in its beak, and tossed it near the nest. Where the gravel was deeper some birds pushed debris from the scoops with their webbed feet.

The antarctic petrels returned to the Mirnyy area October 7. Unlike the pintado petrel, no birds arrived before the main occupation of the nesting sites. The birds arrived in large groups and the whole population arrived within a period of approximately 36 hours. Mass movement was characteristic of the Haswell Island population throughout the season and simplified distinguishing various phases of the breeding cycle.

As the flocks arrived birds went directly to their nesting sites and began their courtship. Pairs or prospective pairs formed immediately after arrival. The following description of courtship behavior is taken from the author's field notes shortly after the species' arrival: "Courtship is quite similar to that of the silver-gray fulmar; perhaps more frequent displays and equally as intense. Two birds observed sitting face to face in the nest display a mutual 'nibbling' of areas about the head and beak. One bird lowers the beak to the breast and slowly raises and extends the head until the beak lies horizontally along the back. Some head wagging observed, but less violent than that of the silver-gray fulmar. Courtship displays are accompanied by a soft clucking sound. Areas near the base of the beak have a definitely puffed or pouched appearance."

Copulation was first observed October 9 and reached a peak on October 17. The species left the area October 26, when only two courting pairs and three single birds remained. The birds left toward the north-northeast and returned later from the same direction. On October 28 five single birds were in the nesting area, which was completely deserted on November 2. On November 18 five single birds returned and claimed nests; 27 single birds and three pairs were on hand November 19, and all nests had at least one member of each breeding pair present on November 20. Usually one member of each pair returned earlier than its mate. These first arrivals re-occupied nests and quickly and efficiently routed any silver-gray fulmars that had moved in during their absence. Battles for nest sites were often as violent as they had been during the initial occupation. As their mates arrived courtship was resumed, and both activity and consequent noise level greatly increased. Prior to egg-laying the birds were reluctant to eject stomach oil and resorted to it only when handled or when escape routes were blocked.

The first egg was found November 21; the next day 20% of the nests had their single eggs, on November 24 approximately 75%. Measurements of 30 eggs showed an average egg size of 69.4 × 47.9 mm; the smallest measured 66.9 × 46.7 and the largest 74.5 × 48.6 mm. As incubating progressed, birds became more reluctant to leave their nests unless they were startled or touched, though more timid individuals sometimes left their nests as soon as the observer approached.

During the incubation period birds not on nests gathered in numbers on the sea ice below the colony and rested on snow banks adjacent to the breeding site. The continued presence of large numbers of birds after the eggs were laid suggested that flights away from the area for possible feeding were of short duration.

A nest count December 2, 1962, showed a breeding population of 1054 ± 50, and 41 fewer nests than the number recorded October 20. Approximately 90% of the December population were breeding birds. The 1962 breeding population was 10.8% greater than that of 1961. The increase was especially noticeable in the less crowded site nearest the southeast corner.

Egg mortality was fairly low; approximately 63% of those produced were still intact December 30. Accidental freezing of eggs that rolled from poorly protected scoops was the most obvious cause of egg loss. The species does not suffer appreciably from skua predation because its nest sites are not easily accessible to the predator.

When observations ended, 45 days had elapsed since the first egg appeared and none had yet hatched.

The antarctic petrel does not nest on other islands in the Haswell group, and none was seen flying south of the breeding sites.

Records for preceding years show the antarctic petrel to be one of the last summer residents to leave the area. In 1962 the last bird was seen May 1; they have previously been recorded during the last week of April [Nudelman, 1962]. Korotkevich [1959] thinks this species leaves somewhat earlier than other petrels. His estimate shows the population in 1956 to be considerably smaller than it was in December 1962. He reports that the colony formed during the first 10 days of October in 1956; many eggs had been laid by November 25, and hatching began 48 days later on January 12, 1957.

SILVER-GRAY FULMAR

Approximately 4,000 pairs of silver-gray fulmars bred on the more precipitous slopes of the north, east, and south sides of Haswell Island during the 1962–1963 summer season. None nested in the interior nor on the lower and less rocky west side of the island. One colony on the northeast corner of the island extended inland approximately 150 meters, where the topography and wind currents made it difficult for the birds to enter and leave their nests. The fulmar is extremely clumsy on land, and its awkwardness was most evident where it had to approach its nest with the wind. In such places the birds fell or tumbled out of the air into the nest site. On the north side of the island, where the birds could usually land into the wind, their movements were much more graceful, and they occupied sites at all elevations from near sea level to the summit of the island.

The silver-gray fulmars returned to the Mirnyy area early the morning of October 7. Unlike the penguins, most of the population arrived in large groups, and the entire breeding population was present by October 8.

Returning birds came from the north-northeast and invariably flew around the end of the island to reach the south side; very few birds ever flew west of the group of islands. Small flocks often landed on the sea ice just east of Haswell and stayed there an hour or more before moving to the island. A few single birds and small groups overflew Haswell to another island nearer the mainland. Most of the returning birds flew directly to nest sites and immediately began territorial defense or courtship. Frequently only a matter of minutes elapsed between the time of arrival and pair formation. Usually the occupation of sites was uncontested, but, where two or more birds claimed the same nest, territorial struggles lasted some time.

Most nests were near the edge of wide ledges or bare rock outcrops where the birds could hobble a few steps and take flight by simply falling into the air.

Fig. 9. Nest site of silver-gray fulmar showing slight elevation of nest.

of the north and south sides of the island nests were tunneled into snow slopes. Birds nesting where gravel or small stones were plentiful frequently picked up pebbles in the beak and tossed them in the general direction of the nest the same way the pintado petrel did. Some nests were carefully repaired, slightly elevated, and lined with stones. Nesting sites of long use were conspicuous by their guano accumulation, and these older sites were the first to be occupied.

During the early stages of territorial establishment and courtship, pairs did not defend nest sites strongly but flew away when an observer approached. After a few days of occupancy the birds became more defensive and met all intrusions with a lifted head and opened beak. The threat display was usually made silently and was only rarely accompanied by the species' raucous call. Where immediate escape was impossible birds turned in the nest with their heads toward the rear and remained silent, watching the intruder with one eye. When touched or handled, the bird grasped fingers or parts of the hand tightly in its surprisingly strong beak. After egg-laying, birds faced the intruder, spread their wings slightly (Figure 10), just as the pintado petrel did, and defended by disgorging a red-orange oil, clear at first but mixed with semi-digested food particles after repeated disgorgement. The silver-gray fulmar seemed more reluctant to disgorge fluids than the pintado or the snow petrel. Observers elsewhere have reported that ejected fluids were sometimes green [*Falla*, 1937], but at

On the more gentle slopes on the north side birds had to push themselves into the air and glide downslope with the wind before turning or sideslipping into full flight. Most nests were shallow scoops in the windblown gravel; many were sheltered from the elements behind large rock outcrops (Figure 9). In some parts

Fig. 10. Threat display by silver-gray fulmar defending nest.

Haswell Island they were red-orange throughout the breeding cycle.

Copulation was first noted October 24; it increased through the last week of October and reached a peak in the first week of November.

Like most other antarctic Procellariiformes, the silver-gray fulmar leaves the nesting ground for a short period prior to egg-laying. On November 7 approximately 60% of the population was absent, and more than 90% had left by November 8. A few birds returned November 11, when nests were occupied by single members of the mated pairs. On November 15 approximately 40% of the population was back, but full repopulation by both members of the pairs was not noted until November 28. Birds usually spent about 3 weeks away from the breeding grounds, but the length of each bird's absence was not so clearly defined as in the antarctic petrel. Single birds occupied and defended nests from November 14 to November 28; it would be interesting to know which member of the pair it was. While the nests were occupied by single birds, the area was extremely quiet and aerial activity was reduced to a minimum.

The first egg was found December 2; laying gradually increased and peaked December 14. In 1912 the discovery party found two eggs November 28 [*Falla*, 1937]. These and data for one other year indicate that the breeding dates may vary as much as one week in different years. Ten eggs collected December 8, 1962, averaged 75.6 × 50.4 mm; the smallest and largest measured 71.8 × 48.2 and 79.7 × 52.2 mm. After the eggs appeared, nest defense increased markedly. One member of the pair usually seemed more eager to defend than was the second member, especially when both were present. Whether or not aggressiveness and timidity are sex characters is not known.

Egg mortality was somewhat higher than in the antarctic petrel and the pintado petrel. On December 20 approximately 57% of the eggs produced were still intact. Thaw flooding was the major cause of destruction, but eggs were lost in other ways. It was impossible for the birds to retrieve eggs that rolled from some of the crude nests or depressions in the bedrock. Practically all of the eggs laid in snow tunnels were spoiled or broken before mid-December. In three sections the skuas fed almost exclusively upon the misplaced and spoiled eggs of this species.

When I left January 5, 1963, no eggs had yet hatched, which was to be expected, for *Prevost* [1958] gives the incubation period as 43 to 44 days.

Fulmars also nest on the second largest island in the group, appropriately called Fulmar Island, which supported a breeding population of 1500 ± 200 birds.

The previous season the fulmars disappeared during the first week of April 1962. According to *Korotkevich* [1959], the species arrived at Haswell during the last days of September 1956. The colony was formed by October 7, and mating was observed on November 8. Eggs had been produced by the beginning of December, and hatching began in mid-January. Chicks completed their molt and began to leave by the first of March. The population estimate for 1956 is approximately 25% of that observed in 1962. According to *Nudelman* [1962], the species left the area during the first 10 days of April and returned October 11.

Although no population counts were made in 1912, Jones reported that silver-gray fulmars were present in very great numbers all over the eastern face of Haswell Island, and "there was a large rookery on a smaller island on which no other birds were present." If the smaller island was Fulmar Island, it is of interest that Jones did not record the presence of Adélie penguins, for 1860 ± 50 Adélies nested on the island in 1962. Possibly Jones was unable to see all parts of Fulmar Island in 1912.

SNOW PETREL

This species has a very wide local distribution and nests at all elevations on the north, east, and south sides of Haswell Island. The greatest concentration of birds was on the southeast corner; none was seen in the western half or inland. Nests on the northeast corner and north side of the island were easily accessible; those high on the southeast corner could not be reached. Although the snow petrel nests only on Haswell Island, flocks were often seen flying over the mainland and as much as 30 km inland.

Nests were characteristically under large flat rocks or in deep crevices in vertical cliff walls. On the south side many nests were scattered intricately in a long talus slope. Most nests had two or more entrances. Birds landed on wide ledges or large flat rocks near the entrances during high winds. The amount of space inside the nests varied greatly. In some parts of the north side the crevices were large enough to permit the nesting birds great freedom of movement; other cliff sites were so crowded and so low the nesting birds could not stand upright. Nests deep in these recesses were well protected from predators and bad weather. Exposed sites near sea level on the northeast corner were frequently inundated by summer snow. The number of addled eggs, feathers, egg shells, and large deposits of guano at some sites suggested that they were old and well established. Quantities of penguin feathers found in many nests on the north side probably were blown in, for the birds were never seen to take them in. Rocks at the nest sites were stained with the oil ejected during the almost constant defense of territories. Although most nests were isolated, a few were near those of Adélie penguins, antarctic petrels, pintado petrels, and silver-gray fulmars.

When the species returned to Haswell October 7, many birds went directly to nest sites on the island, but, unlike most other species, did not immediately start courtship. Others flew to the sea ice west of the island and rested on soft snow banks for long periods. Most nest sites on the north side were then covered with snow, and the birds started immediately to excavate their old nests. Birds often sat in front of holes in the snow banks, which after thawing proved to be the sites of old nests. In some snow-covered areas birds dug tunnels up to 1.5 meters long to their nest sites. Access to these nests was easy after the mid-November thaw.

A number of nests were occupied by single birds October 10. Courtship was first noted October 11, and the noise level increased greatly. Single birds sat in full view near the nest entrances to drive off intruders. Usually an initial warning squeak was enough, but at more persistent intruders the defending bird sometimes had to eject stomach oil. Birds calling from deep within nesting crevices were frequently answered by birds flying by the entrances. Prospective mates cautiously approached the nest entrances and were either accepted or forced to leave. Courting pairs could be heard chattering almost constantly deep in the nest recesses.

During the pre-egg period the nesting grounds were deserted and re-occupied at least twice. Many sites were first abandoned October 16, and for 2 weeks very few birds were present. Birds began to return in numbers October 31. During the first week of November many birds sat on the flat rocks outside the nest entrances in plain sight. The more aggressive member of the mated pair chased intruders from the nest site. The chase often involved intricate aerial maneuvers accompanied by a shrill chattering. In the more violent chases the pursuing bird flew just above the offender, and sometimes the two birds flew short distances with locked beaks. Birds stained with the orange stomach oil flew to soft snow banks on the sea ice and bathed. Making shallow excavations in the snow, they scooted,

rubbed, turned over on their backs, righted themselves, and preened their stained feathers, which proved a most effective way of removing the stains.

Copulation was first seen November 11, and mating reached a peak November 16. Birds usually copulated on ledges outside the nesting crevices. During the act the treading male chattered in a low-pitched but typical call; the female meanwhile uttered a barely audible whine.

A large part of the population left the breeding grounds again on November 19, and activity remained at a minimum until November 28. The island was not fully repopulated until December 3. Overlapping movements made it impossible to determine accurately the numbers of birds and specific dates involved in the second desertion, which was made apparently for feeding just before egg-laying. Nest defense increased greatly during the second re-occupation period; ownership of almost every nest seemed to be contested. Unoccupied birds continued to challenge and harass nesting pairs until mid-December.

The first egg was found November 29, and the number of eggs increased steadily through December 9. Both members of the pair were usually near the nest at this time, but many unoccupied birds congregated on the sea ice nearby. During incubation both birds defended the nest avidly and rarely abandoned the single egg. When an intruder approached, both birds began their shrill chattering call. Frequently the non-incubating bird retired deeper into the nest site, and the incubating bird defended by disgorging oil. Measurement of 10 eggs showed an average size of 53.9 × 39.4 mm, the smallest and largest measuring 51.7 × 37.9 and 60.1 × 41.3 mm.

A count made December 15 showed a breeding population of 700 ± 50 birds. There also appeared to be a considerable number of nonbreeding birds present.

Egg mortality during December was approximately 43% for 100 nests located in accessible parts of the island. Although the birds were most conscientious incubators, a major cause of egg loss appeared to be freezing. Some sites downslope from snow drifts were flooded by the thaw. A very few nests with eggs were abandoned during the latter half of December. No eggs had hatched by January 5, 1963, when I left, 38 days after the first egg was laid.

As part of the USARP Bird Banding Program, 25 incubating birds were banded with metal overlap leg bands. The birds were caught December 26, 1962, at low and easily accessible sites on the southwest corner (Figure 5).

Korotkevich [1959] states snow petrels began to arrive at Haswell October 5, 1956; by October 20 the colony was formed; egg-laying began during the first half of December, and hatching was noted during the first half of January. Birds began to leave about the first of March, but some remained as late as mid-May.

Observations made at Haswell in 1912 were very general. No estimate of numbers is given except for a note suggesting that the petrels were not plentiful [*Falla*, 1937].

WILSON'S PETREL

Its habit of nesting deep in inaccessible crevices made obtaining information on this species very difficult. Many of the data presented here were obtained from 10 nest sites on the north side of Haswell, a very small percentage of the nests present; thus dates and figures are only bare approximates for the summer population of 1962–1963.

The species nests on all parts of the island except the precipitous east- and south-facing cliffs. Nests were most plentiful along the north-facing slope of an east-west saddle running through the center of the island. Others were isolated in otherwise uninhabited sections of the west side.

Nests were characteristically under large flat stones or in deep fissures, and each was cluttered with remains of eggs from previous seasons, feathers, and the desiccated bodies of chicks in various stages of development. Abundant evidence showed most nests were old and long established. Frequently one narrow entrance led to two or more nests. In a few crowded nest sites on the north side, entrance tunnels approximately 0.5 meter in length were lined with Adélie penguin feathers apparently blown in by the wind. Rocks at the nest entrances usually were stained with stomach oil the birds disgorged during territorial defense.

Wilson's petrel was the last migrant to arrive in the Mirnyy area. A few individuals appeared November 10, and a full breeding population was present November 14. Small groups of migrants came in from the north-northeast and split up at once as the birds scattered to individual nest sites. Rough calculations based on the number of birds calling from nests early in the occupation period indicated a breeding population between 500 and 600 in 1962–1963.

The birds' most obvious activities on their return were nest selection and courtship. In two sections on the north side some 30 nests used in previous years were destroyed or made uninhabitable by rock slides. Returning birds searched for other suitable sites in the rubble and established 16 nests in the slide area during the first 2 days. Most of the new nests were poorly positioned and had to be abandoned when flooded by the summer thaw.

Birds searching for prospective mates skimmed along the rocks and appeared to locate nest sites by sound. The calls of birds deep within nesting fissures frequently attracted 4 or 5 prospective mates. These congregated near the entrances, and commonly one of the more aggressive of the group drove the others away by a series of short hopping threats. It then exchanged calls with the bird on the nest and cautiously moved into the nest entrance. If the bird within did not accept the arrival, it drove him away with high-pitched warning squeaks or, finally, by disgorging red-orange stomach oil.

The species was most active between local sunset and sunrise, and activity was reduced to a minimum at midday. The birds courtship chattering increased noticeably at 6 p.m. local time and continued until 2 a.m.

The first egg was found November 27, and laying continued at least to December 6. Measurements of 10 eggs gave average dimensions of 33.9 × 23.4 mm; the smallest and largest measured 32.5 × 22.7 and 35.9 × 24.7 mm.

Wilson's petrel nests on one other island in the area. On December 13, 1962, three pairs had well-protected nests on the Stroiteley Islands closest to the mainland. At least two of the nests had a single egg, and both members of the mated pairs were present. Seven older and unoccupied nests were also found.

In 1962 the last Wilson's petrel was seen March 12.

SOUTH POLAR SKUA

During the 1962–1963 breeding season, 23 pairs of south polar skuas nested on Haswell Island. In addition 17 nonbreeding birds also occupied territories on the island. Three pairs nested on smaller islands in the group, one pair each on Fulmar, Tokarev, and Stroiteley.

On October 12 a single bird was seen circling over the northwest corner of Haswell. This bird remained alone until it was joined by its mate October 19. October 23 one mated pair and two unmated birds were establishing territories; 13 birds were present on October 31. More continued to arrive during early November, and 61 were in residence November 12. Territorial displays were rare during the initial stages of re-occupation, but, as more birds arrived, certain pairs displayed and made reconnaissance flights as soon as the observer approached. Territories were well established and aggressively defended by November 15.

The plumage of early arrivals was generally the color described by *Eklund* [1961], except for one dark pair of birds whose wing feathers were almost chocolate brown and whose heads and necks were much darker than the usual light chamois color. Their color made them easy to distinguish throughout the breeding season. As summer progressed colors of all the skuas became lighter. Bleaching or fading with increased exposure to sunlight was noticeable on all parts of the body and particularly on the head and neck, which became extremely light by the end of December.

Skua territories were confined primarily to the northern half of the island. More than 82% of the population occupied less than 30% of the total space available for nesting. This rather unusual pattern of local distribution seemed to be governed by the equally uneven distribution of Adélie penguin breeding colonies. The skuas were concentrated in three small areas on the north side of the island, where the penguin nests were closest together. They were strangely absent from the higher south side of the island, where Adélies nested in large but widely separated colonies.

Nests were scoops made in gravel or fine pebbles, and birds usually excavated two or more before making a final selection. Two pairs that nested where no gravel was available laid their eggs in shallow depressions in the bedrock. Some nests were in the lee of large rock formations, but most were exposed to the wind on rocky knolls. Often bits of dried alga were found in the nests.

Copulation was first seen October 30. By then territorial defense had increased, and both members of the mated pairs displayed often and challenged all intruders. During the first two weeks of November copulation usually occurred at the nest or near the established territory. Some birds were seen courting and copulating on the sea ice north of the island.

One of the first pairs to arrive, and one of the most persistent defenders of territory throughout the season, laid its first egg November 13 and its second on November 15. The peak period for laying was the last week in November. The egg laid November 13 produced the first chick on December 11. The average incubation period for 14 marked eggs was 31 days; 4 eggs were incubated 30 days; 2 were incubated 29 days; and 1 egg hatched after 32 days of incubation. Most eggs pipped for two days before the chick was free from the shell, but a few emerged one day after pipping.

After the second chick hatched, one of the young birds, usually the older of the pair, appeared more aggressive than the other. It often chased the younger, smaller chick from the immediate vicinity of the nest and by pecking kept it away. Chicks forced away from the nest often died from exposure or starvation, and some were killed by adult skuas. Adults were never seen killing their own chicks, but young that strayed from the protection of their parents frequently fell prey to adults in adjoining territories. The incubating adults ate any eggs that were accidentally displaced and outside the nesting scoop, and also all chicks that died from undernutrition, exposure, or crushing.

Of 40 eggs laid in 23 nests kept under observation, 25 (62%) hatched successfully; 18 of these (72%) were still alive on January 5, 1963. On January 4, 1963, only about 18% of the successful nests were occupied by two chicks; thus the probability of both chicks reared successfully is very low.

Brood mortality (eggs and chicks) up to January 5 was 55%, of which egg loss accounted for about 68%. Exposure during the egg stage was the major cause of failure, and many eggs that did not hatch contained embryos in various stages of development. Strangely, egg loss was lower in exposed nests than in those that appeared to be better protected; of the 15 eggs that failed to hatch, only 4 were in fully exposed nests. Mortality during the chick stage was relatively low, and resulted mainly from exposure and crushing, usually during the first few days after hatching, and the younger chick was usually the victim.

The arrival of the skuas coincided with the birth of Weddell seal pups, and the birds were frequently seen feeding upon seal placenta. Before Adélie penguins started to lay, the skuas made frequent trips to the emperor penguin colony to feed on the frozen bodies of dead emperor chicks, which then seems to be their main food source. After the Adélie eggs appeared, the skuas continued to obtain some food from the emperor colony. At no time did the skua seem to be a determined aggressor; food was plentiful enough throughout the season so that the birds seldom had to resort to active predation. On Haswell alone, the ratio of skuas to Adélies was 1:550, which seemed more than adequate to meet the skua food needs during most of its breeding cycle. In the few cases when the skua did act as a predator, the Adélie suffered the most. At least three pairs of skuas fed almost exclusively on the eggs and young of the silver-gray fulmar; other species nesting on the island contributed little to the skua food supply.

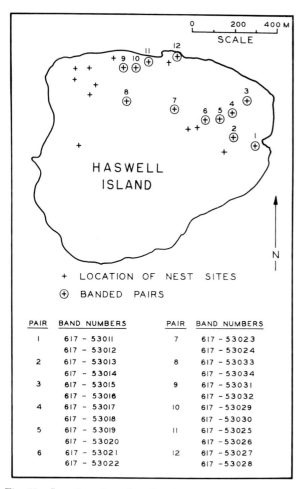

+ LOCATION OF NEST SITES
⊕ BANDED PAIRS

PAIR	BAND NUMBERS	PAIR	BAND NUMBERS
1	617 – 53011	7	617 – 53023
	617 – 53012		617 – 53024
2	617 – 53013	8	617 – 53033
	617 – 53014		617 – 53034
3	617 – 53015	9	617 – 53031
	617 – 53016		617 – 53032
4	617 – 53017	10	617 – 53029
	617 – 53018		617 – 53030
5	617 – 53019	11	617 – 53025
	617 – 53020		617 – 53026
6	617 – 53021	12	617 – 53027
	617 – 53022		617 – 53028

Fig. 11. Location of south polar skua nest sites on Haswell Island in December 1962.

During the 1962–1963 season, 12 breeding pairs of skuas were banded with bands provided by the USARP Bird Banding Program (Figure 11). Three birds that Eklund banded at Wilkes station, one as an adult December 9, 1957, the other two as nestlings in January 1958, were found breeding on Haswell. The bird banded as an adult had a single chick on January 2, 1963; the two banded as nestlings were still incubating eggs January 4, 1963. Of the 37 skuas banded with Russian bands in Mirnyy in 1956–1957, none was seen in the area in 1962–1963; the only one reported to date are three captured at Davis station (68°35′S, 77°58′E) in 1958 and 1959 [Eklund, 1961].

From reports of previous expeditions and the available banding records, the 1962–1963 population was much lower than those of other years. *Korotkevich* [1959] reports that the species was first seen October 7, 1956, and the occupation period had ended by the last of October. His sketch map shows 35 nests on Haswell that year and 15 more on smaller islands in the group. His data show that the breeding population in 1956 was at least 100 as against 52 in 1962. Eggs were produced during the second half of November (many nests with eggs were observed November 25), and hatching occurred during the last days of December. Young birds were flying by the beginning of March, and the species left during the first 10 days of April.

COMPARATIVE CHRONOLOGIES

Chronological data for the summer of 1962–1963 exchanged with Patrick Arnaud at Dumont d'Urville station in the Adélie Coast showed that the emperor penguin, snow petrel, and pintado petrel arrived at Haswell approximately 3 weeks later than at the Géologie Archipelago. The Adélie penguin arrived 4 days earlier at Haswell, and Wilson's petrel appeared at both places during the same 24-hour period.

The timing of the start of egg-laying seemed less variable between the two areas. Although the emperor penguin arrived several weeks earlier at the Géologie Archipelago, the species began to lay in both places at almost the same time, and data available for 1956 show an almost identical pattern. The first egg-laying for both Wilson's and pintado petrels varied between the two areas by a single day in 1962, but the snow petrel started laying 1 week earlier at Haswell.

Data recorded by previous expeditions show the arrival times of both species of penguins at Haswell to be almost identical for 1956, 1958, and 1962, but those for other species to have varied from 1 to 3 weeks between the years. In 1958 the silver-gray fulmar was first observed in mid-September; in 1962 the species returned 3 weeks later on October 7. On the other hand, the south polar skua arrived at Haswell 8 days earlier in 1962 than in 1958. Both penguins and the skua started to lay within the same week in all three seasons, 1956, 1958, and 1962.

Thus arrival dates for some species may vary considerably between areas of equal latitude in a given year, and at the same place from year to year. Dates of first egg-laying seem, however, far less variable between both places and seasons.

Acknowledgments. The author is indebted to Arthur Mirsky and Richard L. Cameron of the Institute of Polar Studies, Ohio State University, for their critical examination of drafts and helpful suggestions during the preparation of the paper. Appreciation is also extended to William J. L. Sladen and to Richard L. Penney, both of the Johns Hopkins University, for their thorough review of the manuscript. The author is also indebted to Jean Prevost of the Paris Museum of Natural History for his examination of data and encouragement. A special note of thanks is extended Patrick Arnaud, Station Marine d'Endoume, Marseille, for the use of data collected on the Géologie Archipelago during 1962. The author also wishes to express his appreciation to many members of the Seventh Soviet Antarctic Expedition who accompanied and assisted him on many trips to the nesting grounds.

Procedures for collecting data in the emperor penguin colony were largely modifications of research methods designed by Prevost. Techniques used for banding and marking birds for observations were those designed and used in other areas by Sladen and Penney. The terminology used in describing behavioral characteristics follows that used by Sladen in various papers on the Adélie penguin.

This study was supported in full by the U.S. Antarctic Research Program, National Science Foundation grant G-18803 (Ohio State University Research Project 1345).

REFERENCES

Arnaud, P., R. Guillard, J. Prevost, and J. Sapin-Jaloustre, Recherches écologiques sur les oiseaux de l'archipel de Pointe Géologie, *Oiseau Revue Française Ornithologie, 34,* 109, 1964.

Budd, G. M., Population studies in rookeries of the emperor penguin *Aptenodytes forsteri, Proc. Zool. Soc. London, 139*(3), 365–388, 1962.

Caughley, G., The Cape Crozier emperor penguin rookery, *Rec. Dominion Mus., Wellington, N.Z., 3,* 251–262, 1960.

Eklund, C. R., Distribution and life history studies of the south polar skua, *Bird Banding, 32*(4), 187–223, 1961.

Eskin, L. I., Concerning the development of shore ice in Antarctica, *Inform. Bull. Soviet Antarctic Exped., 28,* 31–33, 1961.

Falla, R. A., *B.A.N.Z. Antarctic Res. Exped., 1929–1931: Rept.-Ser B, II,* 1–288, 1937.

Korotkevich, Ye. S., Observations on the birds during the first wintering of the Soviet Antarctic Expedition in 1956–1957, *Inform. Bull. Soviet Antarctic Exped., 3,* 83–87, 1958.

Korotkevich, Ye. S., Birds of East Antarctica, *Problems Arctic Antarctic, 1,* 95–108, 1959.

Korotkevich, Ye. S., Habitats of the emperor penguins, *Inform. Bull. Soviet Antarctic Exped., 42,* 37–44, 1963.

Makushok, V. M., Concerning biological collections and observations at Mirnyy Observatory in 1958, *Inform. Bull. Soviet Antarctic Exped., 6,* 40–42, 1959a.

Makushok, V. M., Biological report for 1958, *Rept. Third Continental Exped., 5,* 115–124, Leningrad, 1959b.

Mawson, D., *The Home of the Blizzard,* vol. 2, pp. 108–119, J. B. Lippincott, Philadelphia, 1915.

Murphy, R. C., *Oceanic Birds of South America,* vols. 1–2, American Mus. Nat. Hist., New York, 1936.

Nudelman, A. V., *Soviet Antarctic Expeditions 1959–1961,* pp. 1–147, Publishing House of the Acad. Sci. of the USSR, Moscow, 1962.

Pinder, R., The Cape pigeon, *Daption capensis* Linnaeus, at Signy Island, South Orkney Islands, *Brit. Antarctic Surv. Bull.,* No. 8, 19–47, 1966.

Prevost, J., *Ecologie du manchot empereur,* pp. 1–204, Herman Press, Paris, 1961.

Prevost, J., Influence des facteurs bio-climatiques sur le monadisme des manchots empereurs à la colonie de Pointe Géologie, *Oiseau Revue Française Ornithologie, 33*(2), 89–102, 1963.

Pryor, M. E., Notes on the life history of the emperor penguin, *Aptenodytes forsteri* Gray, at Mirnyy Observatory, Antartica, 1962, *Inform. Bull. Soviet Antarctic Exped.,* in press, 1968.

Sapin-Jaloustre, J., Quelques aspects de la vie du manchot Adélie en Terre Adélie, *Acta XI Congressus Internationalis Ornithologici,* Basel, 231–240, 1955.

Sladen, W. J. L., The Pygoscelid penguins: 1, Methods of study; 2, The Adélie penguin, *Sci. Rept. Falkland Islands Depend. Surv., 17,* 1–97, 1958.

Sladen, W. J. L., and R. L. Penney, Penguin flipper bands used by the USARP Bird Banding Program, 1958–1960, *Bird Banding, 31,* 79–82, 1960.

Stonehouse, B., The emperor penguin *Aptenodytes forsteri,* 1, Breeding behaviour and development, *Sci. Rept. Falkland Islands Depend. Surv., 6,* 1–33, 1953.

Vanhoffen, E., Bericht über die bei Deutschen Sudpolar Expedition beobachten Vogel, *J. Orn. Leipzig, 53,* 500–515, 1905.

TERRITORIAL AND SOCIAL BEHAVIOR IN THE ADÉLIE PENGUIN

RICHARD L. PENNEY[1]

Department of Zoology, University of Wisconsin, Madison

Abstract. The objective of this study is to explore territorial and social behavior in the breeding biology of the Adélie penguin (*Pygoscelis adeliae*). Emphasis is placed on the physical and social aspects of territorial behavior and on the importance of individual recognition in pairing and parent-offspring bonds. A small rookery near Wilkes station, Antarctica, consisting of about 2700 adults grouped in 14 distinct colonies was chosen as the study area. About one-half the occupants were marked with numbered flipper bands. Details on territory location within the rookery, pairing, and breeding success were determined for several hundred birds for portions of three successive breeding seasons. The normal breeding cycle is reviewed as background for analyses of territorial and social behavior.

In 11 colonies nesting territories averaged 0.75 m² in area. Distances between particular nest centers and their two nearest neighbors averaged 69 cm. The spacing of nests was consistent on flat ground but became irregular on uneven surfaces. Territorial defense consisted of three threat displays and overt fighting. Young Adélies visit the breeding colonies prior to becoming reproductively mature at four or five years of age. These birds show a progressive increase in their attachment to particular colonies and territories. Attachment to specific territories becomes strong after first breeding. Males showed a 93 per cent fidelity to their nesting territories in successive seasons; females usually changed territories, but not colony residence, if they did not reunite with their former mates. Return of breeders to their previous colonies was over 99 per cent. Adélies nesting on the periphery of colonies had a 10 per cent lower breeding success than birds nesting in the center during the 1959–1960 season. In the next season peripheral and central nesters had equal breeding success. Young birds tended to nest on the periphery of established colonies. Two small colonies showed a high attraction for new breeders and consequently grew rapidly in size while maintaining a relatively low breeding success.

Four displays used in the establishment and maintenance of pair bonds are described. Three types of sexual association are recognized: (1) trial pairing (no eggs produced), (2) true pairing (eggs produced), and (3) recurrent pairing. Trial pairing was often temporary but led to true pairing in the spring for new breeders and birds whose previous bonds were broken. Temporary sexual associations established in one season rarely influenced choice of mate in succeeding seasons. Fifty per cent of the birds retained their mates in successive years. Annual mortality broke 34 per cent of the pair bonds; the other 16 per cent picked new mates or did not mate. Reunited pairs had a 12 per cent higher breeding success than birds that changed mates. Of birds that lost their mates between years, 11 per cent failed to pair by egg-laying time the second year. Attachment and return to the same nesting territory year after year help maintain the pair bond. Recurrent mating is further encouraged by mate recognition. Experimentally exchanging incubating birds between nests showed that individuals recognize one another by their voices and that relief of the correct mate is more important than relief of the correct nest.

That chicks recognize their own parents' voices was demonstrated by playing back the calls of specific parents to groups of chicks. This recognition helps dispersed creche chicks to find their parents at feeding time. Homing to the nesting territory by parents and their chicks also aids reunions for feeding.

A number of social responses, the effects of social stimulation on reproduction, the components of territorial behavior, and a refined mechanism for individual recognition together contribute toward an efficient system of reproduction in the Adélie penguin.

INTRODUCTION

Objective. This paper examines the roles of territorial behavior and certain social relationships in the breeding biology of a colonial penguin, the Adélie (*Pygoscelis adeliae*), as revealed by observations on banded birds at marked nesting territories during three consecutive breeding seasons. Special attention is given to the physical and social description of territory, to the permanence of pairs between breeding sea-

[1] Present address: Institute for Research in Animal Behavior, New York Zoological Park, Bronx Park, Bronx, New York 10460.

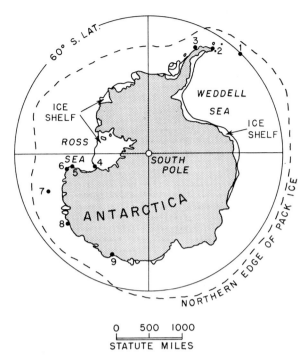

0   500   1000
STATUTE MILES

Fig. 1. Map of Antarctica showing the locations of some important Adélie penguin rookeries: (1) Signy Island, (2) Hope Bay, (3) Petermann Island, (4) Cape Royds and Cape Crozier on Ross Island, (5) Cape Hallett, (6) Cape Adare, (7) Balleny Islands, (8) Phoque Island, and (9) Wilkes station.

sons, and to how paired birds and parents and chicks recognize each other. These facets of Adélie penguin breeding behavior are in turn considered in relation to breeding efficiency.

The Adélie penguin. The southernmost breeding penguin, the Adélie, nests circumpolarly on the coasts of the antarctic continent and on rocky islands south of 60 degrees south latitude (Figure 1). Its winter movements are confined essentially to the pack ice that extends in a belt around Antarctica from a few miles to several hundred miles offshore [*Murphy*, 1936; *Sladen*, 1958].

A migratory species, the Adélie arrives at its breeding grounds in October and early November. It makes brief excursions back to the pack ice to feed during the summer breeding cycle. Though a few birds may remain there to molt during March, most leave the breeding grounds in February to molt and winter in the pack ice.

Study of the behavior and life history of the Adélie began about 50 years ago during early antarctic exploration. Our knowledge of the species up to the 1930's is well summarized by *Murphy* [1936]. *Falla* [1937] reviews earlier accounts and presents observa-

tions of his own made during the early 1930's. *Roberts* [1940] gives additional basic information on pygoscelid penguins. *Sladen* [1953, 1956, and 1958] emphasizes the importance of working with marked birds and uses this approach to resolve a number of questions regarding displays and chick feeding. Many of Sladen's observations are supported by the work of *Sapin-Jaloustre* [1960] who also investigated various aspects of ecology and development. *Taylor* [1962] supplements these basic studies with an excellent account of the population changes and behavior of Adélies in the most southerly rookery of the species, Cape Royds, during the 1959–1960 season.

For information obtained from marked birds, the most important of the early studies is that of *Gain* [1914]. In January 1909 he marked 50 adults and 75 young Adélies on Petermann Island with colored celluloid leg bands. During the next breeding season he found 11 of the adults nesting in the same colony where he had banded them. Some of these adults were also seen by whalers in the 1910–1911 season, but none of the young was recovered [*Gain*, 1913].

In November 1946 Andrew [*Andrew and Roberts, 1952*] banded 10 breeding pairs at marked nests at Hope Bay on the Antarctic Peninsula. During the next October and November Roberts found 17 of the 20 marked birds in the same colony; 6 pairs were back together on their old territories, 2 pairs had new mates with one member of each pair keeping the original territory, and one member of the ninth pair was found at its original site with an unmarked mate. *Sladen* [1958] continued observations on these 10 pairs and summarized their history from 1946 to 1952. He found 5 of the 10 pairs still together on their original territories in 1948. In 1952, six years after banding, 2 birds were still on or very near their original territories.

Study location. My first contact with the Adélie penguin was on December 8, 1958, in the pack ice north of the Ross Sea. From this date until February 2, 1959, I saw many Adélies in the pack ice and at Capes Royds, Crozier, Hallett, and Adare, and on the Balleny Islands while cruising aboard the ice-breaker, U.S.S. *Staten Island.* This cruise provided an excellent introduction to the Adélie and the problems to be investigated during the following two years at Wilkes station.

Wilkes station lies at 66°15′S, 110°32′E on Clark Peninsula, a peninsula which borders the Windmill Islands on the east side of Vincennes Bay. The coast in this vicinity is composed of a number of exposed rocky areas, some of which are connected to the polar

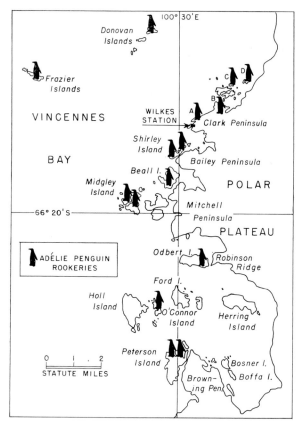

Fig. 2. Map of the Windmill Islands and Wilkes station area showing the locations of the 15 Adélie penguin rookeries.

ice plateau by extensive permanent snow fields. The 16 Adélie rookeries in the area had a combined breeding population estimated at 155,000 by *Eklund* [1961] during the 1957–1958 breeding season. The locations of the major islands and the 16 rookeries are shown in Figure 2.

Rookery A on Clark Peninsula was selected for most intensive study because it was readily accessible during all but the most severe weather. It had a total breeding population of about 2700 birds distributed among 14 colony units of varying size and topography. Exposed metamorphosed bedrock and glacial till composed the rookery substrate. Colony locations, permanent snow slopes (important loafing grounds for Adélies), and landing beaches are shown in Figure 3. Beaches varied with the extent and height of the ice foot (sea and land ice founded on the shoreline) and the tides, but birds rarely landed at points other than the four specified as beaches. Breeding birds in particular were very consistent in their choice of a landing beach.

During the 1959–1960 and 1960–1961 seasons 14 colonies in rookery A produced eggs and chicks. The

size of the colonies ranged from a few birds to several hundred.

Most of the banding, observation, and experimentation was carried out in colonies I through VIII on the west half of the peninsula. Colonies IX through XIV and nearby rookeries were used occasionally for supplementary observation.

Weather and ice conditions. The essential information on weather during the period of this investigation is summarized in Table 1. Although it was often windy at Wilkes during the breeding season, on no occasion was mass mortality of eggs or chicks observed resulting from wind storms or drifting snow. High winds inhibited certain displays (page 113), affected the positioning of birds on their nests (page 93) and caused chicks more than three weeks of age to group into tight huddles (page 123).

The presence or absence of sea ice at Wilkes depends largely on offshore winds, which can be expected to blow away the sea ice at any time of the year [*Tressler*, 1960]. Low temperatures, however, quickly formed new sea ice during the winter and often did so on quiet days in the spring and fall. The existence and form of sea ice during the spring, summer, and fall may be major factors in the control of Adélie movements as well as affecting the timing of the breeding season (see page 90 and *Sladen* [1958].

During the summer pack ice was often visible to the northwest of rookery A at distances of 12 to 16 kilometers. Onshore winds brought pack ice right up to the rookery shoreline. On December 3, 1959, the belt of pack ice extended from 54 to 106 kilometers north and northwest of Wilkes.

Definitions and terminology. A few terms deserve special definition either because of previous confusion in meaning or because of their specialized use in this paper.

A *rookery* is a geographical area, usually a portion of an island or peninsula, that contains one or more colonies (groups) of breeding birds and a landing beach or beaches the birds use to reach the nesting areas from the sea.

A *colony* is a geographically continuous group of breeding birds whose territorial boundaries are contiguous.

A *territory* is any space that an Adélie defends, for which it shows distinct affinity, and within which the breeding birds nest.

An individual is called a *successful breeder* if it is known to have raised one or two chicks up to the

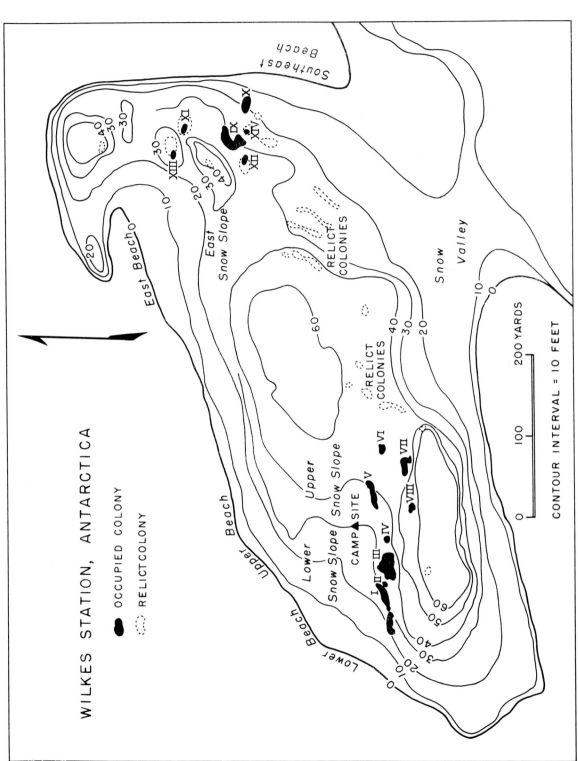

Fig. 3. Map of rookery A showing the locations of occupied colonies, relict colonies, campsite, beaches, and permanent snow slopes, mapped by plane table in October 1960.

TABLE 1. Monthly Weather Summary for the Periods during Which Adélie Penguins
Were Observed at Wilkes Station, Antarctica*

Month and Year	Temperature, °C			Winds			Avg. Sky Cover	Days Vis. <0.42 km	Snow-fall, cm
	Avg.†	High	Low	Avg., m/sec	High,‡ m/sec	Prv. Direction			
Feb. 1959	−3.9	2.8	−12.8	3.9	17.4	NNE	7.8	3	4.1
Mar. 1959	−8.9	1.7	−19.5	4.5	...	SSE	6.1	...	89.2
Oct. 1959	...	1.1	−20.6	6.5	36.6	S	8.7	12	25.2
Nov. 1959	−6.1	0.0	−14.4	7.8	35.7	E	8.1	9	18.1
Dec. 1959	−0.8	6.8	−11.1	3.9	22.8	N	7.6	3	16.5
Jan. 1960	−0.6	5.2	− 7.2	5.0	25.0	E	8.1	1	5.8
Feb. 1960	−2.3	6.1	−12.8	2.9	20.6	S	6.3	0	9.7
Mar. 1960	−10.0	0.6	−20.0	4.8	29.6	NNE	6.0	0	35.6
Oct. 1960	−10.8	−0.6	−22.2	5.3	48.8	E	7.9	8	8.9
Nov. 1960	−4.5	2.2	−13.8	5.5	35.8	E	8.3	3	13.7
Dec. 1960	−1.4	5.6	− 8.9	6.9	33.6	NE	7.8	1	0.0
Jan. 1961	...	5.0	− 5.0	6.0	34.9	E	7.7	1	2.0

* Adapted from monthly summaries of weather by U.S. Weather Bureau personnel at Wilkes station.
† Taken from average daily temperature.
‡ Fastest recorded hour of wind.

beginning of the creche stage (three weeks of age, see page 123). This rather restricted definition ignores chick mortality during the creche stage (see Figure 5) when it is difficult to trace the movements and fates of a large number of chicks.

METHODS AND TECHNIQUES

Observation schedules. I remained in rookery A continuously throughout each breeding season except during persistently high winds, short visits back to base, visits to other rookeries, and for a few days when carrying out homing experiments on the polar plateau. During three seasons I spent more than 1500 hours observing and recording in the rookery.

Camping quarters for living, observing, and recording data were maintained in a small hut in rookery A. Living amidst the Adélies for many months in this manner allowed a more precise feeling for the tempo of the breeding season and provided better opportunities for observing banded birds.

Banding and marking. Methods previously used in banding and marking penguins are summarized in the literature [*Gwynn,* 1955; *Richdale,* 1957; *Sladen,* 1952, 1958; *Austin,* 1957; *Sladen and Tickell,* 1958]. A new type of butt-end aluminum alloy flipper band was developed for this study [*Sladen and Penney,* 1960]. It can be applied quickly with a minimum of disturbance, and the numbers can be read easily with 6× binoculars at distances up to 12 meters or with a 20× scope up to 30 meters. A total of 1528 adults, 66 juveniles (birds hatched the previous season and still

retaining the immature white throat and dark eyelids), and 217 chicks were banded at Wilkes in the present study.

Nearly half the breeding adults in rookery A were banded for individual identification. During the breeding season individuals were banded either while on their nests or after removal to a quiet spot at the edge of the colony. Incubating birds are banded most easily. I approach the nest and with a quick one-hand grab cover the eyes and pull the head down. The bird tries to pull up and back by pushing its flippers against the ground. With the bird in this position it is easy to slip the band on the flipper with my free hand and to close the butt ends neatly. After release the bird usually remains on the nest, shaking its head and adjusting its position on the eggs. This method is quick and causes a minimum of disturbance.

Incubating birds too pugnacious to be banded on the nest, adults without eggs, and large chicks I carried out of the colony to band, placing the bird's head and neck between my knees while affixing the band (Figure 4). Only a very few birds failed to return to their nest or colony immediately after banding. Predation of nest contents by the south polar skua (*Catharacta macormickii*) while banding adults was not the menace at Wilkes that it was by the brown skua (*C. skua lonnbergii*) at Hope Bay and Signy Island [*Sladen,* 1958].

Birds still in the colony after nesting were caught in the open or herded into corrals. Some were run down and caught by the neck or leg, but a slow ap-

Fig. 4. Technique used in holding a penguin for banding.

proach with a small hoop bag on a 6-foot handle was found to be more effective and less disturbing.

Flipper bands of the original design showed a tendency to open at the butt ends. A few bands were known to have been lost from birds within one year of application [*Sladen and Penney*, 1960]. After swimming at sea, banded birds typically showed a small amount of feather wear on the anterior border of the flipper and in the flipper axilla. From the number of unbanded birds showing this wear I estimated that between one and two per cent of the bands were lost per year. That the percentage loss was much greater than this seems unlikely because in a sample of 42 birds banded with both flipper and leg bands, no flipper bands were lost after one year. Additional evidence for low band loss came from colony IV. All the 72 birds in this small colony were banded in one season, and 57 returned to it the next season, 55 of them with their bands intact. One of the two birds without bands showed feather wear that suggested it had lost its band. I saw one more bird in colony IV lose its band during a fight the second season. I have seen bands of this design that still fitted perfectly and showed very little wear 4 years after banding. The time limit of this study did not allow determination of a possible increase in band loss from a weakening of bands with age.

In addition to the aluminum bands I also used various types of paint, colored marking-pen inks, and different types of flipper bands. On the flippers of small chicks I used numbered and colored plastic tapes. The adhesive holding the tape around the flipper gradually came loose as the flipper grew. Males and females in colony IV were given distinctive ink marks on their white breasts so that they could be recognized on the beaches.

Sexing. Adélie penguins show no sexual dimorphism in plumage or body form. Dissection is the only way sex can be told with absolute certainty, but specific behavior patterns and displays indicate sex with near certainty.

Sladen [1958] established several criteria for behavioral sexing and verified them by dissecting a large number of birds. I adopted four of Sladen's criteria on the basis of dissections of 25 key birds observed repeatedly over two breeding seasons. These criteria in order of reliability are: (1) Copulatory position: The male bird takes the top position and treads the back of the female with his feet prior to cloacal contact. Instances of reversed copulatory position, though reported for the Adélie by *Falla* [1937] and *Taylor* [1962], must be very rare. I never observed it at Wilkes, nor did *Sladen* [1958] at Hope Bay or Signy Island. (2) Tread marks: When the ground is wet enough, muddy tread marks show clearly on the backs of females and are easily distinguished from a dirty back attained in fighting. At Wilkes this was a very good clue to sex. (3) Ecstatic display: This display is confined almost entirely to males, especially during the pre-egg stage. This criterion and the next one must be interpreted with caution. (4) Incubation routine: The female departs shortly after laying the second egg, and the male remains behind to incubate. *Sapin-Jaloustre and Bourliere* [1951] claim several cases of the female incubating while the male went to sea, and *Taylor* [1962] reports 6 such cases out of 35 breeding pairs. The birds in these cases were sexed by behavioral criteria.

The consistency of these behavior patterns used in sexing is shown in Table 2. The mates of the 25 dissected birds are assumed to be of the opposite sex and are, on this basis, also included in the table. The dissected birds came from colonies IV and V. Several individuals were chosen for dissection especially because they showed contradictory behavior patterns, which biases Table 2 toward abnormal behavior.

Table 3 shows the consistency of three of the four sexing criteria for 51 birds in colony IV during the 1960–1961 season. Although only 10 birds were dissected, the other 41 birds had associated with them by true or trial pairing (page 117) providing a reliable basis for sexing by association. The 51 birds represent 95 per cent of the colony population and are thought to show more representative behavior than Table 2.

The sex of a bird is not assumed in this report except on the basis of two or more of the four behavior

TABLE 2. The Occurrence of Sexing Clues in 25 Dissected Birds and Their Mates* during Two Seasons, 1959–1960 and 1960–1961

Sexing Criteria	31 Males (11 dissected)		26 Females (14 dissected)	
	No. Instances	No. Birds	No. Instances	No. Birds
Dorsal copulatory position	75	21	0	0
Ventral copulatory position	0	0	73	23
Presence of tread marks on back	0	0	29	19
In ecstatic display	217	25	37†	6
Assumed first incubation fast	42 (46)‡	30	4	4

* Eleven dissected males with 12 mates and 14 dissected females with 20 mates in two seasons.

† Only 3 displays recorded for 2 females before the end of the egg-laying periods. One female gave 31 of the 37 displays.

‡ Number possible. There were 4 cases of non-breeding in the two seasons by the 25 dissected birds and their mates.

criteria discussed above. In nearly all cases paired breeding birds were studied, which provided an additional check on sex by association with another banded bird.

Nest marking and mapping. Nest locations in colonies I through VI were marked with steel welding rods 0.8 centimeter in diameter and 91 centimeters long driven into the "guano-stone" substrate. Color coded with durable plastic tape at the top, 142 rod markers served to locate several hundred nests. In colonies VII and VIII numbers were painted on rocks at the nest sites.

To measure distances between nests, scaled photographs were taken from a camera attached to a long pole. The camera was held either 3.4 or 6.8 meters above a portion of a colony of nesting birds and the photograph taken perpendicular to the plane of the colony.

Colony boundaries, as determined by the distribution of guano, nesting stones, and nest scrapes, were mapped to within 15 centimeters by tape measure transects in early October before colony occupation. Information on the extent of colony area obtained in this way was used in conjunction with population counts to determine territory size and nesting density.

A plane-table map (Figure 3) of the entire peninsula bearing rookery A shows the colonies and campsite.

Population counts. Every four to six days photo-

graphs of each of the colonies were taken from the best available vantage points. The number of adults and/or large chicks was counted on each photograph. *Sapin-Jaloustre and Cendron* [1953] found this an efficient way to maintain population counts on Adélies. This method is timesaving, does not disturb the birds, and provides records for long-term population studies in particular areas.

Systematic population photography does have limitations. Colonies I, III, IX, and X were not easily photographed because of their large size or lack of high vantage points. Direct counts in these colonies during population peaks indicated that the photographic counts were about 10 per cent below the actual numbers present. Periodic checks on the photographic records in other colonies showed only an occasional minor discrepancy.

Observing and recording. One small colony, IV, consisting of 30 breeding pairs in 1959–1960, was chosen for detailed observation on behavior. The activities of all birds in this colony were recorded once or twice daily, and many hours were spent recording behavior in the colony from a high observation point. In addition to this systematic observation I recorded many specific events in the colony while working on other things within view of it, or while recording daily observations at the campsite, which was purposely established overlooking colony IV.

Noel Orton and Frank Soucek made checks on all banded birds in colony IV during the 1961–1962 and

TABLE 3. Incidence of Ecstatic Display, Copulatory Position, and Incubation Routine in Colony IV during the 1960–1961 Season for 10 Dissected Birds and 41 Others Sexually Associated with Them

Sexing Criteria	27 Males (6 dissected)		24 Females (4 dissected)	
	No. Instances	No. Birds	No. Instances	No. Birds
Dorsal copulatory position	119	25	0	0
Ventral copulatory position	0	0	105	23
In ecstatic display	271	25	27[a]	4[b]
Assumed first incubation fast	24[c]	24[d]	0	0

[a] Twenty displays by one female.

[b] Only one female giving the ecstatic display before the end of the egg-laying period.

[c] In four instances the male went to sea for 2 to 4 days between the laying of the first and second egg but relieved the female before or shortly after she laid the second egg.

[d] Three males did not breed.

1962–1963 seasons. Their observations are incorporated with the data on the permanence of territories (page 106).

Once every three to four days a thorough search was made of all the colonies in the rookery to record the presence of banded birds and to note their activities. Binoculars and scope eliminated the necessity of catching a bird to read its band number.

To record daily activity rhythms and seasonal progression of the breeding cycle, a time-lapse camera was installed over a portion of colony I during the 1959–1960 season and over colony IV during the 1960–1961 season. This camera automatically took a 16-mm photograph every 10 minutes throughout each of the two breeding seasons. Postures and displays can be identified in these photographs and related precisely to the time of day and season.

In colony IV many hours were spent tape-recording vocalizations in various situations. A microphone with parabolic reflector, used primarily to shield out noises from other colonies, was suspended permanently 3.5 meters above the colony. A Magnamite model 610 recorder was used at a tape speed of 19 cm per second. These recordings were later analyzed with a Kay Electric "Missilizer" model 675. Tape recordings were used in play-back experiments to study responses showing individual auditory recognition.

Both still and motion-picture photography were used to record and analyze behavior [see *Sladen*, 1958]. Motion pictures were an indispensable aid to verbal description. From about 1500 meters of exposed 16-mm black and white film two short reels of film were prepared to illustrate (*a*) the breeding cycle and basic displays and (*b*) techniques used in the study of individual recognition.

Disturbances. Care was taken during observations to disturb the subjects as little as possible. A densely packed colony of Adélies may at first glance give the impression of indifference to human intrusion, but as I became more familiar with the birds I could perceive that my presence affected their behavior. Some shy birds moved away from me when I was as much as 30 meters distant; others showed no reaction at 6 meters. I lessened my influence on colonies with a slow approach from downhill. The reactions of birds in frequently visited colonies, in comparison with those rarely visited, revealed a limited amount of habituation to my presence.

When I approached birds away from their colonies I generally found that banded birds reacted more strongly to my presence than unbanded ones did. Molting birds were more threatening than non-molting, and juveniles more shy than adults. Molting juveniles often growled threateningly at me or withdrew when I came within 50 meters of them.

Disturbances caused by banding and by experiments were temporary and kept at a minimum. Experimental manipulation of birds used in the study of individual recognition was minor and did not seem to affect breeding success. I used 92 adults in homing experiments. Nearly all these returned to the rookery within a few days and resumed their normal breeding routine (page 106).

On one occasion a husky dog entered the rookery and killed two unbanded birds on the beach. Two injured birds were killed and dissected in addition to the 25 mentioned under sexing (Tables 2 and 3). Twenty molting adults and juveniles from rookery A were temporarily confined for special studies; 13 of these were sacrificed for sexing.

THE BREEDING CYCLE

Most types of social intercourse between Adélies are closely associated with specific stages of the breeding cycle. Several of these social relationships are treated in later chapters. The present sketch of the breeding cycle is presented as background for those particular studies.

The form and chronology of the breeding cycle at Wilkes was very similar to that reported by *Sladen* [1958] for Hope Bay and Signy Island, by *Sapin-Jaloustre* [1960] for Phoque Island, and by *Taylor* [1962] for Cape Royds.

The first Adélies arrived in rookery A for the 1959–1960 season in mid-October. They had to traverse 10 kilometers of relatively fresh sea ice to reach the rookery. Individuals returned walking upright, either singly or in small groups. They distinctly avoided the thin blue ice and favored the older white ice. This avoidance was related to the basic avoidance Adélies show to water and weak ice when threatened by leopard seals (*Hydrurga leptonyx*), humans, or dogs. Crossing the thin blue ice was normally a group activity, as was entry into the water later in the breeding season.

In October of the following season the sea ice was considerably more extensive and covered with a layer of soft snow 15 centimeters thick, which seemed to slow the influx of birds into the rookery. In 1959 rookery A contained more than 1500 Adélies by October 30. At the same date in 1960 only one-fourth of this number was present. Groups of Adélies labored many hours toward the rookery, tobogganing on their bellies through the soft snow. Long lines or "trains" of Adélies could be seen far out from the rookery as birds continued to follow in the tracks broken by a lead bird. During pauses the lead bird was often replaced.

The delay in the influx of Adélies the second season lasted until October 31 when a blizzard blew the sea ice out to sea, and the birds could then swim right up to the rookery. No delay in the timing of the subsequent stages of the breeding cycle was apparent.

An important aspect of the behavior of returning birds was their extreme gregariousness. Adélies on the ice were very responsive to one another's movements. Individuals seldom strayed far from the group, especially when on the thin blue ice, and this appeared in sharp contrast to the behavior of individuals once they entered the boundaries of the rookery.

Upon entering the rookery each individual characteristically went his or her way and would not tolerate another bird within about 0.5 meter. Paired birds did not arrive together. Birds that had previously bred in the rookery made their ways directly to their former colonies and territories (page 103). During the initial occupation of the rookery males tended to outnumber

the females. Recovery of banded birds in colony V in the 1960–1961 season showed 2 males and 3 females present on October 29. On October 31 there were 15 males and 10 females present, and on November 3 the males outnumbered the females 50 to 44. Of 28 birds in colony IV that kept the same mates two successive years, one pair arrived on the same day, the females of two pairs returned one day before the males, and in the other 11 pairs the males arrived first by one to several days.

When a female joined a male on his territory, both birds shared in augmenting the pile of stones that they later hollowed out as a nest for the two eggs. Copulation occurred as early as 2 minutes after male and female met at the territory. It was repeated many times up to egg laying and occasionally between the first and second eggs.

The time between arrival and egg laying varied from 2 to 4 weeks. During this time the rookery population increased rapidly to a peak in mid-November. This period, which *Sladen* [1958] calls the occupation period, overlapped the egg-laying stage and incubation period which began the second week of November (Figures 5 and 6).

In 18 nests in colony IV the time between the laying of the first and second egg averaged 2.2 days and

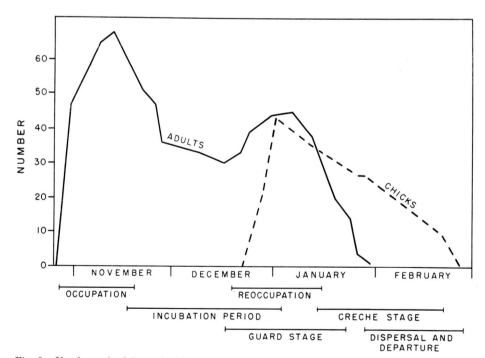

Fig. 5. Numbers of adults and chicks in colony IV of rookery A during the 1959–1960 season and during the various periods and stages of the breeding cycle. The decrease in chick population during January was due to mortality; in February it resulted from dispersal and departure.

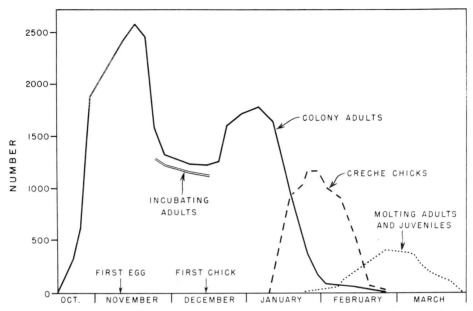

Fig. 6. Numbers of colony adults, creche chicks, and molting adults and juveniles in rookery A during the 1959–1960 season.

ranged from 1 to 4 days. In colony V, where fresh eggs were marked once each day in 23 nests, an average of 3.0 days (range of 1 to 4 days) elapsed between the laying of the two eggs.

Normally within a day after laying their second egg the females left their home colonies and gathered in groups on the ice foot overlooking the sea. At a given time, nearly always coordinated with the presence of other Adélies in the water nearby, the group of females suddenly became very active and belly-flopped into the sea. They swam northwest toward the pack ice, pausing briefly to wash their muddy feathers and preen themselves. Some were generally left behind; these

waited until another group formed before entering the water.

The average length of time ashore for pairing, nest-building, and egg-laying was 17 days for females in the 1960–1961 season (Figure 7). After the females left, the males continued to incubate the eggs uninterruptedly until their partners relieved them 2 weeks later. This occurred the first week of December in 1960, making an average of 5 weeks ashore for the 18 males.

The breeding activities (egg-laying, incubation schedules, and nest reliefs) of the many pairs in rookery A were synchronized to within 2 to 3 weeks. Such syn-

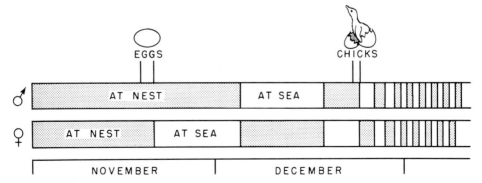

Fig. 7. Average schedule of presence and absence in the colony for breeding males and females in colony IV of rookery A from the occupation of territories to the termination of the guard stage. The movements of 18 pairs with normal incubation routine and success at raising chicks are averaged in this diagram.

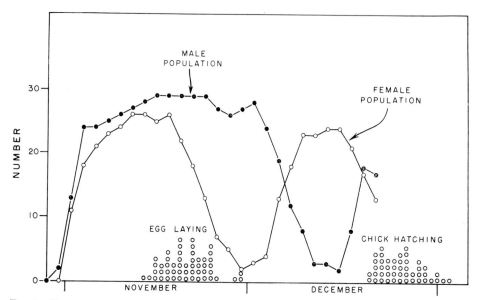

Fig. 8. Total numbers of males and females and the time of egg-laying and chick hatching in colony IV of rookery A during the occupation and incubation periods of the 1960–1961 season.

chrony led to the characteristic drop in colony and rookery populations during the incubation period (Figures 5 and 6) and also led to fluctuations in the percentages of males or females in a colony during the incubation period (Figure 8).

The relief of an incubating bird by its partner took place in a stereotyped fashion. The first interaction between birds was a series of loud vocalizations and mutual head wavings (page 121). After one to several minutes of this the returning bird replaced its mate over the eggs, and the pair again called and waved their heads for several minutes. Before leaving for the sea the relieved bird stood by the nest or made short excursions outside the colony to collect stones for the nest, sometimes for several hours. This stone gathering insures a well-built nest, keeps the eggs high and dry from melt water, and reduces the likelihood of the eggs rolling away. Adélies were never observed to retrieve an egg once it was completely outside the nest.

During the incubation period the Adélie colonies were quiet and inactive. Incubating birds lying prone on their nests changed their positions every few hours (see Figure 9). The only major activities to disturb this quiet were nest-relief ceremonies. During warm weather brooding birds often just crouched over the eggs, with one egg on and between the two tarsi and the other egg on and between the two feet. High winds and blowing snow induced the birds to face into the wind and to lie close on the nest. Interestingly enough

this orientation to the wind reversed after the eggs hatched.

After being relieved by their partners and going to sea, breeding males remained away for an average of two weeks before returning and relieving the incubating females. This second period of incubation by the males usually lasted less than a week and closely approached hatching time (Figure 7).

In the 23 nests in which the eggs were marked daily, the incubation time averaged 37.2 days for the first egg and 34.6 days for the second egg. In the 14 nests where both eggs hatched, they hatched on the same day in 12 and on successive days in 2. In 7 nests only one egg hatched; the fertile egg was the first in 2 cases and the second in 5. Two nests lost both eggs before hatching.

Either the male or female of a pair might be on the nest when the eggs hatched, producing downy chicks weighing about 80 grams. At this time the frequency of nest reliefs increased, and one parent guarded the chicks while the other went off to collect food at sea. Cursory observation of food spilled during feeding showed it to consist primarily of euphausid shrimps plus various species of amphipods. This period of guarding and food collecting, called the guard stage, began the last two weeks of December and continued into the second week of January.

The increase in nest-relief frequency after hatching was essential to chick survival because a fasting incubating parent cannot feed a chick. In 39 of 45 nests

Fig. 9. Variations in colony topography. *A, (top)* colony XI during the incubation period. *B, (bottom)* colony VII during the incubation period.

the first nest relief took place within 2 days after hatching. Three chicks died after 5, 6, and 7 days, respectively, apparently from starvation, while being guarded by a fasting parent.

In at least three cases parents whose eggs had hatched during their absence at sea were able to feed chicks immediately upon returning to the nest. The food demands of small chicks were not heavy and were characteristically met with well-digested food (more juices than solids) regurgitated in small amounts throughout the day. Larger chicks, of course, demanded greater amounts of food. Parents returned to late guard-stage chicks heavily laden with food.

The time between the first two nest reliefs of the guard stage was 1 to 3 days, and within a week nest reliefs generally occurred every day (Figure 7). These frequent nest reliefs contributed toward an increase in social activity within colonies.

Non-breeding adults that had been present in the colonies during the occupation period, adults that had lost their eggs, and some newcomers occupied territories within the colonies at this stage of the breeding cycle. This new surge of activity produced a second peak in colony and rookery population called the reoccupation period (Figures 5 and 6).

Chicks were left alone by their parents for the first time at about three weeks of age. When so deserted the chicks gather in groups called creches. This creche stage of the breeding cycle began in the second and third weeks of January and continued to the end of February. In colony IV the two youngest chicks were left alone when between 18 and 23 days of age. *Taylor* [1962] found that early chicks were guarded longer than later chicks, and the average age at desertion was 22.4 days. *Sladen* [1958] found considerable variation in the age of chicks at desertion. At Hope Bay the average desertion age for 6 chicks was 30 days, whereas for 10 chicks at Signy Island it was 19 days.

The first chicks to be left alone by their parents were harshly pecked by neighboring adults when they wandered over territory lines. A chick in this situation is often buffeted about by a series of adult territorial defenders until it returns to its own territory, is ousted from the colony, or finds shelter with the guard-stage chicks of a less pugnacious adult. In colony VII the same pair of adults produced the earliest chicks two successive years. Both years the chicks, after being left alone for the first time, were so severely treated by neighboring guarding adults that they were ousted from the colony and died within five days.

The number of chicks in a fully formed creche varies from a few individuals to several dozen. Small colonies usually have one creche, and large colonies several. The tightness with which the chicks huddle together varies with the temperature, threatening situations, and age of the chicks. On cold windy days the chicks huddle close together, sometimes one on top of another. On warm days they usually disperse over the colony, some of them, usually older chicks, returning to their natal territories. In the bright warm sun, creche chicks often lie prostrate with head, flippers, and feet extended. If a skua threatens to attack, the creche reforms quickly into a tight group of alert standing birds jostling each other. They respond similarly to human disturbance.

At approximately seven weeks of age creche chicks lose most of their down and acquire the juvenile white-throated plumage. Chicks of this age are more independent and do not form creches as often or as tightly as younger chicks.

The daily feeding routine established by adults during the guard stage continues into the creche stage. My observations were not consistent enough to determine whether the feeding continued daily when the chicks were six to seven weeks of age.

The oldest chicks began dispersing from their creches and colonies during February and gradually assembled on the beaches. The chicks often moved up and down the beaches exercising their flippers and calling. Some chicks were fed on the beaches almost to the moment of departure. The chicks go to sea independently, unaccompanied by their parents. When the last had left by the end of February, it marked the end of the breeding season. A small population of adults and juveniles remained behind for the molt during February and March (Figure 6).

TERRITORY AND TERRITORIAL RELATIONSHIPS

Size and Spacing of Nesting Territories

In Adélie penguin colonies the nests are spaced closely and regularly. A ratio of the number of occupied nests to the colony area expresses the average nesting density and territory size. Nest counts are best made during the early incubation period so as to include the maximum number of occupied sites.

The nesting densities of 11 colonies of rookery A during the 1959–1960 season were determined from area maps and a visual count of occupied nests on December 11, 1959 (Table 4). The average area per nest in all 11 colonies was 0.75 m². Circular colonies

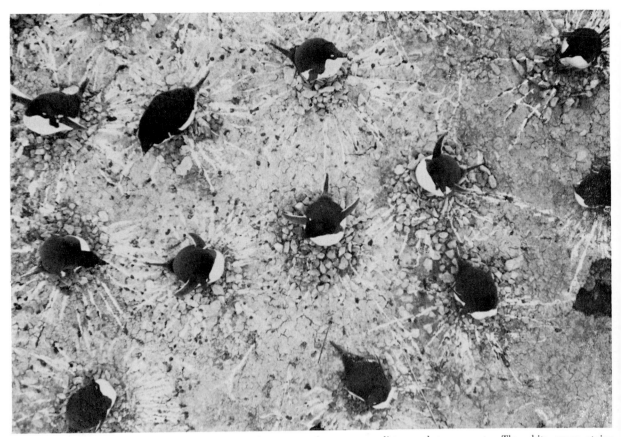

Fig. 10. Spacing of nests. Such vertical photographs were used to measure distances between nests. The white guano stains distributed radially around the nests reflect the positioning of the incubating birds on their nests.

with uniform flat topography, such as XI, reflect territory size (0.49 m²) more accurately than do colonies such as VII that are broken up by uneven ground (0.93 m²). The extremes in colony topography in rookery A are shown in Figure 9a and b.

I was repeatedly impressed by the consistency of

TABLE 4. Area and Nesting Density for 11 Colonies in Rookery A during the 1959–1960 Season

Colony	Area, m²	Number occupied nests on Dec. 11	Area per Nest, m²
I	300.2	352	0.85
II	14.1	20	0.70
III	281.7	306	0.92
IV	20.2	30	0.67
V	73.6	93	0.79
VI	21.3	29	0.73
VII	51.9	56	0.93
VIII	15.7	25	0.63
IX	138.0	184	0.75
X	38.8	48	0.81
XI	35.0	72	0.49

nest spacing by birds in the center and on the periphery of colonies. Measurements showed the average distances between nest centers to be only 7 centimeters greater for peripheral birds (Table 5). A nest was considered peripheral if on the colony edge and central if it had at least one other nest between it and the edge. Only five nests in rookery A during the 1959–1960 season had no neighbor within 2.3 meters, yet there were innumerable apparently suitable nesting sites beyond the colony boundaries.

The sizes of nesting territories in Table 4 vary because the areas of colonies, as measured, include varying numbers of rocks, water pools, and drainage gulleys unsuited for nests. Also, some potential nesting sites were never occupied or were deserted by December 11th. Measurements of distances between nests (Table 5) reflect the possible degree of crowding in a colony. Distances measured arbitrarily to the two nearest neighbors failed to show consistent variation in nest-spacing because of different types of surface in the nine areas measured. A hypothetical nesting territory constructed on the basis of distances between nests

TABLE 5. Distances between Nest Centers for Peripheral and Central Nests in 9 Locations in 7 Colonies as Measured from Scaled Photographs Taken on December 9, 1960

Location	Peripheral Nests		Central Nests	
	No. of Nests Measured	Average Distance to Two Nearest Neighbors, cm	No. of Nests Measured	Average Distance to Two Nearest Neighbors, cm
1.	12	67	4	63
2.	4	75	6	64
3.	. . .	. . .	4	66
4.	5	78	. . .	. . .
5.	12	69	7	58
6.	. . .	. . .	5	70
7.	5	74	2	67
8.	4	82	. . .	. . .
9.	7	63	2	66
Total	49	Avg. 72	Total 29	Avg. 65

averaged 65 to 72 cm in diameter and between 0.4 and 0.5 m² in area.

Territory limits can best be defined behaviorally. Nesting Adélies frequently pecked each other across territory boundaries. Birds at central nests could generally reach four to six neighbors in such disputes, but this varied with their position on the nest (Figure 10). A territory with boundaries contiguous to six others would represent maximum use of available nesting area.

Displays Used in Territorial Defense

To defend and maintain its territory the Adélie uses a number of different postures and movements. These vary from simple threat displays to prolonged combats between two or more birds. On the basis of the positioning and movements of the head and flippers, three displays, termed here the alternate stare, the direct stare, and the bill-to-axilla display, are encountered regularly in territorial defense. Attack, pecking, and outright fighting (page 104) are considered extreme variations of the displays described in this section.

The alternate stare. This display is one of two threat attitudes *Sladen* [1958] describes as pictured in an earlier illustration by *Roberts* [1940]; he did not name it specifically. Unfortunately Robert's illustration of what he terms "aposematic display" lacks sufficient description to differentiate it from intermediate positions of two different displays recognized in this study.

Standing, prone, or crouching birds may give the alternate stare (Figure 11). In this display the bird depresses the bill and arches the neck up. The neck arching is accentuated by erecting the feathers on the neck and back of the head (crest). The bird keeps the bill closed, turns the eyes down to expose the white sclerae, and usually extends the flippers laterally and waves them slowly and irregularly. The bird presents alternate sides of the face toward the adversary (another penguin, a skua, or a human) by turning the head in irregular movements from side to side, while trying to keep the adversary in front. Although usually given silently, the alternate stare is sometimes accompanied by a low growl (Figure 12a).

The alternate stare was confined to birds on their territories except when approached by a human in a position from which they could not escape easily. Females, being generally shier than males, were less inclined to this display during the pre-egg stage of the breeding cycle. While incubating or guarding chicks, however, both males and females frequently assumed this display in response to intruding penguins, skuas, or me.

There can be little question of the threatening nature of this display. On many occasions it delayed and discouraged the approach of territorial intruders, both avian and human. If the intruder continued to advance, this display often developed into either flight (more frequent with birds on newly acquired territories) or a more intense threat followed by pecking and attack.

The characteristic jerky movements of the flippers in this display lead me to believe that it is a low-level threat display. Such flipper movements, as noted first by *Sladen* [1958], suggest nervous apprehension and often precede flight.

Fig. 11. Incubating birds giving alternate stare to a bird seeking nesting stones.

The direct stare. The direct stare, also referred to as a threat attitude by *Sladen* [1958], is given by a standing bird facing its opponent. The crest feathers and eyes are the same as in the alternate stare. The bird keeps its bill tightly closed except while vocalizing, usually holds the flippers close to the sides, and presents a frontal view of the face to the threatened object (Figure 13). When directed toward a flying skua, a giant petrel, or to a tall human, the bill is pointed upward. When disturbed by another penguin, the bill is held nearly straight out. The direct stare may be silent or accompanied by a harsh growl or

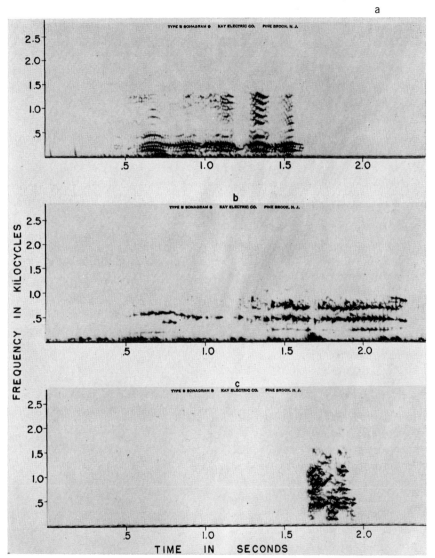

Fig. 12. Sound spectrograms of the sounds given during the alternate and direct stare threat displays: (a) alternate stare growl, (b) direct stare growl, (c) squawk given by a bird lunging at a skua. (All spectrograms presented in this report were produced on a Kay Electric model 675 sound spectrograph. Using a single taped pattern, highly variant sound spectrograms can be produced by varying the signal amplitude input, gain, and speed adjustments of the spectrograph. Absolute equivalence in the amplitude of field recorded sounds is impossible to obtain. For all comparisons within displays, however, the sounds in this report were analyzed with standard amplitude input from the recorder to the spectrograph and standard gain adjustments. The amplitude range of these and following spectrograms is about 22 decibels from a very dark marking to very light grey. Time and frequency are as indicated on the ordinates. The presence of sound components below but rarely above 100 cps is due to background noise.)

Fig. 13.   A creche stage chick presents a direct stare display
to an attacking Skua.

squawk (Figure 12*b* and *c*) and is often the precursor
to attack.

Adults and creche stage chicks gave direct stares
both on and off their territories. In birds on their
territory this threat display occurred most often just
before attacking an intruding penguin or in response
to a diving skua. Off the territory the display was
given by pugnacious birds on the loafing grounds prior
to attacking my trouser leg, before rushing a resting
skua, and before retreating. On rare occasions a loafing bird gave the direct stare when approached closely
by another Adélie.

As a prelude to overt attack, the direct stare is
clearly meant to repel individuals. In this way it also
functions to prevent territorial violation.

When accompanied by jerky movement of the extended flippers, this display nearly always indicates
that the bird intends to retreat and not attack. In this
form the direct stare coincides with *Sladen*'s [1958]
"nervous" response and *Sapin-Jaloustre*'s [1960] "la
position angélique."

The bill-to-axilla display. This display was first described and named by *Sladen* [1953]. At first he considered it a subsidiary of the ecstatic display, but he
later [1958] described it separately. During this display the body is bent forward 45 degrees and the
closed bill points into the axilla of the flipper (Figure
14). The feathers on the crest and neck and the eyes
appear as in the previous threat displays. As the bill
points into the right or left flipper axilla, the head is
slowly swiveled or rotated through 50 to 60 degrees
of arc, and the flippers are waved up and down irregu-

larly. A low growling sound is emitted with each roll
of the head.

The posturing during this display is quite variable.
In about 10 per cent of the displays, birds lean forward and roll the head while pointing their bills toward the ground instead of to the axilla. In a few
cases within a single sequence the bird may display
first on one side of the body, then on the other. It is
not unusual to see a bird perform the display while
standing upright or while lying on a nest.

Sonographic analysis of five normal display vocalizations revealed considerable variation in structure.
Figure 15*a* illustrates a harsh sound while 15*b* illustrates a soft whirring with easily recognized harmonics.
The sounds show little variation in their lower frequencies, which begin at 140 to 160 cps. The upper
frequency limit ranges from 490 to 1410 cps, a span
of three to seven octaves. The duration of the sound
varies from 0.8 to 4.1 seconds during which 3 to 8
pulses are usually recognizable with each roll of the
head.

The intensity of the display varies from posturing
with no sound to very rigorous movement accompanied by relatively high amplitude growls not necessarily proportional to the duration of the display or its
tonal qualities.

Because it often followed the ecstatic display (page
109), the bill-to-axilla display was seen most frequently
during the occupation and reoccupation periods. It
was given predominantly by unpaired lone males, but

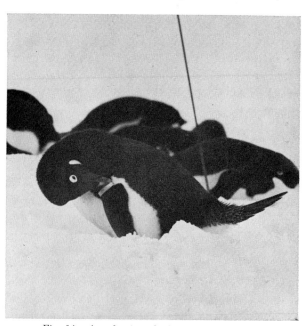

Fig. 14.   A male gives the bill-to-axilla display.

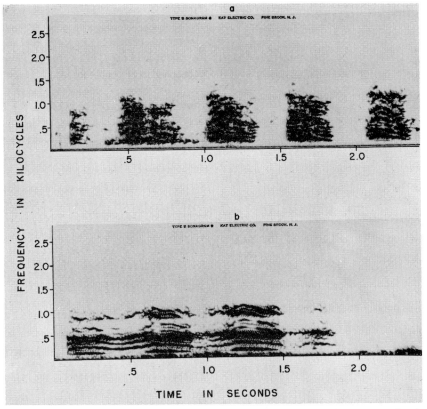

Fig. 15. Sound spectrograms of two types of bill-to-axilla vocalizations: (*a*) a harsh sound lacking clear harmonics, and (*b*) a soft "whirring" sound with clear harmonics.

also frequently by males incubating eggs or guarding chicks.

A count was made from 1400 meters of unselected motion picture film to determine the frequency with which the bill-to-axilla display followed the ecstatic display (Table 6). Lone males without eggs or chicks gave the bill-to-axilla display after the ecstatic 54 per cent of the time. The incidence for incubating or guarding males was 86 per cent. The frequency dropped to zero for males giving the ecstatic display when the mate was present or when a female approached a male in response to his ecstatic display. Bill-to-axilla displays given independently of ecstatic displays occurred most frequently in incubating or guarding males. In this situation the display was generally more intense.

The bill-to-axilla display was a normal response by males in defensive situations. A male driven off a territory by another male during the occupation or reoccupation period, or one threatened by the intrusion of a wandering bird while incubating or guarding, often gave the bill-to-axilla display independently of the ecstatic display or after an interrupted one.

The four females in colony IV that gave ecstatic displays in threatening situations (see Table 3 and page 88) followed the ecstatic with bill-to-axilla displays in 14 of the 27 cases. Two of the four females were incubating and gave bill-to-axilla displays without the ecstatic in response to the nearness of wandering birds.

Sladen [1958] suggests that the bill-to-axilla display may play a role in pair-formation by selectively repulsing birds of the same sex that are attracted by the ecstatic display. The only positive instance of repulsion by the bill-to-axilla display observed at Wilkes involved an incubating male and another bird of uncertain sex. The male, while giving an ecstatic display, was approached by an unbanded bird with the appeasing oblique stare bow (page 114). The incubating male performed an intense bill-to-axilla display, which suddenly halted the approach and oblique stare bow of the unbanded bird. On no occasion was the bill-to-axilla display observed as part of a normal pair-formation sequence (page 118).

In light of the meager evidence indicating the function of this display, my conclusions must remain ten-

TABLE 6. Temporal Association of the Bill-to-Axilla Display
and the Ecstatic Display

Status of the Displaying Bird	No. of Ecstatics with Bill-to-Axilla Following	No. of Ecstatics without Bill-to-Axilla Following	No. of Bill-to-Axilla Given Alone
Lone males without eggs or chicks	26	22	4
Incubating or guarding males	23	5	9
Males with or approached by female	0	20	0

tative. The posturing of the bill-to-axilla display and
its vocalizations are similar to those of the alternate
threat display. On a very few occasions the alternate
stare was observed to merge directly into the bill-to-
axilla display. When given independently of the ecstatic display, the bill-to-axilla often occurred when the
displaying bird's territory or position on a territory
was threatened. Upon this basis the bill-to-axilla appears to be an exaggerated threat display functional in
territorial declaration and defense.

Establishment of Territory

Because the ages of adult birds at Wilkes unfortunately
were unknown, I could not distinguish with certainty
between the territorial behavior of experienced breeders and inexperienced birds occupying territories for
the first time. *Sladen* [1958] presents behavioral observation and morphological data to show the existence of several age groups in communities of Adélie
penguins. His provisional age groups are:

1. *Established (experienced) breeders* three to five years old
   and over, mostly four years old and over.
2. *Unestablished (inexperienced) breeders* two to four years,
   mostly three years old.
3. *Non-breeders ("wanderers") in adult plumage* two to
   three years old, mostly two years old.
4. *Non-breeders ("wanderers") in immature plumage (yearlings)*; from leaving the rookery until they moult into
   adult plumage when about fifteen months old.
5. *Nestlings (chicks)*; up to the time when they leave the
   rookery, nearly two months after hatching.

The suitability of this age classification has not been
fully tested on a population of birds of known age.

One record has been published of an Adélie banded
as a chick and known to have nested (with a chick)
in its natal colony 4 years later [*Sladen*, 1958; *Sladen
and Tickell*, 1958]. John Cranfield (personal communication) reports the recovery of ten Adélies in
their natal rookery at Cape Hallett 4 years after being

banded as chicks; five of these were recovered at territories in their natal colonies. Two of the ten birds
were females that laid one egg apiece (see page 103),
but neither raised a chick successfully. None of the
other eight bred at 4 years of age. These birds generally arrived late at the rookery, and at least six
changed territories frequently or wandered about the
rookery to various colonies; five of the eight that occupied territories did so on the periphery of colonies.

On December 15, 1962, in rookery A at Wilkes,
Frank Soucek (personal communication) recovered a
single Adélie nesting in colony V 4 years after being
banded as a chick in the same colony. In colony I he
also recovered a bird 2 years after it was banded as a
molting juvenile. Studies by the U.S. Bird-Banding
Program at Cape Crozier (Sladen and Penney, unpublished data) recorded both 2- and 3-year-old Adélies
present in their natal rookery in 1964 and 1965. A few
of the 3-year-olds were breeding.

Young birds. The behavior of non-breeding males,
assumed to be young birds, was watched closely in
colony IV for comparison with breeding birds. The
presence and activities of the seven non-breeding males
are summarized in Figure 16. The major activities of
these birds were ecstatic displays and non-reproductive
pairing during the occupation and reoccupation periods. Late in the breeding season they collected nesting
stones and pecked at creche chicks to drive them from
their territories. Non-breeding males performed these
activities more often than the unsuccessful breeders
that also reoccupied territories late in the breeding
season. Non-breeding males changed territories rather
often; the unsuccessful breeding males consistently
reoccupied the same territories.

Four of the seven non-breeding males returned to
colony IV the following season. Two of them (58195
and 58118) bred successfully; one (58754) lost its
egg, and another (57768) failed to get a mate. The
two successful breeders kept the territories they had
occupied late the previous season. The fifth male
(58260) was observed unpaired in rookery B the second season, and the two remaining birds were never
seen again.

The non-breeding males in colony IV during the
1959–1960 season associated with five different females
during the reoccupation period. These five females, all
non-breeders, differed most noticeably from breeding
females in that they spent time with a greater number
of different males at different sites (page 117). All five
females returned the second season, mated, and laid

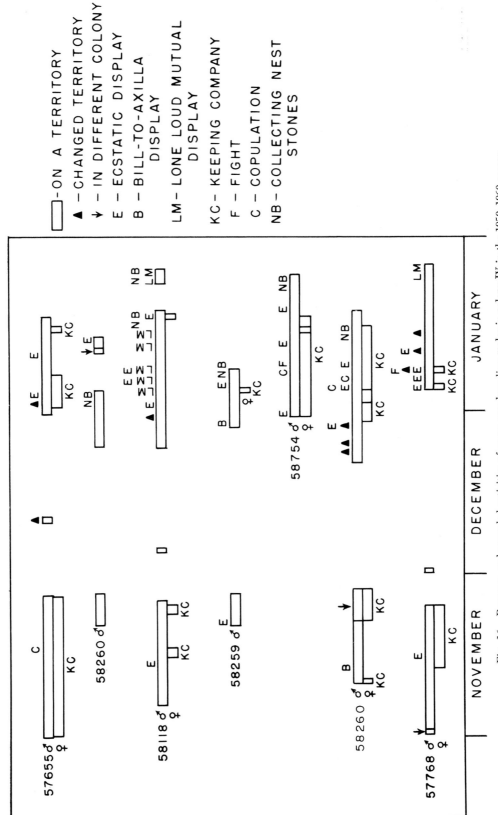

Fig. 16.   Presence and recorded activities of seven non-breeding males in colony IV in the 1959–1960 season.

TABLE 7. Behavior of 24 Breeding Males within the First Ten Minutes of Arrival in Their Previous Breeding Colony

Number Observed to:	Of 18 That Returned before Their Former Mate	Of 6 That Returned after Their Former Mate
1. Return to their former territory	18	4
2. Give loud mutual display toward their former territory	8	3
3. Give loud mutual display to a female on their former territory	3	6
4. Give ecstatic display	9	2
5. Give bill-to-axilla display	7	0
6. Quarrel or fight with another male	10	3
7. Collect nest stones	4	0
8. Copulate with a female on their former territory	0	2

eggs; three of the five were successful breeders while the other two laid only one egg each and failed to raise a chick.

The non-breeding males and females in colony IV seem to fall somewhere between the "non-breeding 'wanderers' in adult plumage" and the "unestablished breeders" of Sladen's [1958] classification. They showed high preference for colony IV and a limited attachment to certain territories within the colony. They did not move aimlessly about to other colonies as did wandering adults.

Initial reactions of experienced breeders to former territories. Former nesting territories hold great significance for experienced breeders. This was most obvious in the way the birds acted during the first few minutes of their return to their former breeding colonies in late October and early November 1960 (Tables 7 and 8). The previous season's nest locations and mates were known for all the birds in these tables.

All the 24 males but 2 returned to their former territories at once (Table 7). None of the sites was covered by snow. The two males that did not return in the first 10 minutes did so later on. A loud mutual display (page 115) with head initially directed at the ground was generally the first reaction upon arrival. If the former mate was present the display was done with the female; otherwise it was alone. The ecstatic (page 109) and bill-to-axilla displays were performed fairly frequently by males that returned before their mates, especially if the colonies were already crowded. When a male found another male already on his former territory, a fight ensued.

All but 2 of the 22 females also returned to their former territories within 10 minutes of arrival (Table 8). If the former mate was present, they stayed on the territory. If the mate was not present, the females stayed on the territory, sometimes only briefly, or for as long as two days before joining a new male at a different site. Females that joined new males in the absence of their former mates nearly always rejoined their former mate if and when he returned later (page 121). The females did not respond as often as males with loud mutual displays directed toward the territory. If the former mate was present and alone, the pair normally gave the loud mutual display upon reunion. If the former mate had another female on the territory, the newly returned female always attacked and drove off the strange female.

The loud mutual displays given between paired birds when first reunited at their former territories are considered greeting ceremonies. When given by a lone bird toward the nesting territory, the stimulus may be a psychological association between the site and memory of past reunions (nest reliefs in which the loud mutual display is such an intimate part), or a territorial declaration. The ecstatic and bill-to-axilla displays by males are interpreted as territorial displays functional in pair-formation and in declaration and defense of territory.

These distinct reactions to former nesting territories by experienced breeders are in sharp contrast to the territorial behavior of non-breeding birds when they

TABLE 8. Behavior of 22 Breeding Females within the First Ten Minutes of Arrival in Their Previous Breeding Colony

Number Observed to:	Of 10 That Returned before Their Former Mate	Of 12 That Returned after Their Former Mate
1. Return to their former territory	8	12
2. Give loud mutual display toward their former territory	4	1
3. Give loud mutual display with their former mate	0	10
4. Fight and defeat female with their former mate	0	7
5. Keep company with a new male at a new territory	4	0
6. Wander about colony	3	0
7. Later reunite with their former mate	5	0
8. Collect nest stones	2	2

first establish a territory during the occupation or re-occupation periods. Non-breeders from the previous season were never observed to give the loud mutual display toward their territories and usually did not give the ecstatic display until they had occupied a site for several days. These birds appeared rather casual as they assumed a site, and they did not seek many nesting stones early in the season as did experienced breeders when they returned to their former territories. I watched two males, non-breeders in the 1959–1960 season, as they arrived back at their previous territories in early November 1960. Neither bird gave the loud mutual display toward the territory, but one did give an ecstatic display within the first 10 minutes. Non-breeding females, when they returned the second season, were not seen to show attachment to any territory where they had associated with males the previous season.

Territorial fighting. Active defense of territory took the form of threat displays and open attack. In attack the bird rushes toward its opponent, keeping the flippers at its sides until contact is made. The attacking bird, while pecking or holding onto the opponent with the bill, uses its flippers to pummel the adversary rapidly and forcibly. Fighting ranged from more or less continuous bouts lasting 10 to 15 minutes down to disputes involving quick pecking across territorial boundaries. The latter is the commonest activity between Adélies of both sexes in a dense colony.

Major fights often involved more than two birds. To determine the cause of a dispute one needed knowledge of sex and circumstance. Severe fights over territory were observed four times in colony IV. Territorial fighting was well illustrated when male 58741 (A) returned late to colony IV to find his former territory and mate taken over by another male (B). The incident, as detailed in my notes, exemplifies a number of individual and group reactions:

On November 15, 1960, at 2045 hours male A came from the beach and entered colony IV. He gave the loud mutual display to his former territory and mate. Male B, who had taken over the previous territory and mate of A four days earlier, immediately attacked A, driving him through the colony. Two neighboring males disturbed by the fighting also attacked A, but not B. After intermittant hard fighting A returned to his former territory and gave the loud mutual display toward the site and up in the air. His former mate replied, but B attacked again, and others joined in the fight amidst many loud mutual displays. A was forced to one side of the colony where he was also attacked by a female upon whose territory he had been forced during the fighting. The severe fighting lasted 10

minutes before A retired to the edge of the colony, breathing heavily and shaking his head. After 5 minutes he moved 40 feet away where he remained for 15 minutes. He then spent another 5 minutes standing on the edge of the colony before entering. Attacked by several birds including B, A was ousted again.

Throughout the next morning and early afternoon A remained on the edge of the colony giving many low-intensity ecstatic displays followed by intense bill-to-axilla displays. At 1520 hours he neared his former territory and the territory of another male. This male responded by a series of bill-to-axilla displays directed toward A over the right shoulder, but relaxed when A moved back further away.

At 1215 hours on the second day A again entered his former territory, but was defeated by B. After the fight B assumed the direct stare display toward A, who stood outside the colony. This was followed by intense bill-to-axilla and ecstatic displays. A series of quiet and loud mutual displays were given by paired birds, and several males joined in with ecstatic and intense bill-to-axilla displays.

Male A never regained his former site or mate but remained on the edge of the colony for various periods the rest of the season.

Three other major territorial fights were recorded in colony IV, all but one during the occupation period. Two of the fights were between non-breeding males. The longest and most severe fights seemed to occur when two males were both strongly attached to a particular site. Usually this happened in a situation such as that described above, but it also occurred between close neighbors during the occupation and reoccupation periods when the colonies were densely populated. An experienced breeding male just returning to his former colony was prone to fight at the least provocation (Table 7). The severity of these fights (blood was often drawn when a flipper hit a rock or the metal band of an opponent) was apparently limited only by the combatants' stamina.

Fights between individuals of opposite sex were unusual. Females commonly pecked at strange birds that approached their territory, but I saw only two females actually chase males away. Unaccountably a banded male was once observed to fight a female with which it had associated for four days. In another instance a male and female were observed in a serious fight which was suddenly interrupted when the female gave the loud mutual display. The male returned the display, and the two birds returned to the territory, bowed at the ground, and performed the quiet mutual display. Similar situations were observed in unbanded birds. Some of these may have started as mistakes in individual or sex recognition.

TABLE 9. Permanence of Territories in Two Consecutive Seasons in Colony IV

Breeding Status and Number 1959–1960	Number Returned 1960–1961	Number Keeping Same Territory 1960–1961
Breeding males	30	25	22
Breeding females	30	20	13
Nonbreeding males	7	5	2
Nonbreeding females	5	5	0

Finally, there were a few rare cases of males found beating the backs of females to force them off the nest and eggs. Two banded males thus forced their broody mates off the eggs, and two unbanded males also did so to females other than their mates. The banded males, but not the unbanded, took over incubation in these abnormal nest reliefs. As the unbanded birds ignored the eggs, they were apparently trying to take over the nesting territory illegitimately.

Permanence of rookery and colony residence. Transfers between rookeries in successive seasons were rare. Of the 1081 banded adults maintaining territories in rookery A in the 1959–1960 season, 792 (72.8 per cent) returned to A the next year and 9 (0.8 per cent) were found in rookery B, 1.2 kilometers east. None was found in cursory checks of the more distant rookeries of the archipelago (Figure 2). Of the 9 recovered in B, none was known to have produced eggs during its association with rookery A the previous season; therefore all are thought to have been young birds.

Within the rookery the tendency to return to the same colony in successive years is strong. Of 428 breeding adults banded at marked nests during the 1959–1960 season and recovered the following season,

TABLE 10. Territory Permanence for 60 Breeding Males and Females in Colony IV in Four Consecutive Seasons

Males Banded 1959–1960	Present 1960–1961	Present 1960–1962*	Present 1962–1963†	Females Banded 1959–1960	Present 1960–1961	Present 1961–1962*	Present 1962–1963†
57636	x			57688	x	x	y
57778	y	x		57676	y		
57682	x	x‡	y	57696	x		
57693	x	x	x	57792	x		
57698	x		x	58044	y		
57770	x	x		58072	x		
57794	x	x	x	58073	x	x	
57910	x	x		58074	x	y	
58076	x			58085	x	y	y
58117	x	x	x	58088	x	x	x
58119	x	x	y	58114	y		
58143	x	x		58150	y		y
58144	x		y	58252	x		
58198	y			58256	x	y	x
58254	x			58263	x	y	
58261	x			58572	y	x	
58262	x	x‡		58574	x	x	y
58274	x			58576	x		
58573	x	x	x	58104	y§		
58575	x	x		58196	y§		
58577	x		x	58223	x§		
57641	x§			58273	x§		
57681	x§			57638			
57802	x§			57675			
58741	y§			57678			
57639				57701			
57684				57713			
58071				58116			
58103				58258			
58199				58578			

x, Same territory as previous season.
y, Different territory from previous season.
* Colony checked on three visits by Noel Orton.
† Colony checked on ten visits by Frank Soucek.
‡ Band number not read with certainty, but probable.
§ Sacrificed at end of season for sexing.

TABLE 11. Permanence of Territory for 277 Breeding Pairs between the 1959–1960 and 1960–1961 Seasons

Sex	Total No. 1959–1960	No. and % Returning 1960–1961	No. and % Keeping Same Territory	No. and % Changing Territory
M	277	206 (73%)	191 (93%)	15 (7%)
F	277	222 (79%)	138 (62%)	84 (38%)
Total	554	428 (77%)	332 (78%)	96 (22%)

only 2 (0.5 per cent) changed colonies the second year; both were females. Five non-breeding females and two birds of uncertain sex and breeding history were also known to have changed colonies between two seasons.

Permanence of individual territories. Previous studies on the faithfulness of Adélie penguins to their territories are reviewed in the introduction. I kept a complete record of all territories in colony IV and their residents for two seasons. Table 9 shows the permanence of territory and the breeding status of the individuals during the 1959–1960 season. Territory permanence was high for breeding males (22 of 25), intermediate for breeding females (13 of 20), and low for non-breeding birds of both sexes (2 of 10). The lower territory fidelity in breeding females was due to relatively weaker territorial bonds in females who changed mates.

Colony IV was checked in four successive seasons for permanence of individual territories (Table 10). After four years, 6 of 9 males were within 3 feet of their original territories. Only 1 of 6 females was within 3 feet of her original territory.

Further information on fidelity to territory was obtained in other colonies. Table 11 shows the numbers and percentages of territory permanence between two seasons for 277 breeding pairs in various colonies in rookery A. The results are similar to those obtained in colony IV. The few shifts in territory sites by males (15) were all between 0.6 and 2.4 meters from the original site. Females changed territories much more often. Except for four females that changed territory together with their former mates, all changes accompanied a change in mate the second season. In 18 per cent (15) of the cases the females changed to neighboring territories between 0.6 and 0.9 meters from their original sites. The remaining changes (69) were to territories farther from the original site but always within the same colony. Only 2 females of 222 kept their original territories in the absence of their former mates. Obviously males choose the territory.

The high degree of attachment of male Adélies to their colonies and nesting territories was further illustrated by homing experiments. In experiments at Wilkes, sex was determined on the basis of the ecstatic display and/or copulatory position before the birds were taken from the rookery. Early in the 1959–1960 and 1960–1961 seasons 92 males were released 8 to 57 kilometers inland from Wilkes. Most returned within a few days, and all but seven returned within one month to the colony from which they were taken.

A more spectacular result was obtained from five males taken from their territories on December 3, 1959, and flown to McMurdo Sound near Cape Royds (Figure 1) for release. None of the five was paired, yet two and probably three (an individual with a worn spot at the band position on the flipper was found back at a territory from which one of the birds had been taken) returned to their original territories early the next season. The homing distance was at least 3800 kilometers along the coast past many Adélie rookeries. The average speed was about 13 kilometers per day.

Territory in Relation to Colony Form and Breeding Experience

Access to territory. Breeding colonies of Adélies take many shapes and sizes, but the distance a bird will travel through dense nesting to reach its territory seems to be limited. I do not recall seeing a colony anywhere in Antarctica in which this distance exceeded 12 meters.

Individuals with nests deep within colonies often hesitate to run the gantlet of aggressive territorial defenders. The females usually assume a docile, nonaggressive look while traversing other birds' territories to reach their nests. Sometimes a bird is so deflected from its route that it has to try again later.

In colony III, a large compact colony, birds followed the paths of least disturbance to their central nests. These paths were often along drainage gulleys lacking nests. The departure of unsuccessful breeders also opened new paths. Figure 17 shows a large colony with an empty center that contained the remains of many old nesting territories, still apparently quite suitable. This suggests that the colony grew so large it limited access to the center. I believe that access to territory is an important factor in limiting the size of dense circular colonies.

Choice of territory in relation to breeding experience and success. Sladen [1958] suggests that "unestablished breeders" tend to occupy peripheral nests and have low breeding success.

Fig. 17. A large colony in rookery B during the occupation period showing the absence of nesting at the center (see p. 106).

I tested this theory at Wilkes with 725 banded birds whose territories were noted as either central or peripheral during both the 1959–1960 and 1960–1961 seasons (Table 12). Birds that changed from central to peripheral nests or vice versa are not included in this analysis. In the first season the peripheral birds had a significantly lower (9.8 per cent) breeding success than the central birds ($\chi^2 = 6.7$, $p < 0.01$). The same birds showed an over-all increase in breeding success the second season. The peripheral birds showed a 14.3 per cent increase, and the central birds a 5.3 per cent increase, which made breeding success by nest location nearly equal the second year.

Sladen based his observation on birds that had produced at least one egg. Included in my analysis are 44 non-breeding adults (rated as unsuccessful in Table 12), 26 at peripheral nests, and 18 at central nests during the 1959–1960 season. These 44 birds were included in the comparison because they consistently occupied territories during the occupation and reoccupation periods. As mentioned before, nonbreeders that show persistent attachment to territories within a colony are not easily fitted into Sladen's age groups. Behaviorally they most nearly resemble his "unestablished (inexperienced) breeders." Exclusion of these non-breeders from the comparison still leaves peripheral birds with a 6.6 per cent lower breeding success, a significant difference from the central birds ($\chi^2 = 4.5$, $p < 0.04$) for the 1959–1960 season.

The general increase in breeding success by the birds analyzed in Table 12 could have resulted from a number of factors. Food may have been more plentiful, ice conditions more favorable, weather less severe, and loss of eggs and chicks lower, owing to

TABLE 12. Breeding Success of Central and Peripheral Birds in Two Seasons

	1959–1960				1960–1961			
	Successful breeders,		Unsuccessful breeders,		Successful breeders,		Unsuccessful breeders,	
	No.	(%)	No.	(%)	No.	(%)	No.	(%)
Central birds	294	(73.3)	107*	(26.7)	225	(78.6)	61	(21.4)
Peripheral birds	209	(64.5)	115†	(35.5)	193	(78.8)	52	(21.2)

* Includes 18 non-breeders (see text for explanation).
† Includes 26 non-breeders (see text for explanation).

the smaller partially predatory skua population. Food and ice conditions on the feeding grounds were impossible to measure. In both seasons no mass mortality of eggs or chicks resulted from winds or drifting snow, yet the winds were generally higher the second season (Table 1). The rookery skua population dropped from five to four pairs the second season.

The marked increase in breeding success by peripheral over central birds in successive seasons is not easily explained by external causes. The drop in skua population may have been a partial factor because skuas, when not scavenging, tend to harass peripheral nests more than central nests for eggs and chicks. No other external factors were noted that might increase the breeding success of peripheral over that of central nests. It seems reasonable to suppose that subtle changes within the birds through experience and maturity could be crucial factors, accounting for the higher increase in success in peripheral nests, if they tend to be younger.

In discussing non-breeding and presumably young birds in colony IV, it was shown that 8 of the 9 that returned the second year paired successfully and produced eggs, and 4 of the 8 successfully raised their chicks on their first try. This is a 44 per cent increase in breeding success for the 9 birds. All these birds were at peripheral nests. The 44 non-breeders in Table 12 showed similar results. Of 31 that returned the second season, 27 paired successfully and produced eggs; 15 of these raised chicks to the beginning of the creche stage, 9 of them at peripheral nests.

Age composition of a colony may also influence the choice of territories. Colonies XII and XIII showed high increases in breeding success in successive seasons. Both these colonies expanded rapidly in size (Figures 18 and 19). The number of pairs breeding successfully in colony XII increased fivefold, and those in colony XIII increased sixfold between the 1959–1960 and 1960–1961 seasons. Photographic counts of the maximum occupation period populations and minimum incubation period populations show that this increase continued into the 1961–1962 season (Table 13). Other colonies remained more or less stable in size in the three consecutive seasons, and another colony was added to the rookery during the third season.

Considering the high fidelity of breeding Adélies to their former territories and colonies, the rapid growth of colonies XII and XIII must have been due to influxes of young birds. These colonies reached their peak populations later in the occupation period than stable colonies did and had a higher percentage of non-breeding pairs and lone males during both the occupation and reoccupation periods. While most birds in the more stable colonies were intent on the normal breeding routine, many birds in colonies XII and XIII were busily establishing territories, displaying (ecstatic displays were very common), and pairing (Figures 20a and b). This higher level of activity also seemed to attract wandering adults that were passing through the rookery.

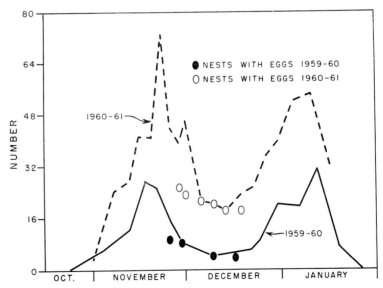

Fig. 18. Numbers of adults and nests with eggs in colony XII during the 1959–1960 and 1960–1961 seasons.

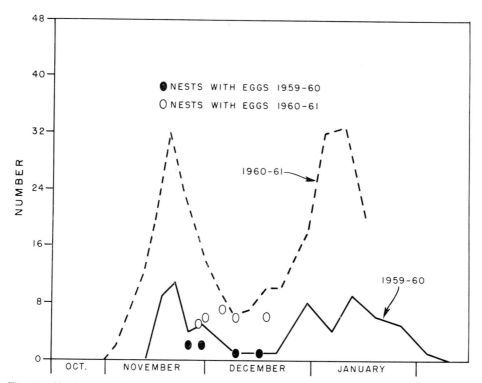

Fig. 19. Numbers of adults and nests with eggs in colony XIII during the 1959–1960 and 1960–1961 seasons.

ESTABLISHMENT AND MAINTENANCE OF THE PAIR BONDS

Sexual Displays

Four displays are recognized in this study as fundamental to the establishment and maintenance of pair bonds. These are discussed below together with their possible significance in non-sexual situations.

Ecstatic display. The ecstatic display was first described and named by *Wilson* [1907]. During this display the breast thrusts out, the bill points up vertically, and the feathers on the occipital crest and down the back of the neck are erect (Figure 21). The bird exposes its conspicuous white eyelids by rolling the eyes ventrally to show the white sclera. The bill is kept closed except for a slight opening at the climax of the display. The flippers are waved backward and forward in a slow rhythm while the bird delivers a series of prolonged rasping sounds. In full display the flippers move about 30 degrees backward and 45 degrees forward. The rate of waving ranged from 0.51 to 0.66 beats per second and averaged 0.61 beats per second in ten full displays.

The sounds accompanying five normal displays were analyzed by sound spectrograph. Figure 22a is a spec-

trogram of the sound produced during a complete ecstatic display consisting of a pulsating pumping followed by the hoarse rasps of the climax. During the pre-climax the breast heaves as air is pumped in and

TABLE 13. Maximum Numbers (during Occupation Period) and Minimum Numbers (during Incubation Period) in All Colonies in Rookery A in Three Successive Seasons (Counted from Photographs)

Colony	1959–1960		1960–1961		1961–1962†	
	Max.	Min.	Max.	Min.	Max.	Min.
I*	673	322	790	388	. . .	301
II	41	17	43	19	49	17
III	676	311	758	377	642	275
IV	68	29	52	26	59	27
V	157	73	158	69	155	61
VI	61	23	61	26	52	19
VII*	105	40	123	54	98	51
VIII	48	24	48	22	55	22
IX*	355	175	411	178	344	190
X*	70	31	122	49	103	46
XI	123	69	170	88	196	91
XII	27	5	75	18	118	44
XIII	11	1	32	7	38	8
XIV	4	2	6	2	7	3
XV	1	0	1	0	18	5

* Limited photographic coverage in 1959–1960 and 1961–1962. Number probably 10 per cent lower than actual.

† From photographs taken by Noel Orton on November 15 and December 15.

Fig. 20.   Colonies VI and XII on the same day. (*Top*) Colony VI with nearly all occupants lying over eggs. (*Bottom*) Colony XII with a number of unpaired birds standing at territories at the right or growing portion of the colony.

Fig. 21. Ecstatic display given by a male during the occupation period.

out. Figure 22*b* gives a more detailed analysis of the pumping sound. Rather uniform in rate, it ranges from 7.0 to 7.7 and averages 7.4 pulses per second; its duration ranges from 3.2 to 6.9 seconds and averages 5.3 seconds. The sound is harsh and discordant, lacking in purity and clear harmonics, with overtones ranging over 2½ octaves.

During the climax of the display the body is contracted, and the air expelled produces another harsh pulsating sound ranging a little over 2 octaves (Figure 22*c*). The duration of the climax averages 3.0 seconds (ranging from 1.9 to 4.3 seconds).

The intensity of the display varies from brief posturing with the bill up and flippers only slightly out to an intense display that tends to abbreviate posturing and flipper movement in favor of the climax vocalization uttered with the mouth fully open. These highly intense displays (called "imploring" by *Sladen*, 1958) often have a repetitive climax that sounds similar to the intense loud mutual vocalizations illustrated in Figure 26*c*.

The ecstatic display given by incubating or guarding males crouching over eggs or chicks (Figure 23) generally sounds more intense. Ecstatic displays by females (rarely given) do not sound or appear different from those of males (one was analyzed by sound spectrograph).

The ecstatic display, although seen and heard throughout the breeding season, was most common during the occupation and reoccupation periods (Table 14; also see Figure 23 in *Sapin-Jaloustre*, 1960). Reference has already been made to the general restriction of this display to males and its use as a sexing criterion. It may be given by either paired or unpaired males. In colony IV during the occupation period of the 1960–1961 season 209 ecstatic displays were recorded for particular males, 92 by males alone on their territories and 117 by males with a female present on the site. During the incubation period and guard stage of the same season an additional 101 displays were recorded for the 30 males resident in the colony. Of the total 310 displays recorded for specific males in colony IV, 134 displays (43 per cent) were given by four neighboring males.

The ecstatic display generally occurred in spasmodic waves involving several birds at a time, which, in large colonies, may pass from group to group of neighboring males. One such wave, lasting 10 minutes, had 84 displays given by only 12 males, 10 of them incubat-

TABLE 14. Number of Birds in Ecstatic Display Posture at Various Periods of the Breeding Season*

Period	Photographs Covering About 50 Nests in Colony I, 1959–1960, No.	%	Photographs Covering About 30 Nests in Colony IV, 1960–1961, No.	%
1. Occupation Period				
a. Pre-egg stage	No record		87	(36)
b. Laying stage	113	(27)	73	(30)
2. Incubation Period				
a. Early, mostly incubating males	25	(6)	19	(8)
b. Middle, mostly incubating females	No record	...	13	(5)
3. Reoccupation Period and hatching stage				
a. Late incubation	63	(15)	49	(20)
b. Guard stage	225	(53)	No record	
Total	426		241	

* Data recorded by time-lapse camera between 0300 and 1800 hours (LST) during good weather and for three days per period of counting.

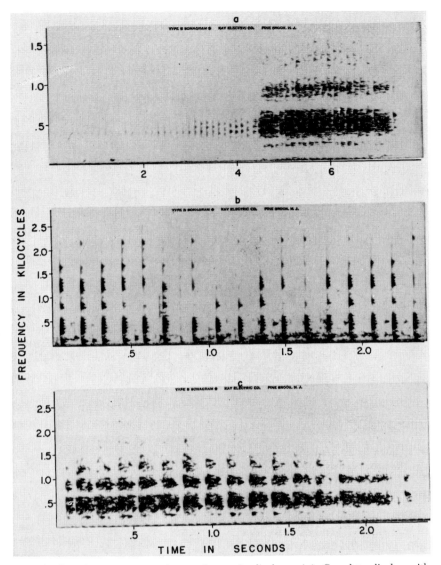

Fig. 22. Sound spectrograms of normal ecstatic displays: (a) Complete display with 4.2 seconds of the preliminary "pumping" followed by the climax. (b) A more detailed analysis of the preliminary pumping sound. (c) A more detailed analysis of the climax vocalization.

ing. Another time an outburst of displays by a group of incubating males lasted 1 hour and 40 minutes without interruption.

Ecstatic display outbursts by males over eggs or chicks were usually initiated by the displaying of a lone bird or by some other disturbance in the colony. Ecstatic outbursts often followed fighting or nest-relief ceremonies. Even a human's nearness to a colony induced ecstatic displays in some individuals; during 2 successive breeding seasons male 57664 in colony V usually went into ecstatic display whenever I neared the colony for routine observations or photography.

Threat displays, the alternate stare, and bill-to-axilla displays usually preceded this bird's ecstatic postures, and, once started, he was usually joined by one or more neighbors. Males in defensive situations, such as immediately after being defeated in territorial disputes, sometimes gave the ecstatic display, but they never gave it when acting aggressively toward another bird.

Sladen [1953] states that the ecstatic display was seen ". . . more among the unpaired 'inexperienced' and 'wanderer' males, particularly when attempting to establish themselves, than among the paired 'experi-

Fig. 23. Two males giving ecstatic displays while crouched over guard stage chicks. A third male (in left background) with no chick is just beginning the ecstatic posture.

enced' breeders." Males at Wilkes known to have bred successfully in the two preceding seasons (certainly they were "experienced") gave the ecstatic display before reuniting with their former mates, before gaining a new mate, when accompanied by their mates, and while incubating or guarding in the absence of their mates the third season. In colony IV no good correlation was found between the breeding status of individuals and the frequency with which they performed the ecstatic display.

This discrepancy in observations is probably due to the contagious nature of the display. Excluding the displays given in the spasmodic waves described above, my observations support Sladen's conclusion that the display is more common in the unpaired males. High individual variation in the incidence of the ecstatic display (it ranged from 0 to 53 displays for males in colony IV during 1960–1961) plus incidence variation in different colonies points to the need of further investigation into its causes.

The time-lapse camera records fail to show any particular daily rhythm in the occurrence of the ecstatic display (Table 15). I had the impression that the displays were most common during the late evening, but unfortunately the exposures were not clear after 1800 hours [Müller-Schwarze, this volume].

"Imploring" ecstatic displays of high intensity were seen and heard in only 9 banded birds, 8 males and 1 female. One of these males was recorded in this extreme display 7 times and another male 5 times. Female 58223 gave the "imploring" ecstatic and was also the only female repeatedly seen giving the normal display. Case histories of these birds suggest all were

particularly reactive to disturbance and could be described as having nervous temperaments.

All 27 ecstatic displays recorded for four females in colony IV during the 1960–1961 season (Table 3) occurred while the birds were reacting to threatening situations. One female displayed early in the season when another female threatened to evict her from her territory. After eviction early the next day, the same female gave the ecstatic display while standing near the site from which she had been evicted. The other 25 displays were given by three females, all of them while threatening wandering birds that frequented the nesting areas during the incubation period and guard stage. Other females were known to give the ecstatic display when no disturbing influence was manifest.

Ecstatic displays given by males while their mates were present on the territory invariably elicited quiet mutual displays (page 116) from the females. Such responses, given while standing next to the males or lying next to them on the nest, inhibited the ecstatic displays and brought them to a nearly immediate halt.

The climax of the ecstatic display was also inhibited by the approach of a female to a lone male during pair-formation (page 117) or by any nearby disturbance directly affecting the displaying bird. High winds generally prevented the ecstatic display. After a windstorm lasting a few hours or more the incidence of this display increased.

Ordinarily the ecstatic display was confined to the nesting territory. Rare instances of it were observed

TABLE 15. Number of Birds Recorded in Ecstatic Display Posture during Successive Hours of the Day*

Hour, LST	12 Days in Colony I, 1959–1960	15 Days in Colony IV, 1960–1961
0300	28	12
0400	32	16
0500	14	13
0600	34	13
0700	13	25
0800	26	14
0900	16	16
1000	29	23
1100	27	6
1200	30	10
1300	29	8
1400	24	11
1500	35	18
1600	17	23
1700	29	9
1800	34	24

* Data recorded by time-lapse camera on clear days during the occupation and reoccupation periods.

when birds were in awkward situations, as when stranded on the edge of a precipitous ice foot or stranded alone on an ice floe. It was also observed in several of the experimental birds released far inland during navigation experiments.

Twice I saw molting birds I disturbed give the ecstatic display. Unbanded birds were occasionally seen in ecstatic display on the loitering grounds.

As *Sladen* [1958] and *Sapin-Jaloustre* [1960] note, the ecstatic display definitely functions to attract lone females to the nesting territories of lone males and thereby facilitates pair-formation. This response to the display is seen most commonly during the occupation and reoccupation periods.

Persistent outbursts of displays by groups of incubating or guarding males are interpreted as mimetic behavior [*Armstrong*, 1951]. The contagious nature of the ecstatic display undoubtedly also contributes toward repetitious displays in groups of unpaired males or males already accompanied by females.

The occurrence of the ecstatic display in response to certain disturbed, defensive, or artificial situations is assumed to be an inappropriate response. Under these circumstances it may be regarded as a displacement activity as defined by *Tinbergen* [1952]. This interpretation coincides with that of *Sladen* [1958] when he noted an association between "nervous strain" and the incidence of the very intense "imploring" ecstatic displays by males and the normal display by females.

Oblique stare bow. *Sladen* [1958] gives 3 examples of ritualized bowing when male and female meet at a territory. This was seen repeatedly at Wilkes and is named the oblique stare bow display to differentiate it from bowing related to the species' upright stance. An upright bird giving the oblique stare bow lowers its bill 30 to 60 degrees below the horizontal and presents one side of the face toward a bird of opposite sex (Figure 24). The bill is kept closed. Males sometimes turn the eyes down, but not the females. The crest feathers on a male are always erected, but the female relaxes or only partially erects the crest. Both sexes keep their flippers relaxed next to their bodies. Movements are very slow, with the positions held for as long as 30 seconds. Although the head position is similar to that in the alternate stare, the head is never moved rapidly.

Very little variation occurs in this display, other than an exaggerated bowing of the head which brings the bill closer to the ground. A male giving the ob-lique stare bow to a female approaching from behind looks over his shoulder to present the side of his face.

The oblique stare bow display occurs between two previously unpaired birds of opposite sex during their initial meeting on the male's territory. This is most common during the occupation and reoccupation periods when lone males in ecstatic display attract females to their territories. The display was never observed between banded birds of the same sex, and was only rarely seen given by males away from their territories. Twice I saw lone males leave their territories and approach lone females with this display. Unbanded birds occasionally approached other birds away from the colonies with the oblique stare bow display. The response to such an approach was a quick peck, a direct stare threat, a walking away or, in one case, a return of the oblique stare bow followed by copulation.

Sladen [1958] concludes that the form of bowing described here as the oblique stare bow is a display and suggests that it may be an appeasement ceremony. Another form of bowing, before or after copulation, Sladen also considers a display. My observations support these conclusions, except that bowing before or after copulation is not known to convey any particular information and might best be considered a part of the copulatory act. The oblique stare bow, however, clearly plays an important role in the process of pair-formation and may serve as an aid in sex recognition. To a strange female, this bow appears to be an essential signal for acceptance on a male's territory and probably communicates sexual acceptance and appease-

Fig. 24. Oblique stare bows by a male and female during the pair-formation process.

Fig. 25. A pair just reunited after separation during the winter months giving the loud mutual display.

ment. To a male, the oblique stare bow appears to communicate receptivity and the sex of the approaching bird.

Loud mutual display. The names of this and the quiet mutual display are taken from *Sladen* [1958], who shows them to be forms of *Roberts'* [1940] "mutual epigamic display." The loud mutual display is characteristically given between two birds of the opposite sex. The birds stand facing each other and point their bills upward together. The feathers on the crest and neck are erect, and the eyes turn down as in the threat and ecstatic displays. Both birds hold their flippers pressed to their sides (Figure 25), wave their heads from side to side out of phase, and utter harsh, sonorous notes through the open bill.

This sequence is often repeated many times. Such series begin very loudly and diminish gradually, each phrase of the call becoming shorter and covering a decreasing frequency range. Five analyzed phrases varied in length from 0.9 to 1.8 seconds and averaged 1.2 seconds. The sound sometimes pulsates throughout the phrase and often consists of several pulses followed by a tone of constant or changing frequency with discreet harmonics (Figure 26a). The sound in low intensity displays covers a limited frequency range (Figure 26b), and at higher intensity it covers up to 10 octaves (Figure 26c).

The loud mutual display is used in many social situations and occasionally is given by lone birds.

Circumstances for the loud mutual display recognized in the present study are listed in their approximate descending order of frequency:

1. Between a male and female at a territory during their reunion at the beginning of the season and during nest reliefs.
2. Between a parent and its late guard stage or creche chicks prior to and during feeding.
3. Between two or more adults, not necessarily of opposite sex, involved in fighting.
4. By a lone adult toward the territory upon arrival during the occupation period.
5. By a lone or paired adult at a territory in response to a nearby nest relief ceremony or fight.
6. By a parent toward eggs or nestlings.
7. By a reoccupying bird toward creche chicks.
8. Between creche chicks over four weeks of age.
9. Between two or more adults when confined and disturbed.
10. Between two birds away from the nesting territory (seen only once).

The frequency of the loud mutual display throughout the breeding season was related to changes in the incidence of the circumstances given above, i.e. initial reunions, fights, nest reliefs, and parent-offspring relations.

Sladen [1958] and *Sapin-Jaloustre* [1960] consider the loud mutual display primarily a greeting ceremony that plays an important role in confirming and reassuring individual recognition. My observations at Wilkes support this interpretation.

The wide diversity of circumstances that elicit the loud mutual display suggests the need of caution in assigning a single function or cause. *Sladen* [1958] discusses and rejects the proposition that the mutual display possesses elements of incipient threat and appeasement [*Roberts*, 1940]. Yet the obvious usefulness of the display as a means of recognition between paired birds and parents and chicks does not in the least preclude its having aggressive elements. Loud mutual displays by lone or paired adults in response to and directed toward a disturbance (such as nearby nest-reliefs and fights), by non-parents toward creche chicks, and between creche chicks are best viewed as agonistic, though in certain cases the display may be imitative. The same interpretation can apply to adults giving the display toward their former territory upon first arrival in the spring. The same display given between paired birds with an established relationship, although still having aggressive qualities, has taken on

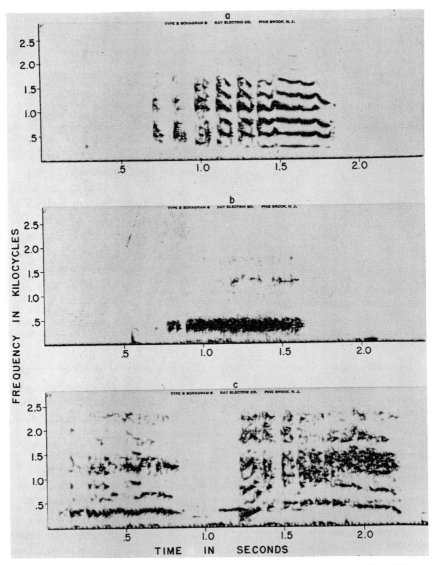

Fig. 26. Sound spectrograms of loud mutual vocalization phrases given at three different intensities: (a) normal, (b) low, and (c) two phrases of high intensity.

additional meaning and communicates identity. Certainly in those cases where identification was not confirmed (page 121) open aggression resulted.

The spontaneous but unusual occurrence of the loud mutual display by a parent towards its eggs or nestlings seemed to occur most often when the parent was "nervous," had been incubating a long time, or was unable to feed its begging chicks. The displays in these situations may possibly be misplaced responses through low response thresholds.

Quiet mutual display. In full display the birds face each other holding their bodies upright. They point their bills up and toward each other, but keep them closed instead of open as in the loud mutual display. The feathers on the crest and neck are erected, and the eyes are turned down in intense displays. The birds hold their flippers pressed to the sides (Figure 27) and wave their heads slowly from side to side and out of phase as one or both birds make a low humming sound (Figures 28a, b, and c). The five vocalizations analyzed showed distinct harmonics with individual variation extending over 3 to 5 octaves. The duration of the five displays varies from 0.7 to 1.9 seconds and averages 1.3 seconds.

It is not uncommon for the quiet mutual to be given between a standing bird and its mate lying on the nest.

Fig. 27. Quiet mutual display given by a pair shortly after reunited and after giving a series of loud mutual displays.

The posturing sometimes lacks sound and involves only a brief pointing of the bill.

With few exceptions the quiet mutual display was observed only between paired birds. It was the normal response of a female to her mate's ecstatic display. A bird that had been away from its territory collecting nesting stones or eating snow often gave the quiet mutual display to its mate upon return.

This display was frequently given by one member of a pair in response to the loud mutual display by the other member. Pairs that had first given a series of "greeting" loud mutual displays often ended by giving the quiet mutual.

It was easy for me to induce paired birds to perform quiet mutual displays by approaching a colony slowly. Occasionally I observed this display among penned birds when I disturbed them.

The general restriction of the quiet mutual display to paired birds, its inhibitory effect on the ecstatic display (page 113), and its frequent occurrence after birds were briefly separated or slightly disturbed leads me to believe that it functions primarily to ensure established pair-bonds.

Sladen [1958] and *Sapin-Jaloustre* [1960] treat the quiet mutual as a lower intensity form of the loud mutual display. I feel that the two displays are sufficiently different in form, occurrence, and function to warrant separation. I did not note the quiet mutual occurring in nearly so many circumstances or so often

as the loud mutual display. As the quiet mutual sometimes merely involved bill pointing, it was often difficult to differentiate it from intentional movements of parents and chicks preceding food begging or regurgitation. The quiet mutual display, although giving reassurance to the pair-bond, was never known to act as a test for individual identity as did the loud mutual display (page 115).

Types of Sexual Association

I recognize three types of association between birds of opposite sex. The first type, trial pairing, is often of brief duration and need not lead to egg production. The second type, true pairing, is stable for at least one season and leads to egg production. A third type is recurrent pairing, which always led to the production of eggs during my study. All three types of association are formed (with rare exceptions) on a territory chosen and maintained by the male.

In the following descriptions no birds are designated as paired or mated unless they were known to have produced eggs. Birds involved in recurrent pairing are called reunited pairs, those failing to reunite are called disunited, and when pair bonds have been broken by mortality the pair bond is designated as split.

Trial pairing. *Sladen* [1958] uses *Richdale's* [1951] term "keeping company" to describe temporary associations between birds of opposite sex on a territory. I observed this behavior repeatedly at Wilkes. In colony IV five non-breeding females from the previous year each associated with an average of 3.2 different males before pairing reproductively in the 1960–1961 season. Four non-breeding males from the preceding season each associated with 2.6 females before obtaining a mate that laid eggs the second year. One male had five different females on his territory for various periods of time, none of which stayed with him. Of the birds that kept the same mates two successive seasons in colony IV, 61 per cent were known to have associated with other birds of opposite sex during the absence of their mates. Successful breeders did this only during the occupation period; unsuccessful breeders and non-breeders did so during both the occupation and reoccupation periods. Of the birds in this colony that changed mates between seasons, 25 per cent were known to have made pairings with other birds before mating the second season.

Beyond initial reactions, the behavior of birds during trial pairing and recurrent pairing is not easily differentiated. In trial pairings the defense of territory

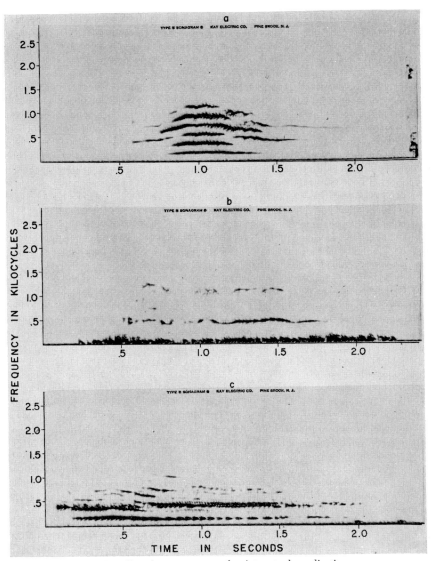

Fig. 28.   Sound spectrograms of quiet mutual vocalizations.

and mate, nest building, mutual displays, and copulation occur as with reunited birds, but at a somewhat lower intensity. While a general receptivity of birds to others of opposite sex may facilitate true pairing when previous bonds are broken, trial pairing does not seem to influence mate choice in subsequent seasons (see pages 101–120). I recorded only one instance of birds mating after a non-reproductive association the previous year. These two birds spent considerable time together during the reoccupation period the first season. The male was a non-breeder and the female a breeder that deserted her eggs when her mate failed to return.

True pairing. The behavior of previously unacquainted birds when meeting at a territory for the first time is the same whether the two birds finally produce eggs or dissociate before egg production. By present definition the process of pair-formation is consummated when eggs are produced.

Pair-formation is initiated by a female approaching a male's territory in response to his ecstatic display. Rarely does a strange female approach a male who has not given the ecstatic display within the previous few minutes. The slowly approaching female characteristically gives the oblique stare bow when 0.6 to 1.5 meters away from a displaying male. The male, sometimes interrupting his ecstatic display, returns the stare bow. Both birds hold this posture for as long as 30 seconds while the female approaches until she is within about 0.3 m of the male. At this point the male usually turns his head to the resting position and bows to the nest site. Often he lies down slowly and scrapes the ground or his pile of stones with one foot at a time. This scraping, which sometimes appears to be a ritualized part of the pair-formation sequence, is also done at other times to accentuate the nest form and to make a depression in the pile of stones for the eggs.

While standing next to the scraping male the female usually bows to the site. After bowing she sometimes walks away, picks up a stone in her bill, and returns to drop it next to the male. The male picks up the stone and places it on the site either while still prone or after standing up. Sometimes the male stands upright after briefly scraping the site and allows the female to occupy it. Mutual bowing to the nest site is common at this point. Subsequently mutual collecting of stones, scraping of the site by either bird, and copulation can occur in almost any sequence and within a few minutes.

Associations begun in this slow ritualized manner may be broken any time after the initial approach of the female or continue through reproductive pairing. In some cases two birds may repeat the sequence from ecstatic display through scraping by the male several times before the association becomes well established.

Recurrent pairing. The behavior of birds reuniting at their former territory after separation during the seven winter months is described in part on page 115. The characteristic immediate and intense loud mutual displays by the reunited birds is in sharp contrast to the slow, quiet behavior of birds initiating an association.

The frequency with which paired breeding Adélies from the 1959–1960 season reunited in the 1960–1961 season was studied for 277 pairs. In 165 cases (60 per cent) both members of the pair returned the second season; 138 (84 per cent) reunited, and 27 (16 per cent) disunited. From the original 277 pairs 98 (35 per cent) were split by one member missing the second season, and 14 (5 per cent) had both birds missing. These disappearances are presumably due to mortality; losses were 26 and 20 per cent, respectively, for males and females.

In about half the disunited pairs, return to the rookery was widely asynchronous. Seven of the males and eight females arrived in the rookery 7 or more days later than their previous mates. As shown on page 104, a late-arriving male can lose his previous territory and mate. Late returning females probably run the risk of losing their former mates to new females (see below).

Sexual Fighting

Male and female Adélies fight with birds of the same sex in somewhat different situations. Males fight hard for territories whether or not a female is present (Table 7). During fights between two females the males, with rare exceptions (page 104), are bystanders and show no preference for either female. Females fight hard with each other only when a male occupies the territory in question (Table 8). The differences, if any, between territorial and sexual fighting are vague and are based only on the relatively weaker territorial bonds of females.

Ten major fights were observed between banded females. Seven of these began when newly arrived females returned to their former territories to find their former mates in residence with new partners (Table 8). In each case the newly returned bird attacked, drove off the strange female, and regained the former territory and mate. One fight involved a bird that had lost her mate and then deserted her eggs. During the reoccupation period this bird drove off the female and joined the male of a pair that had occupied her old nest site. Early next season she mated after defeating another partner of the same male. The final case of fighting between females occurred during the reoccupation period, but the cause was uncertain.

The severity of sexual fighting is in proportion to the length of time the two females have spent with the male on his territory. Where a male had associated with a new female for only a day or two, I could predict a brief fight and an easy victory for the male's former mate. In contrast, a late returning female would probably lose the fight if her former mate had progressed far enough into pair-formation with a new female. I never observed this directly but presume it happened in some of the failed rematings mentioned earlier. Two blood-smeared victims of such fights were found at the edge of their colonies; the victorious females showed blood on flippers and breasts as evidence of recent fighting.

Pairing in Relation to Breeding Success

Breeding success was determined for 427 birds whose pairing relationships were known for two successive seasons (Table 16). Successes at raising chicks to the creche stage during the 1959–1960 season were not significantly different between birds that were reunited, had their previous bonds split by mortality, or were disunited. During the next season, however, the reunited birds showed a significantly higher breeding success than either the split or disunited birds.

TABLE 16.   Breeding Success in Relation to Mate Stability in Rookery A

	Breeding Status							
	1959–1960 Season				1960–1961 Season			
	Successful		Unsuccessful		Successful		Unsuccessful	
	No.	(%)	No.	(%)	No.	(%)	No.	(%)
Reunited birds	224	(81)	52	(19)	240	(86)	36	(14)
Newly paired birds	111	(74)	40	(26)	112	(74)	39	(26)
a. Split birds	74	(76)	24	(24)	74	(76)	24	(24)
b. Disunited birds	37	(70)	16	(30)	38	(72)	15	(28)

χ^2 *and Probability Values*

1959–1960 Season		1960–1961 Season	
Reunited vs. split	$\chi^2 = 2.2$, $p = 0.12$	Reunited vs. split	$\chi^2 = 6.5$, $p = 0.01$
Reunited vs. disunited	$\chi^2 = 3.4$, $p = 0.06$	Reunited vs. disunited	$\chi^2 = 8.6$, $p = 0.004$
Split vs. disunited	$\chi^2 = 0.6$, $p = 0.4$	Split vs. disunited	$\chi^2 = 0.2$, $p = 0.7$

The most obvious reason for this lower breeding success for the split and disunited birds was the failure of some to gain mates in time for egg laying the second season. Eight per cent of the split birds and 17 per cent of the disunited birds failed to gain true mates by egg-laying time during the 1960–1961 season.

Other factors, dependent and independent of delayed pairing, were also examined. (1) Some of the split and disunited birds paired with unbanded newcomers to the colonies in the 1960–1961 season. These unbanded birds were possibly new breeders, which generally have a lower breeding success. (2) There was no difference in the average egg-laying times between 28 reunited birds and 26 newly paired birds in the 1960–1961 season. (3) Nest locations, central or peripheral, were randomly distributed between the reunited, split, and disunited birds. (4) Split males and split females showed no significant differences ($p < 0.01$) in breeding success. Split males nearly always kept the same territories, whereas split females nearly always changed territories (page 106). This suggests that a change in nesting territory by females is not a factor in reducing breeding success.

Individual Recognition between Adults

Sladen [1958] raises the question of whether memory of the territory or of the mate keeps pairs together year after year. Both Sladen and *Sapin-Jaloustre* [1960] feel that Adélie pairs recognize each other by sight and sound. Sladen considers the mutual display a greeting ceremony that acts in "confirmation, reinforcement, and 'reassurance' of individual recognition" and that individual recognition between previously paired birds is more important in keeping pairs together than is faithfulness to territory.

I collected evidence for individual recognition by direct observation and experimentation including data on reunions of paired birds after the long winter separation. Banded birds meeting their former mates at their previous territory for the first time in October and November 1960 gave clear evidence of individual recognition. The vocal and behavioral exchanges between a pair just reunited were spontaneous and intense (pages 103 and 115).

I have already shown that females returning later than their former mates nearly always attacked and defeated any female temporarily paired with their former mate. The consistency of these responses can best be explained in terms of immediate recognition of mate and territory. This type of evidence is circumstantial, but worth illustrating from my field notes:

At 1131 hours ♀C, wet from sea, came running into colony V and gave the loud mutual display to a lone unbanded bird on a territory next to ♀C's former site. This unbanded bird returned the display, but ♀C about-faced and began the same display toward the pair at her former site. Present on this site was ♀C's former mate (♂D) who had been associating and copulating intermittently with another female (♀E) for seven days. This pair returned the loud mutual display of ♀C with all three birds directing their heads together. ♂D then attacked ♀C and beat her hard for a few seconds, but stopped suddenly as all three birds began the loud mutual display again. At this point an all-out fight started between ♀C and ♀E and lasted for four minutes. During the fighting ♂D attacked ♀E once, but not ♀C. At the end of the bout ♀E remained three feet from the site as ♀C and ♂D went through intense and repeated loud mutual displays. Ten minutes after the fighting began ♀C and ♂D began to copulate, but were interrupted by the return of ♀E who was first pecked at by the male and then severely beaten again by ♀C. (♀E later bred successfully in another colony.)

Though this set of events can be explained on the basis of territorial recognition, I prefer the following interpretation: The initial exchange of calls between the returning female, first with the unbanded bird and then with the pair, told her which bird was her previous mate. The aggressive behavior of her former mate was at first a territorial defensive response which halted once the former mate was recognized. The behavior, appearance, and/or voice of ♀ E must have allowed the newly arrived ♀ C to recognize her as a strange female. This led to the big fight. The aggression of ♂ D toward ♀ E may have indicated some preference for his former mate, or may have again been a territorial response.

When females return to their colonies before their mates they nearly always return first to their former territory (Table 8) and then move off to pair temporarily with another male at a different site. Some begin trial pairing within a few minutes, and others stay alone at their former territories or wander about the colony for up to two days before joining another male. When the original mate of such a female returns to the colony, the female generally leaves her temporary mate and joins her former mate on his territory. Both birds give the spontaneous and intense loud mutual displays at their reunion. This sequence was observed three times in colony IV, and must have occurred many times more, judging by the females that later reunited with former mates they preceded back to the colonies.

Exceptions to this situation were observed in the two cases of disunited pairs in colony IV. One male returned late and never regained his former mate or territory after another male usurped them (page 104). The second male did not return until the guard stage. His former mate was raising chicks at the time and no reunion occurred. Similar exceptions must have occurred with other disunited pairs.

Auditory clues are important in individual recognition. On November 9, 1960, I played tape recordings of two separate nest relief ceremonies from the previous year outside colony IV. Both tapes contained loud mutual calls of males whose former mates were associating with new males in the colony. The response of one female (asleep at the time the recording was started), after hearing her own and her former mate's voice for one half a minute, was to look about the colony intently and to take three steps toward her former territory, now vacant. Four minutes later the same tape was replayed, and at her former mate's very first call she quickly moved to her previous territory

and remained there looking about for one minute. The other tape recording brought no observable response from the female whose previous mate's voice was played.

Incubating birds repelled all birds that approached the nest other than their own mates. A bird relieving its mate had merely to give the loud mutual vocalization to be accepted in relief. Occasionally a bird returning to relieve its mate approached the wrong nest and gave the greeting vocalization. The bird incubating on the nest immediately pecked the approaching bird instead of replying to the greeting calls. The relieving bird then moved on to its correct nest and mate for a normal nest relief.

Similar situations were created experimentally by exchanging incubating birds between neighboring nests while their mates were away. Males and females so displaced remained on the new site in six cases (three exchanges), and the subsequent nest reliefs were watched and filmed in two cases. Both the relieving birds first returned to the correct nest, gave the greeting calls, and were pecked away by the illegitimate birds on the nests. The initial calls of both relieving birds were answered by their correct mates on the wrong nests a meter away. These replies by their proper mates redirected the attention of the relieving birds, which turned and displayed with their correct mates, and normal nest reliefs ensued at the wrong nests.

Shortly after the conclusion of one of these experiments one of the exchanged birds and the mate that relieved the other exchanged bird each returned to its correct nest. This left three nests with non-relieved but exchanged birds on them. The subsequent nest reliefs at these three nests were not observed, but checks on the nests each day showed that partners in each case relieved the correct mate and not the correct nest. All six pairs involved in this experiment bred successfully, four of the six raising their foster chicks.

The behavior of reunited birds and the tape recording and switching experiments demonstrate the existence of individual recognition and suggest the importance of auditory means of identification. Normally local orientation to territory and visual and auditory clues must all work together to insure the reunion of mates during and between breeding seasons.

PARENT-OFFSPRING RELATIONSHIPS

Guard Stage

During the guard stage contacts between mated birds

gradually lessen as the developing chicks demand more and more attention. This increases parent-chick intercourse and aids the development of individual recognition between parents and their chicks so essential during the next (creche) stage of the breeding cycle.

During the early guard stage the relatively short nest relief ceremonies of the incubation period are extended. A single relief often takes 15 to 20 minutes. Stone collecting is reduced or is totally neglected. The loud mutual displaying and calling, still generally directed toward the mate, increasingly turns toward the small begging chicks.

In the guard stage a brooding parent sometimes spontaneously gives the loud mutual call with its head directed down to the chicks. This occurs more frequently when the chicks beg for food and the parent has exhausted its food supply. Small chicks show no apparent response to this calling, but large chicks often beg more persistently after hearing it.

By late guard stage when the chicks are too large for their parents to cover easily, the nest relief ceremonies are brief. The calls of the relieving bird are now directed increasingly toward the insistently begging chicks, while the bird being relieved still directs its displaying to the mate. When the arriving parent calls to the chicks, they point their bills toward it and beg for food. The begging chicks vibrate their bills next to or against the bill of the parent. They call

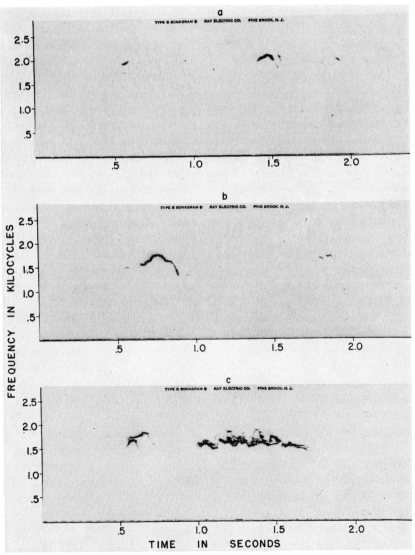

Fig. 29. Food begging calls by chicks at various ages: (a) 2 days old, (b) 13 days old, and (c) 20 days old. Frequency scale should be doubled.

Fig. 30. Food begging and feeding position of 1-week old guard-stage chicks.

with very brief "peeps" when one week old. At three weeks the chicks' "peeps" are extended and take on a wavering quality with a falling pitch (Figure 29a, b, and c).

The feeding position of small chicks (Figure 30) is constant; that of late guard stage chicks changes constantly. Before and between feedings the parent bird shuffles around within the territory as the chicks constantly vie for a position in front of it (Figure 31).

Creche Stage

With the suspension of parental stone collecting during the guard stage, nests are gradually obliterated by chick excreta, and the stones are scattered about the territories. For the first few days after parents leave their chicks unguarded during the creche stage, the chicks tend to stay on their natal territories. A parent returning to feed creche chicks still on the natal territory characteristically prefaces the feeding with loud mutual calling directed toward the territory and chicks. This is followed by chick begging and feeding. Some adults also call loudly between feedings. The feeding of young creche chicks is normally accomplished in about an hour, although the parent may remain on the territory with the chicks several hours more. Parents feeding small creche chicks lead their charges on short feeding chases within or very near the colony. Only rarely do paired parents meet on their territories during the creche stage, and then probably fortuitously.

As more and more chicks are left alone, they leave their individual territories and gather in close-knit groups or creches within the colony boundaries. In large colonies some chicks may be 9 to 12 meters from their natal territories. Creches from colonies near each other sometimes mix. This situation presents a

problem for parents if they are to feed only their own chicks during the creche stage.

Parent-Offspring Auditory Recognition

By marking a number of chicks and adults Sladen [1958] discovered that parents, contrary to earlier accounts [Murphy, 1936; Roberts, 1940], feed only their own creche chicks. These observations are supported by Taylor [1962] and by my own studies. In 51 feedings of marked chicks during the creche stage at Wilkes, only one instance occurred in which a parent fed other than its own offspring.

In the early creche stage a food-laden parent on reaching the colony approaches the site of its former nest and gives the loud mutual call toward the ground, or sometimes in the air if a creche occupies the site. Almost immediately its chicks, and often one or two others, run up and begin begging for food.

The parent bird seldom feeds the chicks right away, but pecks lightly at all begging chicks before running away with the chicks in pursuit. The feeding chase is an important part of the feeding procedure. Parents will run, pause, and sometimes call to the chicks again. Strange chicks are now pecked more severely as more chasing ensues. Feeding by regurgitation takes place during brief pauses in the chasing. The duration and distance of feeding chases generally increase with the age of the chicks and the number of chicks in pursuit, especially when these include strange chicks. As strange chicks are eliminated, chasing is reduced.

The total time for feeding is usually only 10 to 15 minutes, and appears to be governed by the amount of

Fig. 31. Food begging and feeding position of late guard-stage chicks about 3 weeks old.

food regurgitated and the amount of interference. A parent that has sated its single chick sometimes returns to the colony with it and remains there for hours. The chick is usually fed again before the parent leaves. Parents that appear to have exhausted their food supply usually outrun their chicks. Sated chicks stop chasing before hungry ones do. The chicks return to the colony and the creche at the end of the feeding while the parent heads back to sea or to the snow slopes above the beaches where it may loaf for several hours before returning to sea.

It is not unusual for a parent to return to its colony and former territory and remain silent for some time, and the absence of chick response is striking. After several minutes the parent may call from its territory, or it may first approach a nearby creche before calling. In either case the chicks rally to the call, and the usual chasing and feeding ensue. In the late creche stage the parent often does not return to its territory, but calls from the colony edge or after approaching a creche.

The greatest distance at which I observed a chick respond to a parent's feeding call was about 11 meters.* Chicks may wander farther than this, especially during feeding chases, but they usually return to their natal colony within an hour or two.

I watched the actions of parents returning to feed creche chicks closely on many occasions. Only twice did I see a parent return from sea, approach its colony, and feed its chicks without giving any calls. I often watched chicks beg from reoccupying adults without being fed. The response of a reoccupying adult was to peck away the chicks, call toward them, or occasionally go through partial regurgitation movements.

To test the importance of auditory identification between parents and creche chicks, I set up a loud speaker and amplifier outside colony IV and played back certain parent's voices to creche chicks. Playback experiments were conducted while the experimental chicks were tightly huddled in the creche 1.2 meters or more away from their natal territories.

Seven different recordings were played back, each for 15 to 60 seconds. One recording was played three

separate times, and another was used twice for a total of ten separate playings. Five of the seven recordings were of different parents calling loudly to their creche-stage chicks just prior to and during feeding; the other two recordings were of two pairs of birds in loud mutual display during nest-relief ceremonies.

In nine out of the ten playbacks the chick or chicks of the parent whose voice was played stood up, looked around the colony, and then moved to the edge of the creche in an alert posture. None of the other creche chicks changed position, although several raised their heads and looked briefly about. In seven of the nine cases the chicks kept moving and returned to their natal territory within one minute after the playback had started. These chicks remained on the territory for three to nine minutes after the sound stopped. On the territory they assumed an appearance best described as alert and anticipatory.

The two recordings of nest-relief ceremonies brought responses equal to those of parents calling loudly to the chicks. One recording had two females calling their chicks simultaneously. The two chicks of the one parent and the single chick of the other responded by returning to their natal territories. The tenth playback, which brought no distinctive reaction, was of a parent whose chick had recently been fed. To one tape played backwards by mistake, no chicks in the creche responded except to lift their heads.

These results demonstrate three important points. First, creche chicks recognize the voices of their own parents. Second, the chicks know accurately the location of their natal territory. Third, on hearing their parents' voices, they act as if they anticipate reunion with their parents on that territory.

Conditions during the guard stage are favorable for learning parental calls. The chicks from hatching to three weeks of age hear their parent's voices at each nest relief. The gradual transfer of a parent's calls from mate to chicks in the late guard stage and the chicks' accurate local orientation facilitate parental finding and feeding of their own young during the creche stage. The chicks' response to individual voices and nest locations is undoubtedly strongly reinforced during the early creche stage by the regular calling from the territory that precedes each feeding. Feeding chases, which first occur in the late guard stage and continue through the creche stage, could also be considered reinforcement for specific responses.

Sladen [1958] reports that the parental mutual displaying to chicks decreases in intensity as the chicks grow older. He also notes that recognition is more im-

---

* In 1962–1963 and 1964–1965 I visited several Adélie rookeries on Ross Island during a series of experiments on Adélie sun-orientation [*Emlen and Penney*, 1964]. For several weeks I watched parent-chick relations with the aim of testing earlier conclusions. All my observations and impressions confirmed conclusions regarding individual recognition presented here. At Cape Crozier I saw parents call their chicks from greater distances than observed at Wilkes; twice chicks responded to parental calls from 25 to 30 meters away.

mediate with older than with younger chicks. Possibly visual clues supplement or replace auditory ones later on. Another possibility is that as recognition increases in efficiency, chicks beg from the wrong parents less often. This would decrease the confusion of creche chick feeding and eliminate the need for elaborate and intense displaying.

Auditory recognition of parents by chicks combined with local orientation toward the territory facilitates but does not insure that parents feed their own chicks. The parents must also have recognition marks, both auditory and visual, by which to sort out their own chicks from strange chicks that also beg them for food.

Individuality in the Loud Mutual Vocalization

The use of the loud mutual and possibly other calls as a basis for individual recognition implies distinctive individuality of voice in penguins. In the field I was able to identify only two birds by voice. Later I analyzed a number of loud mutual calls by sound spectrograph. Figures 32, 33, and 34 present spectrograms of three loud mutual calls given by each of six birds. Each call given by the same bird was given in a different display series.

A quantitative analysis of spectrograms to determine whether different calls of one individual show less variation than do those of different birds reveals no single measure of differentiation. Studying the spectrograms as a group, however, one can easily sort out individual voices by the general pattern of each display. Such selection becomes much more difficult if the calls are of low intensity. When of the same strong intensity, the individuality remains evident regardless of the situation in which the call was given, i.e. nest-relief ceremony or calling to creche chicks.

DISCUSSION

Territory. Adélie penguins are highly territorial during the breeding season. Each individual concentrates its activities on a small plot of land within the colony. The territory location is normally chosen by the male (page 106), but both members of the pair may defend it (page 121). Fighting appears to serve two principle functions, to drive intruders from the territory and to protect the mate from competitors. Unlike yellow-eyed penguin behavior [*Richdale*, 1951], the mate is not defended when outside the territory.

The function of Adélie territorial behavior seems to be to "secure objects or situations that are indispensable for reproduction" [*Tinbergen*, 1936]. Within this broad context five specific functions may be recognized:

1. For subadult birds the territory is a place for gaining sexual and social experience. An Adélie's first

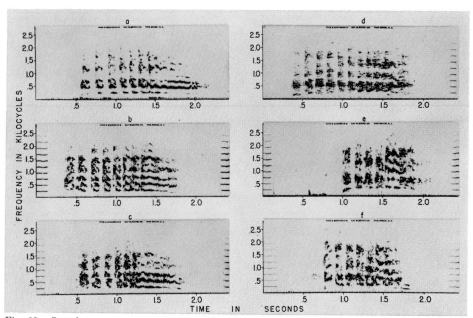

Fig. 32. Sound spectrograms of phrases of the loud mutual vocalizations of two birds to illustrate the distinctive and consistent characters of voice (see text).

(*a*), (*b*), and (*c*) Male 57636.

(*d*), (*e*), and (*f*) Female 58085.

(The regularly spaced lines on the right and left of *b*, *c*, *d*, *e*, and *f* are 240 cps calibration marks.)

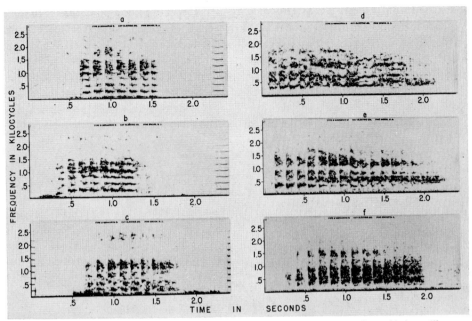

Fig. 33. Sound spectrograms of phrases of the loud mutual vocalizations of two birds to illustrate the distinctive and consistent characters of voice (see text).

(a), (b), and (c) Male 58117.

(d), (e), and (f) Male 58118.

(The regularly spaced lines on the right and left of a, b, and c are 240 cps calibration marks.)

successful attempt to reproduce is preceded by brief, then extended attachments to territory and to birds of the opposite sex during several seasons [*Sladen*, 1958 and pages 101–108]. For reproductively mature Adélies the territory has a special significance that elicits spontaneous displays from the birds when they arrive each spring (page 115). This relationship to territory based on and strengthened by previous experience serves to enhance successful reproduction.

2. A territory provides a relatively undisturbed site for reproductive display, courtship, copulation, and pairing. All major displays except threat are restricted to individual territories. During mutual displays between birds of opposite sex, at least one and generally both birds are on territory.

3. The maintenance of a territory gives protection for eggs and small chicks. Site protection is both inter- and intraspecific. In the former case the eggs and chicks are protected against skuas, dogs, and humans, and in the latter against aggressive pecking by neighboring territorial Adélies (page 97).

4. A territory serves to facilitate reunion of previously paired adults (pages 119 and 121). Orientation to a territory during a breeding season is retained until the next season and, combined with individual recognition, encourages the reunion of experienced pairs for more efficient reproduction.

5. The reunion of parents and their creche chicks at feeding time is also enhanced by an orientation to territory (pages 123–125) in combination with individual recognition.

The actuality of these subsidiary functions depends on demonstration of their contribution to increased reproductive success. Function 1 seems well established for experienced breeders, but needs additional support from observations on young birds. Functions 2 and 3 are amply supported in this report and are well known in many birds, both colonial and solitary [*Hinde*, 1956; *Nice*, 1941].

The role of territory in the reunion of previously paired birds (function 4) is not well documented. It has been suggested for the yellow-eyed penguin by *Richdale* [1951], for Wilson's petrel (*Oceanites oceanicus*) by *Roberts* [1940], and for the white stork (*Ciconia c. ciconia*) by *Schüz* [1938]. This function of territory may very likely be of importance in many species of migratory birds that show local orientation to territory year after year. In the Adélie penguin the perennial reestablishment of pair bonds between the same individuals, aided by site tenacity, appears to contribute toward successful breeding (page 119–120).

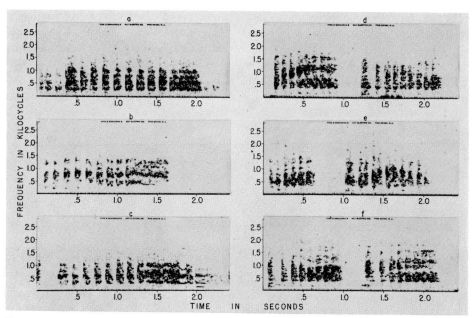

Fig. 34.   Sound spectrograms of phrases of the loud mutual vocalizations of two birds to illustrate the distinctive and consistent characters of voice (see text).
(a), (b), and (c) Male 58274.
(d), (e), and (f) Female 58783.

Site tenacity in successive seasons is widespread among birds. The classic documentation of this in a colonial bird is that for the common tern (*Sterna hirundo*) [*Austin*, 1949]. Austin's conclusion that site tenacity functions in the maintenance and stability of colonies seems applicable to the Adélie penguin. Faithfulness to particular colonies and to territories within colonies by established breeders inevitably perpetuates nesting in places whose geography and ecology favor a high survival. This could be further enhanced by offspring tending to return to their natal colonies. There is some indication of this for the Adélie (page 101), and it is hoped that future returns of birds banded as chicks will provide more specific information on the dispersal of young and the nature of recruitment within Adélie rookeries.

The role of territory in the reunion of parents and chicks during feeding in the creche stage (function 5 above) is an important but rather specialized function related to the creche phenomenon in the Adélie.

Social stimulation. *Darling* [1938] emphasizes the necessity of social stimulation, synchrony, and threshold numbers for successful breeding by colonial animals. While I lack quantitative data, several of my observations seem to support this thesis.

Weather and ice conditions imposed a 2-week delay in the mass influx of Adélies into rookery A during the 1960–1961 season. This delay did not appreciably influence the timing of the subsequent breeding season (compared with the previous year), yet the relative frequency and intensity of displays, nest building, and other activities seemed a response to colony crowding and not to time of year. If the tempo of the breeding season depended on season or gonadal state alone, I would have expected early arrivals the second year to be as active as those in crowded colonies the first year.

The contagious nature of the ecstatic display was discussed earlier. Social stimulation of a display that is so essential in the process of pair-formation argues strongly that this display is a synchronizer of breeding activities. Synchrony is important once the eggs have hatched. *Taylor* [1960] shows a variation in the guard-stage length for individual pairs that is independent of chick age, and I have shown the low chance of survival of the earliest chicks to be left alone in a colony (page 95). Survival of chicks from late breeders is endangered by the return of low temperatures and a refreezing of the sea. The benefits of creche behavior, such as protection from skuas and cold, can only be attained by coincident communal breeding.

A corollary of synchronization of breeding activities by social stimulation is Darling's theory of threshold numbers. *Richdale* [1951] criticizes this theory by pointing out the difficulty of demonstrating it without knowledge of the age and breeding status of the individuals composing the group. In colonies XII and XIII at Wilkes low breeding success was correlated with substantial colony growth and the attendant mean youthfulness of the birds (page 108). The high level of sexual activity by non-breeding birds in these colonies, while behind schedule, was important in attracting new potential breeders. Rookery A contained many unoccupied relict colonies (Figure 3) with distinct signs of past occupation such as nesting stones, guano, and chick remains. Why they were deserted is unknown, for they still seem to offer excellent nesting sites. The presence of these suitable vacant places so near to closely spaced nests of occupied colonies (pages 96 and 107) bespeaks the social and colonial nature of the Adélie penguin.

Pairing. The type of pairing relationship shown by the Adélie seems to fall somewhere between pairing for a single season and pairing for life. *Lack* [1940] is careful to differentiate between birds that pair for life and remain together throughout the year and those that "re-mate" in successive seasons. Within Lack's classification the Adélie is best labeled as one that mates for an entire breeding season and that generally mates recurrently in successive seasons (page 118).

In the yellow-eyed penguin the number of pairs that reunite in successive years averages 60 per cent [*Richdale,* 1957]. For the two years covered by my study on the Adélie the comparable figure was 50 per cent, with 84 per cent rejoining their previous mate when both survived to return to the rookery. The retention of mates year after year is reported as common for the erect-crested penguin (*Eudyptes sclateri*) [*Richdale,* 1941], the little blue penguin (*Eudyptula minor*) [*Richdale,* 1951; *Kinsky,* 1960], and the rockhopper penguin (*Eudyptes chrysocome*) [*Warham,* 1963].

Pair formation in the Adélie centers on the male's territory during the occupation period. In the yellow-eyed penguin, where the earlier history of individuals was known in detail, *Richdale* [1951] reports that pair formation can occur almost any time of the year as long as the birds are at their nesting stations. In the king penguin [*Stonehouse,* 1960] and the emperor penguin [*Prevost,* 1961] pair formation takes place in the absence of a nesting site.

True pairing in the Adélie seems to be restricted to the period between arrival at the rookery and the egg-laying period. At Wilkes only one productive pair formed after the egg-laying period (page 118), whereas many new productive pairs were formed during the

2-week pre-egg stage by birds that had not associated the previous year. This result is contrary to the suggestion of *Richdale* [1951] that the formation of new mated pairs in penguins during the short pre-egg stage would be unusual.

Adélies are essentially monogamous. This characteristic is violated only by the temporary associations designated "trial pairing" (page 117) and by a few rare instances of promiscuity (on two occasions a paired male was observed to leave the side of his mate and copulate with a prone female four feet distant). In a species whose food sources are remote from the breeding grounds, monogamy and a precise coordination of the mates in the breeding routine is essential to successful reproduction. Fighting, when restricted to the vicinity of the territory and between birds of the same sex, also helps to promote monogamy [*Tinbergen,* 1936].

Sex recognition. My studies confirm *Sladen's* [1958] rejection of the trial-and-error explanation of sex recognition in the Adélie. The birds appear to be immediately aware of the sex of every other bird they meet on territory. This awareness is undoubtedly founded upon a number of subtle differences in appearance and action. *Richdale* [1951] also rejects trial-and-error sex recognition for the yellow-eyed penguin and states that he can differentiate between male and female with near certainty by appearance and behavior. I feel similarly about the Adélie when I can watch two active birds at a nest site for a few minutes. Surely the task is much simpler for a penguin. *Prevost* [1961] shows a sexual dimorphism in emperor penguin voices; my analyses of Adélie calls showed none.

Individual recognition. Adélies quickly discriminate between their mates and strangers after the 7-month winter separation (pages 120–121) and at the time of nest reliefs (pages 114–115). Auditory clues are essential in this recognition. Clear evidence of what visual clues may be used was not found in my study, but they must also be important [*Sladen,* 1958].

Mate recognition has been noted as assisting rematings in the yellow-eyed [*Richdale,* 1951] and the rockhopper penguins [*Warham,* 1963] and is probably important in other penguins.

Mutual recognition between parents and chicks is common in colonial penguins. It has been shown for the rockhopper [*Pettingill,* 1960; *Warham,* 1963], yellow-eyed [*Richdale,* 1951], king [*Stonehouse,* 1960], and emperor penguins [*Prevost,* 1961]. *Sla-*

den [1953] first demonstrated parent-chick recognition for the Adélie. My own studies confirm this and point again to the essential role of auditory clues. Evidence of how parents recognize their chicks is still lacking. The situation Stonehouse [1960] describes for the king penguin appears to be very similar to that in the Adélie. Warham [1963] believes that vision and voice are both used in adult and adult-chick recognition in the rockhopper penguin. Recognition between adults in the yellow-eyed penguin is thought to be visual by Richdale [1951], but he thinks recognition between parents and their chicks relies on both appearance and voice. Experimental evidence is needed on all these species.

Individual recognition is very important in the biology of the Adélie. The mechanisms by which it is accomplished entail mutual responses to specific locations in the rookery (territories) and the ability to discriminate among individual variations of physiognomy and voice. Auditory and visual clues confirm identities of adults and chicks after the birds have been brought together by their ties to a particular territory. These behavioral phenomena are essential components of the Adélies' reproductive system.

SUMMARY

Territorial and social behavior in the colonial Adélie penguin was studied in detail at Wilkes Station, Antarctica, during portions of three consecutive breeding seasons. Much has been written about the behavior and life history of the Adélie during the past half-century, but rarely has one investigator had the opportunity to observe the same marked birds in successive seasons. To evaluate breeding behavior properly, prolonged periods of observations on marked birds are necessary because behavior varies with age and individual temperament. The present paper is based on the study of more than 1000 banded breeding Adélies in 14 distinct colonies within a rookery of approximately 2700 birds. The conclusions in this report are based largely on birds of known sex, breeding history, and territory.

As background for analyzing territorial and social behavior, the breeding cycle of the Adélie is reviewed briefly. Successful breeding is seen to depend on close synchronization and coordination between members of a pair during incubation and the guarding and feeding of chicks.

Adélies defend a small space around the established or potential nesting site. Defense is both intra- and interspecific. The average area defended per nest was 0.75 m² on the basis of a ratio of the colony areas to the number of occupied nests. The range in territory area per colony was 0.49 to 0.93 m², depending on variations in colony topography. Distances between particular nest centers and their two nearest neighbors averaged 69 cm, a measure of Adélie social tolerance.

Limited evidence from observations on birds of known age indicates that Adélies normally do not breed until 3 to 5 years of age. Before reaching breeding age young birds visit the breeding grounds for short periods. During these visits distinct attachments are created for specific colonies and territories.

Experienced breeders give immediate and spontaneous territorial displays upon arrival at their colonies in the spring, showing that nesting territories hold their special significance from one year to the next. An accurate homing to territory by both males and females assists in the reunion of previously paired birds.

Territorial defense takes the form of threat displays and outright fighting. Fighting ranges from brief bickering across territorial boundaries to long and continuous bouts ended only by exhaustion of one of the combatants. Fights over territory between birds of opposite sex are rare, and females repel intruders only in the presence of their mate or when incubating or guarding chicks.

The rate of return to specific territories in successive years was low (20 per cent) for the small sample of birds that occupied territories but did not produce eggs, intermediate for experienced breeding females (62 per cent), and high for experienced breeding males (93 per cent). Among recaptures in all parts of the rookery more than 99 per cent of the birds banded while occupying territories returned to the same colony the following year. Site and colony tenacity was further demonstrated by both short- and long-distance homing experiments.

Nests in Adélie colonies are so close together that central nests can only be reached by "running a gantlet" of aggressive territorial defenders. The difficulties imposed by this phenomenon appear to limit the dense circular colonies to a diameter of about 24 meters.

Certain correlations were noted among territory location, breeding success, and the age and breeding experience of individuals. Birds at central nests showed a significantly higher breeding success (10 per cent) than those at peripheral nests in the 1959–1960 season but not in the 1960–1961 season. Young birds tended to nest on the periphery of established colonies and also to concentrate in small colonies. Thus the ages of the birds in a colony must be known before generalizing about the effects of colony size, form, or social stimulation on reproduction.

Four displays are described as fundamental to the establishment and maintenance of pair bonds. Three types of sexual association are described and discussed. The first, trial pairing, is a temporary mutual receptivity between birds of opposite sex. It does not influence the choice of mates in succeeding seasons, but does lead to the second type of association, true pairing, early in the season when former mates do not return. Both trial and true pairing begin with a slow ritualized pair-formation process. This process is initiated by the approach of an unmated female to a territorial male in response to his ecstatic display. Pairing is consummated by the production of eggs. The third association is recurrent pairing by the same birds in subsequent years.

Former mates greet each other in the spring with mutual displays the moment the second member arrives. Fifty per cent of the pair bonds remain the same between two seasons. The greatest single factor in the dissolution of pair bonds is mortality. Widely asynchronous arrival of members of a previous pair early in the breeding season works against recurrent mating.

Females fight severely with each other when one female returns to find her former mate associating with another female. During these fights the males, with rare exceptions, show no clear preference for either female. The newly returned former mates won each of seven such fights observed.

Pairs reuniting in successive seasons had a significantly higher breeding success (12 per cent) than those that changed their mates the second season. Failure in consummating reproductive pairing by the end of the rookery egg-laying stage accounts for 11 per cent of the lower breeding success by birds that were separated from their mates.

Observations on birds returning in the spring showed that both local orientation to territory and recognition of mate play a part in the reunion of pairs. Tape-recording experiments and the exchange of incubating birds between nests just prior to nest relief revealed that, after approaching the vicinity of their territories, birds recognize and relieve their correct mates rather than their correct nests. Auditory clues were shown to be an essential part of individual recognition.

A system for reunion and recognition between parents and chicks develops during the guard stage of the breeding cycle. This system is necessary for chick feeding during the creche stage when parents no longer guard their individual chicks and the chicks gather in groups away from their natal territories. Observations and experiments using tape-recording playback again showed the importance of orientation to territory by both parents and chicks and the essential role of voice in the recognition of parents by chicks. Sound spectrograms of the adult loud mutual vocalizations between mated birds and by parents calling to chicks showed individualistic characteristics.

Finally, the effects of territorial behavior, social stimulation, stable pairing relationships, sex and individual recognition, and other aspects of Adélie reproductive behavior are discussed in relation to the breeding biology and success of the species and are compared with those of other penguins. The survival value of territorial behavior and individual recognition is emphasized.

Acknowledgments. This study was made possible by grants from the National Science Foundation administered by the University of Wisconsin, Department of Zoology. Logistics concerned with field work and administrative assistance were given by the United States Antarctic Research Program, the U. S. Navy's Task Force 43, and the Australian National Antarctic Research Expeditions. Personal thanks are due many members of these organizations.

For guidance in the planning of this project I am indebted to W. J. L. Sladen, George A. Llano, and the late Carl Eklund. I owe thanks to many of my Australian and American comrades in Antarctica for help in many different phases of this project. Special assistance in field work was given by John Dearborn, John Denholm, Keith Jones, Jan Lunde, Noel Orton, Frank Soucek, Ian Tod, Robert Underwood, and Dave Ward.

Technical assistance for the modification and operation of the sound spectrograph was given by Martin Joos. The maps and line illustrations were done by John Dallman.

I am grateful to John T. Emlen for his supervision of this project and wish to thank him, Oliver L. Austin, Jr., and W. J. L. Sladen for helpful suggestions in the preparation of this report.

REFERENCES

Andrew, J. D., and J. M. Roberts, The pairing of Adélie penguins, *Ibis, 94,* 540–541, 1952.

Armstrong, E. A., The nature and function of animal mimesis, *Bull. Anim. Behav., 9,* 46–58, 1951.

Austin, Oliver L., Site tenacity, a behaviour trait of the common tern (*Sterna hirundo* Linn.), *Bird Banding, 20,* 1–39, 1949.

Austin, Oliver L., Jr., Notes on banding birds in Antarctica and on the Adélie penguin colonies of the Ross Sea sector, *Bird Banding, 28,* 1–25, 1957.

Darling, F. Fraser, *Bird Flocks and the Breeding Cycle,* University Press, Cambridge, 124 pp., 1938.

Eklund, Carl R., Distribution and life history studies of the south-polar skua, *Bird Banding, 32,* 187–223, 1961.

Emlen, J. T., and R. L. Penney, Distance navigation in the Adélie penguin, *Ibis, 106,* 417–431, 1964.

Falla, R. A., *B. A. N. Z. Antarctic Res. Exped. Rept., 1929–31* (*Adelaide*), Birds, Series B, 2, 1–288, 1937.

Gain, L., The penguins of the antarctic regions, *Smithsonian Rept., 1912*, 475–482, 1913.

Gain, L., *Oiseaux Antarctiques, Deuxieme Exped. Antarctique Française, 1908–10*, (Paris), 2, 5–46, 1914.

Gwynn, A. M., Penguin marking at Heard Island, 1953, *Australian Natl. Antarctic Res. Exped. Interim Rept. 8*, 8–12, 1955.

Hinde, R. A., The biological significance of the territories of birds, *Ibis, 98*, 340–369, 1956.

Kinsky, F. C., The yearly cycle of the northern blue penguin (*Eudyptula minor novaehollandiae*) in the Wellington Harbor area, *Rec. Dominion Mus., 3*, 145–218, 1960.

Lack, D., Pair formation in birds, *Condor, 42*, 269–286, 1940.

Müller-Schwarze, Dietland, Circadian rhythms of activity in the Adélie penguin (*Pygoscelis adeliae*) during the austral summer, *this volume*, 1968.

Murphy, R. C., *Oceanic Birds of South America*, 1245 pp., Macmillan, New York, 1936.

Nice, M. M., The role of territory in bird life, *Am. Midland Naturalist, 26*, 441–487, 1941.

Pettingill, Olin Sewall, Jr., Creche behavior and individual recognition in a colony of rockhopper penguins, *Wilson Bull., 72*, 213–221, 1960.

Prevost, Jean, *Écologie du Manchot Empereur*, 204 pp., Hermann, Paris, 1961.

Richdale, L. E., The erect-crested penguin (*Eudyptes sclateri*) Buller, *Emu, 41*, 25–53, 1941.

Richdale, L. E., *Sexual Behavior in Penguins*, 316 pp., University of Kansas, Kansas, 1951.

Richdale, L. E., *A Population Study of Penguins*, 195 pp., Clarendon Press, Oxford, 1957.

Roberts, B., The breeding behaviour of penguins with special reference to *Pygoscelis papua* (Forster), *Sci. Rept. Brit. Graham Land Exped., 1934–37*, (London), *1*(3), 195–254, 1940.

Sapin-Jaloustre, J., *Écologie du Manchot Adélie*, 211 pp., Hermann, Paris, 1960.

Sapin-Jaloustre, J., and F. Bourliere, Incubation et développement du poisson chez le Manchot Adélie, *Pygoscelis adeliae, Alauda, 19*, 65–83, 1951.

Sapin-Jaloustre, J., and J. Cendron, Une Technique de Dénombrement et d'Etude d'une Rookery des Manchots Adélies par la Photographie Systématique, *La Terre et la Vie, 1*, 1–27, 1953.

Schüz, Ernst, Über Biologie and Ökologie des Weisen Storchs (*Ciconia c. ciconia*), in *Proc. Intern. Ornith. Cong., 1934*, pp. 577–591, University Press, Oxford, 1938.

Sladen, W. J. L., Notes on methods of marking penguins, *Ibis, 94*, 541, 1952.

Sladen, W. J. L., The Adélie penguin, *Nature, 171*, 952–955, 1953.

Sladen, W. J. L., Bird study in the Antarctic, in *The Ornithologist's Guide*, pp. 52–59, B. O. U., London, 1956.

Sladen, W. J. L., The Pygoscelid penguins, 1, Methods of study, 2, The Adélie penguin *Pygoscelis adeliae* (Hombron & Jacquinot), *Falkland Is. Depend. Survey Sci. Rept. 17*, 97 pp., 1958.

Sladen, W. J. L., and R. L. Penney, Penguin flipper-bands used by the U. S. A. R. P. bird-banding program, 1958–1960, *Bird Banding, 31*, 79–82, 1960.

Sladen, W. J. L., and W. L. N. Tickell, Antarctic bird-banding by F. I. D. S., 1945–47, *Bird Banding, 29*, 1–26, 1958.

Stonehouse, B., The king penguin, *Aptenodytes patagonica*, of South Georgia, 1, Breeding behaviour and development, *Falkland Is. Depend. Surv. Sci. Rept. 23*, 81 pp., 1960.

Taylor, Rowland, The Adélie penguin at Cape Royds, *Ibis, 104*, 176–204, 1962.

Tinbergen, N., The function of sexual fighting in birds, and the problem of the origin of "territory", *Bird Banding, 7*, 1–8, 1936.

Tinbergen, N., The behavior of the snow bunting in spring, *Trans. Linn. Soc. N.Y., 5*, 1–95, 1939.

Tinbergen, N., "Derived" activities: their causation, biological significance, and emacipation during evolution, *Quart. Rev. Biol., 29*, 1–32, 1952.

Tressler, Willis L., Oceanographic and hydrographic observations at Wilkes IGY Station, Antarctica, *J. Wash. Acad. Sci., 50*(5), 1–13, 1960.

Warham, John, The rockhopper penguin, *Eudyptes chrysocome*, at Macquarie Island, *Auk, 80*, 229–256, 1963.

Wilson, E. A. Aves, *British National Antarctic Exp. Rept. 1901–4*, Nat. Hist., 2, 1–121, 1907.

CIRCADIAN RHYTHMS OF ACTIVITY IN THE ADÉLIE PENGUIN (*PYGOSCELIS ADELIAE*) DURING THE AUSTRAL SUMMER

DIETLAND MÜLLER-SCHWARZE

San Francisco State College, San Francisco, California

Abstract. The distribution of activity through the day of the 100,000 members of the Adélie penguin colony at Cape Hallett at 72°19'S was investigated. The light amplitude was on the average 200 times greater at midday than at midnight. Notes were made on the following activities: general activity prior to mating; the ecstatic display of the males; adults walking over the ice to and from their feeding grounds; adults swimming to the feeding grounds; and the general activities of the chicks. A midday minimum is common to all forms of activity. This seems connected with the adaptation to cold. The high summer noon temperatures are apparently near the birds' upper limit of tolerance. The ecstatic display of the males showed morning and evening maxima in both the number of males participating and the number of performances per male. The ratio of complete to incomplete performances remained constant throughout the day.

The number of penguins walking across the ice to their feeding grounds reached the maximum at 0900 hours. I encountered the fewest individuals resting on the ice at 1200 hours. The percentage of walking birds to the total number of birds present on the ice resembles a sine curve; the maximum occurred at 1000 hours, the minimum at 2200 hours. The largest number of swimming birds was observed at 0800 hours, the minimum at 1700 hours. The morning peak is due to birds swimming outward, the number of which rises sharply between 0400 and 0500 hours, the time of local sunrise. This activity resembles the morning promenade of the king penguin. The Adélie penguins returning with food show no clear daily rhythm; perhaps they respond to the stimuli of well-filled stomachs.

Chicks 2 to 4 weeks old were observed at intervals of 3 hours. The observations made at 0900 and 1800 hours showed the greatest activity, in regard to both preening and general activity.

INTRODUCTION

The Adélie penguin (*Pygoscelis adeliae*) has a circumpolar distribution, and like the emperor penguin (*Aptenodytes forsteri*) breeds in the coastal regions of Antarctica and on the South Orkney, Bouvet, and South Sandwich islands. The two species are thus the southernmost ranging penguins. The southernmost penguin colony in the world is the Adélie colony at Cape Royds on Ross Island.

The Adélie penguin arrives on the coasts of Antarctica in October to breed in traditional colonies. Most of the colonies are on such level surfaces and so exposed to wind that they are seldom snow-covered, even in winter. The same birds return to the same colony and usually to the same nest site and the same mate year after year [*Sladen*, 1958; *Penney*, 1964]. They build nests of stones 1 to 5 centimeters in diameter and occasionally use bits of bone and wood as well. Immediately upon arrival the males begin to display. In the display *Roberts* [1940] and *Sladen* [1953] call the "ecstatic display," and *Sapin-Jaloustre* [1960] the "position extatique," the bird waves its flippers slowly and rhythmically, raises its head, and utters a series of soft clacking noises, followed by sudden loud guttural sounds with the beak raised to a vertical position (Figure 1). Immediately after an ecstatic display the male often puts its head under a flipper with a croaking "orr-orr-orr," which *Sladen* [1958] calls the "bill-to-axilla" movement. After the chicks hatch the females may also exhibit the ecstatic display [*Sladen*, 1953].

Common to both sexes is the "mutual display" (*Sladen* [1953], the "parade mutuelle" of *Sapin-Jaloustre* [1960]), in which the pair face one another, move their necks from side to side with loud cries and open beaks, and finally bow deeply to each other simultaneously.

The Adélie penguin lays two bright blue-green eggs, which are incubated for 31 to 37 days. The male incubates first for about 14 days, then the female for about

Fig. 1.   The ecstatic display of the male.

the same length of time, and finally the male again for a shorter period.

The first down of the newly hatched chicks is thick and dark grey in color. The head is more darkly colored than the body, which ranges from anthracite ash to black. If a chick falls out of the nest, the parent bird rolls it back in, orienting itself with its beak to the black head and the neck of the chick.

At the age of 14 to 20 days, the chicks grow a second coat of dark gray down and begin to collect in "creches." On January 4, 1965, 24 days after most of the chicks hatched, some creches had as many as 29 chicks (Figure 2). *Sladen* [1953, 1958] refutes the classic descriptions by Wilson [1907], *Gain* [1914], and *Levick* [1915] of some adults watching over the groups of young while the others procure food. The so-called creche-guards, according to Sladen, are merely old birds that have lost their eggs or young but still maintain and defend a territory at or near the creche.

Each Adélie penguin feeds only its own young, which it recognizes acoustically [*Penney*, 1964]. At the age of 8 weeks the young molt into their first juvenile plumage, which differs from that of the adults in having a white throat and in lacking the white eye ring (Figure 3), and they are then able to swim. The old birds feed their young mainly with regurgitated krill—shrimp of the family Euphausiidae (Fig. 4).

In early February both the young and the older birds gather at the shore and enter the water, the young birds following the old. All swim northward or let themselves be carried on ice floes with the current. They spend the winter in the pack ice, where open water is available for feeding and ice floes for resting and sleeping. Other conditions under which they live during the winter are unknown.

Enemies of the Adélie penguin are the leopard seal (*Hydrurga leptonyx*) in the water and the south polar skua (*Catharacta skua maccormicki*), which preys on their eggs and their young on land.

The Problem

Comparatively little is known about the daily rhythms of activity in the immense colonies of breeding birds in either the Arctic or the Antarctic. Although the daily rhythms of many species of other animals in the Arctic have been investigated (birds by *Palmgren* [1935, 1943], *Paatela* [1938], *Karplus* [1952], *Armstrong* [1954], *Cullen* [1954], *Hoffmann* [1959], *Remmert* [1965], and others; rodents by *Swade* [1963] and others; insects by *Remmert* [1965]), no systematic investigation has been carried out on the daily rhythms of antarctic animals. The assertion by *Armstrong* [1954] to the effect that "in the Antarctic so little study has been given to the reaction of birds in continuous daylight the discussion has to be limited to high northern latitudes," is still valid.

Aside from its purely academic interest, an investigation of the daily period of activity of penguins during the polar day might clarify two problems:

1. It should contribute to our understanding of the "biological clock" that informs the individual bird of the time of day, and which *Emlen and Penney* [1964] found plays a basic role in penguin orientation.

2. In extreme latitudes human beings often have difficulty adjusting themselves to the polar day and night. Insomnia, the so-called "big eye," occurs often. It is thus of interest to determine how higher vertebrates that are adjusted to the higher southern latitudes distribute their activities through the day.

Their lack of timidity, their flightlessness, and their high-density concentrations make the penguins excellently suited for behavioral studies. A penguin colony represents an ideal natural laboratory.

Location of Study Site

I lived at the New Zealand-United States station at Cape Hallett from November 6, 1964, to February 25, 1965. The station is located at 72°19'S, 170°13'E on

Fig. 2. A creche of 3-week-old Adélie penguins.

a gravel-covered peninsula about 800 meters long (Figure 5), the remains of a former moraine extending westward from a rocky ridge, the northern point of which is Cape Hallett. Until 1956, the peninsula was completely covered by a breeding colony of Adélie penguins. Constructing the station dislodged some thousands of the penguins, but these still nest immediately behind the buildings. Its topographical situation simulates a night situation during the polar day, for when the sun is in the southeast it casts the entire peninsula in shadow. On Midsummer Day, December 21, the noon altitude of the sun at Hallett is 41°11'; at midnight on the same day the sun is 5°49' above the horizon. The amount of light present, which I measured with a Gossen Trilux light meter, is lowered by a factor of 200 at midnight compared with midday. As most of the Adélie penguin colonies lie at the foot of coastal slopes, they are shadowed part of the time during the polar day.

The day I arrived, November 6, 1964, Moubray Bay, into which the peninsula extends, was still entirely frozen over. The colony, which according to a count in 1963 by B. Reid and J. Cranfield [*Reid*, 1964] numbers about 100,000 birds, was already fully occupied, and only a few stragglers were still coming in over the sea ice.

The Climate

Temperatures. Between November 6, 1964, and February 20, 1965, air temperatures ranged between −19°C and +3°C. Table 1 gives temperature ranges in degrees Centigrade for individual days at 10-day intervals. For November the average temperature was −7.7°C; for December, −1.4°C; and for January, −0.8°C. Additional meteorological data are given in Table 2.

In 1964 Hallett experienced the warmest and cloudiest December and the highest barometric pressure of any December since the station was established in 1956.

Sapin-Jaloustre [1960] has shown that strong blizzards, because of the ice they carry with them, have a much greater cooling effect than winds of the same strengths that carry no ice. In water-filled metal cylin-

Fig. 3. Young Adélie penguins about to leave the colony in February at the age of 8 weeks. Note the white throats, lack of eye rings, and wisps of down still present, mainly on the head.

ders he found for a wind of 3 m/sec a cooling effect of 2.1 cal/g/sec; for a wind of 10 m/sec, twice that; for a wind of 40 m/sec, 7.7 cal/g/sec; but in a blizzard of the same wind speed of 40 m/sec the cooling effect was 19.2 cal/g/sec. At a temperature of −20°C, the cooling effect of a blizzard of 110 km/hr, which occurs commonly on the stormy Adélie Coast, is equal to that of a temperature of −180°C in still air. Thus, blizzards produce the severest physical conditions to which antarctic penguins are exposed. Blizzards are one of the main causes of the high mortality young Adélie penguins suffer, which amounts to 70% from the time the eggs are laid until the young are independent [*Sladen*, 1958; *Sapin-Jaloustre*, 1960] and may even rise to 100% on occasion. Most of the time, of course, the microclimates at the penguin nests are somewhat more favorable than the general climatic conditions.

Method

Three conditions dictated the method:

  1.   The cold, often accompanied by strong blizzards,

Fig. 4. Feeding the young.

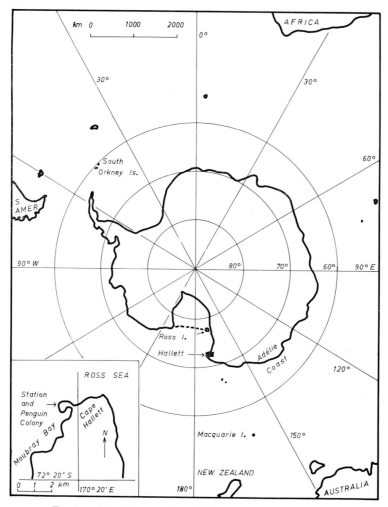

Fig. 5a. Map of Antarctica and the region of Hallett station.

prevented long periods of outdoor observations. I soon discovered that under the worst possible conditions I could spend 15 continuous minutes motionless in observation, and so selected this as the standard observation time span. Counting regularly once every full hour, up to 15 counts per day were possible.

2. In the short antarctic summer the breeding season is much accelerated. The birds proceed rapidly from one stage to the next. The period for the young Adélies from hatching to departure to sea averages 53 days, whereas young rockhopper penguins (*Eudyptes chrysocome*) of Macquarie Island at 54°30'S do not leave the colony for 70 days after hatching [*Warham*, 1963]. The physical conditions also change; for example, when the sea ice breaks up in summer, the penguins must swim instead of walk. To be uniformly evaluated, observations should therefore not extend over longer periods than about two weeks. For ex-

ample, the chicks may gain about 77% in weight from the sixth to the seventh day of life and about 112% from the twelfth to the fourteenth day [*Sapin-Jaloustre*, 1960].

3. Finally, the method was prescribed by the fact that I worked alone at Hallett.

For these reasons I observed the various activities for 15 minutes out of each hour, with the full hour of true local time always in the middle of the observation period. After I obtained five samples for each hour, I combined them to arrive at an average daily action. Aside from the fact that this was the only practical method under the circumstances, it also permitted elimination of random peaks of activity—if these peaks occur at different times on different days they eliminate each other in the combining and averaging. Obviously this method can differentiate only between a 24-hour rhythm and random distribution; rhythms of

Fig. 5b. Hallett station and the Adélie penguin colony.

a period other than 24 hours cannot be established by this method.

I observed the frequency of the following activities at various times of the day:

1. The general activity of a group of three pairs of penguins that nested apart from the colony.

2. The ecstatic display.

3. Wandering over the sea ice by birds in quest of food.

4. Swimming of the food-gathering birds.

5. General activities of the chicks.

In addition, I kept pairs of male Adélie penguins in a partly buried hut under constant light of 100 lux (= 10 foot-candles) and recorded on tape the sounds they uttered and the sound of their movements on the floor of the hut.

TABLE 1. Daily Temperatures (°C) at Hallett, Polar Summer, 1964/1965

Date	Time							
	3	6	9	12	15	18	21	24
Nov. 6	−18	−19	−14	−12	−11	−11	−14	−18
16	−13	−11	− 8	− 7	− 4	− 6	− 8	− 2
26	−10	− 7	− 4	− 3	− 2	− 2	− 6	− 8
Dec. 6	− 2	− 2	− 2	− 2	− 1	− 2	− 3	− 3
16	− 6	− 3	− 2	− 1	− 1	0	− 2	− 3
26	− 3	− 3	− 1	0	0	− 1	− 1	− 1
Jan. 5	− 3	− 3	− 2	+ 1	+ 3	+ 2	0	− 2
15	− 1	− 1	− 1	− 1	− 1	− 1	− 1	− 1
25	− 3	− 3	− 1	+ 2	+ 1	+ 2	− 1	− 3
Feb. 4	− 4	− 4	− 3	− 1	− 1	− 2	− 2	− 2
14	− 5	− 4	− 2	− 1	+ 1	− 1	− 3	− 4

The prevailing wind direction was south-southeast. Further meteorological data are given in Table 2.

TABLE 2. Further Meteorological Data from Hallett, Polar Summer, 1964/1965

	Nov.	Dec.	Jan.
Average wind speed (km/h)	9.26	10.9	11.5
Highest wind speed (km/h)	83.4	90.8	68.6
Average cloud cover	4/10	7/10	6/10
Number of clear days	13	7	8
Number of partially cloudy days	12	5	10
Number of days with complete cloud cover	5	19	13
Average barometric pressure (mm Hg)	744.0	745.4	745.2
Precipitation (cm)	0.08	0.74	0.56
Drifting snow deposit (cm)	0.76	10.7	7.4

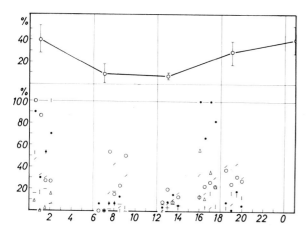

Fig. 6. General activity. *Top:* average of all individuals in four two-hour periods. *Bottom:* activity of single individuals during 30-minute periods. Six individuals represented by open circle, solid circle, vertical line, slanted line, horizontal line, and triangle.

RESULTS

General Activity

During a visit of 30 hours to the 300,000-member Adélie penguin colony at Cape Crozier on Ross Island on November 2, 1964, I noted that around noon the birds were quieter and that more of them lay asleep in or beside the nests than at other times of the day.

At Hallett I marked with colored flipper bands three Adélie pairs that had built their nests at the periphery of the colony about 20 meters from the next group (this was the group designated "Ak 18" on *Reid's* [1964] map of the Hallett penguin colony) and watched them between November 12 and 16. From a tent 15 meters away that served both as concealment and protection from the wind, I recorded the activities of the three pairs, deviating from the usual method by observing for periods of 2 hours, beginning at 0000, 0600, 1200, and 1800, local time. This was at the beginning of the incubation period, and in one case I was able to follow the laying of the second egg closely. When the three females left on November 15, I discontinued the observations.

During each observation period I classed as activity: interruptions in incubation such as turning the eggs over, preening, gathering stones, walking, nest building, making sounds, and courtship displays. Designated as rest were: incubating, lying down, sleeping while lying down, standing, sleeping while standing.

Figure 6 shows the percentages of time each individual bird spent in activities during 30-minute observation periods. The top graph combining the aver-

ages for the four 2-hour periods shows the six individual birds were active 40% of the time at midnight and only 7% at midday. The values are of high significance as of non-random distribution ($p < 0.01$). Figure 7 shows the locomotor activities of the individual birds, which at this time were largely for the purpose of gathering pebbles. Here again the nighttime activity and the midday inactivity are strongly manifest.

The Ecstatic Display of the Males

Between November 14 and 23 I counted the frequency of the ecstatic display in an isolated Adélie penguin group of 49 nests, 5 of which were occupied by unmated males. The males were recognized easily, because the ecstatic display is restricted to them up to the time the chicks hatch [*Sladen*, 1958]. Among the few paired birds that did not display it was impossible to determine the sex. Marking had to be omitted because it would have taken too much time in a group of this size and disturbed their normal activities unduly. During the period of observation about 70 individuals were constantly present, of which about 40 were males. I made three series of counts for 16 minutes out of each hour. During 9 observation days three full 24-hour cycles were recorded.

Total number of displays. Figure 8, curve a, shows the frequency of ecstatic displays, combined for each 2 hours. Each point represents 6 displays in each 16 minutes. Regardless of the number of males present, the pattern of activity shows morning and evening peaks and a minimum around noon (difference between morning maximum and noon minimum statistically established at $p < 0.05$; t test). In the incomplete ecstatic display the Adélie penguin raises its beak only to an oblique angle and utters only the first soft clacking sounds. Elevation of the beak to the vertical position and the explosive guttural sounds do not occur. Curve b, Figure 8, shows the frequency of incomplete displays at various times of the day; the ratio of incomplete to complete displays remains much the same throughout the day (Figure 8, curve c).

Display frequency. Plotting the greatest frequency of displays in the group per minute yields a curve with maxima for complete displays per minute at 0900 and 2100 (Figure 9), but the data do not prove significant by t test.

Number of displaying males. The number of males participating in the group display is difficult to determine because some birds perform only intermittently.

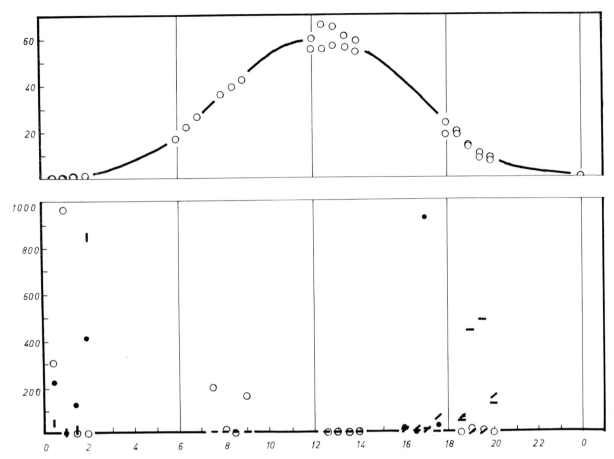

Fig. 7. Locomotion. *Bottom:* number of steps of single individuals during 30-minute periods; five individuals represented by open circle, solid circle, vertical line, slanted line, and horizontal line. *Top:* the light intensities at that time in units of 1000 lux.

Many of them even limit their activity to one or two displays, while others repeat the ecstatic display for periods of 10 minutes or longer. Therefore, it is best to express the number of participating males as the greatest number observed displaying at the same time.

This value is shown in Figure 10. The peaks of activity lie at 0900 and 1900 hours with 18 and 17 males performing simultaneously ($p < 0.05$).

Displays per male. If the number of displays per male is plotted against the time of day, morning and evening maxima again appear (Figure 11), but the values

Fig. 8. Frequency of ecstatic display. a = total number of complete displays; each point represents the sum of six 16-minute periods. b = number of incomplete displays; same as a. c = percentage of complete displays for all hours. Ordinate shows per cent scale for c.

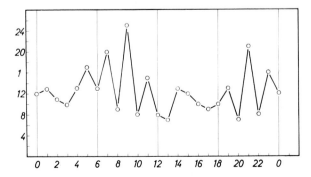

Fig. 9. Display intensity, i.e., number of displays per minute in a group of 40 males.

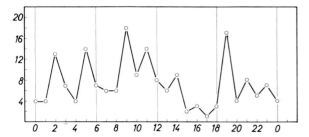

Fig. 10. Highest number of males displaying at the same time.

are not statistically significant. The frequency peak around 0900 hours shown in Figures 8, 9, and 10 is masked by the greater number of males participating from 0100 through 0900.

Frequency of ecstatic display during the reoccupation period. In the period between the hatching of the chicks in the middle of December and the establishment of the creches at the beginning of January the number of birds in the Adélie breeding colony reaches a second maximum [*Sladen*, 1955]. Not only do the adults return from the sea to feed their young, but the unsuccessful breeders who lost their nests also return to occupy new territories, and non-breeding itinerant 2- to 3-year-old young penguins return to claim territories. This period *Sladen* [1955] calls the reoccupation period. During this period both sexes perform the ecstatic display, and both the ecstatic and the mutual display occur with increasing frequency for a time.

In a group of 10 nests with a total of 17 chicks I noted the frequency of the ecstatic display of the males and females. Between December 27, 1964, and January 15, 1965, I counted for a 15-minute period 8 times a day, starting at midnight and every three hours thereafter. This was repeated 5 times (i.e. for five full daily cycles), so that a total of 8 × 5 = 40 values was obtained. Here the number of displaying birds fluctuated irregularly, because the adult birds, both male and female, were away part of the time in search of food for the young (Figure 12, curve b).

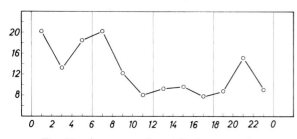

Fig. 11. Average number of displays per male.

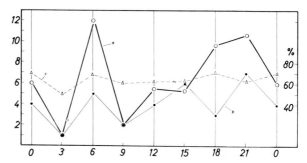

Fig. 12. Frequency of ecstatic displays during the reoccupation period. *a* = number of displays per male during five 15-minute periods for each point (see text). *b* = number of males displaying. *c* = percentage of complete displays, with scale at right.

The number of performances per bird fluctuated during the day in the same way as in November, with peaks of activity at 0600 and 2100 (Figure 12, curve a), but there is no statistical significance.

The relationship of complete to incomplete performances is as during the pairing period, about the same throughout the day, and is not significantly different from that of the pairing period (Figure 12, curve c). There seemed to be little synchronization in the colony. When the observation group was silent, other groups often displayed intensively.

Walking over the Sea Ice

To determine the extent to which adult Adélie penguins wander across the sea ice to open water to gather food at various times of the day (Figure 13), I made counts of the penguins walking across a nearby stretch of shore ice one-kilometer long from December 5 to 15. I made hourly counts on the hour for a total of 5 daily cycles and so obtained 120 values. On December 15 I had to end these counts because the ice broke up and the ice shelf receded closer and closer to the shore. On December 20 no continuous ice surface was left, and the penguins had to swim across short stretches of open water.

Total numbers. The total number of penguins walking on the ice was greatest around 0900 local time, the minimum occurred around noon (Figure 14, lower curve).

Outward-bound and returning birds. If the numbers of outward-bound and returning birds are evaluated separately, a morning maximum is evident for both groups (Figure 14, upper curves a and b). Most of the birds left around 0500, but some continued to depart through the morning up until 1100 hours. Most

Fig. 13. Adélie penguins walking over the sea ice.

of the penguins returning with food arrived between 0800 and 0900. The minima occurred at noon and 1300 hours, respectively. The difference between maxima and minima is highly significant ($p < 0.01$) for the birds leaving and significant ($p < 0.05$) for those returning. The greater number of outward-bound birds is in part due to an unavoidable observation defect: the outward bound birds are easily recognizable from the shore by their black backs; the white bellies of those returning are hard to distinguish in the fissured sea ice, especially when the birds are still at a distance. Also, departing non-breeders will stay at sea and not return [*Sladen*, 1958].

Number of resting birds. The Adélie penguin often interrupts its traveling while whole groups remain motionless from one to several hours on the ice. They may lie down or sleep standing (Figure 15). As might be gathered from the walking minimum, few birds rest on the ice around noon. The greatest number of resting or sleeping groups was found around 0300 (Figure 16). Thus the midday rest period is passed pre-

dominantly at the colony or out near open water, in any case not in motion.

Relationship between active and resting penguins on the sea ice. If the percentage of birds walking is selected as the standard of activity for the birds present on the ice, the activity maximum occurs around 1100 and the minimum around 2200. The whole curve (Figure 17) resembles a sine curve and runs parallel to the amount of light present (linear measurement correlation, 0.915; highly significant, $p = < 0.001$). However, the birds walking on the ice are only partly subject to the light conditions I measured at the colony. They walk northward past Cape Hallett through an area that is in shadow for at most two hours before midnight. The curves show that the birds on the ice increase their walking and general activity as the light increases. In this respect the Adélie penguin differs from the Weddell seal (*Leptonychotes weddelli*), which inhabits the same biome. The Weddell seal becomes more active as the light

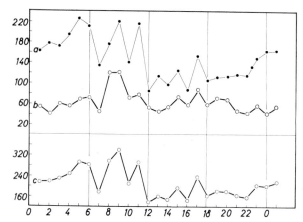

Fig. 14. Numbers of penguins traversing the sea ice. Each point represents the average of 5 counts. a = number of birds walking out. b = number of birds returning. c = total number of walking birds.

diminishes; its activity maximum occurs between midnight and 0200 [*Müller-Schwarze*, 1965].

Swimming. In early January the ice in Moubray Bay was so broken up that the penguins had to swim all the way to their feeding grounds. Between January 20 and February 2 I made five daily counts of the birds swimming out and in along one kilometer of coastline for 15 minutes on the hour every hour. As in crossing the ice, the maximum occurred around 0800, but the minimum was at 1700 (Figure 18).

If the birds are classified according to whether they are swimming outward or inward, a considerable difference becomes apparent; toward 0500 the number swimming outward increased suddenly (Figure 18), which is highly significant, $p < 0.01$. This is the time of day when the increase in light is greatest; within an hour the daylight doubles in strength (Figure 19). As in the case of the birds walking outward, the greatest activity continued from 0500 to 1100 (cf. Figure 14, upper curve a).

This morning departure of the Adélie penguin corresponds to the morning parade of the king penguin (*Aptenodytes patagonica*), described by *Stonehouse* [1960]. In winter up to 500 adult birds of the South Georgia colony gather in the early morning and walk slowly toward the coast. At the shore they divide into groups of about 50 individuals, pressed as closely together as possible without fighting. They enter the water and swim completely submerged. Suddenly, with the sound of flippers beating upon the water, they return to shore. Half a minute later, they again walk slowly into the water; this may be repeated several times.

Adélie penguins about to enter the water in the early morning also form groups numbering 10 to 20 or more.

At all times of the day, about the same number of penguins swim back to the colony. These birds have full stomachs, and their activity is probably deter-

Fig. 15. A group of penguins resting during their departure.

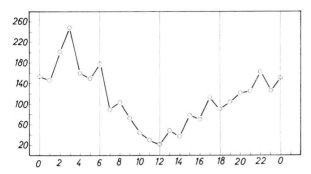

Fig. 16. Number of birds resting during departure; averages from five observations for each hour.

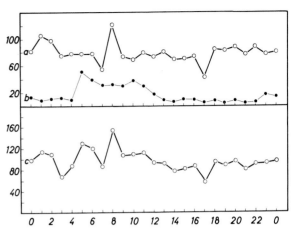

Fig. 18. a = numbers of swimming birds returning with food. b = numbers of birds swimming out for food. c = total numbers of birds swimming to and from the colony. Each point represents the average of five fifteen-minute periods.

mined to a greater extent by stimuli from the filled stomach than by external influences.

The returning Adélie penguins swim porpoise fashion in groups of 2 to 20 birds. At intervals of about 5 seconds they leap 10 to 60 centimeters out of the water for ½ to 2 meters and submerge again completely. Between jumps they swim for 100 meters or more fully submerged. (All values are estimates.) *Stonehouse* [1960] described the same swimming style in the king penguin, which, like the Adélie, employs this style for slow long-distance swimming.

Correlation with amount of light. The total number of swimming birds is correlated neither to the absolute amount of light ($r = +0.289$, not significant), nor to changes in the amount of light per time unit ($r = -0.092$).

Activity of Chicks

On December 26 I observed the first creches, which grew in size from day to day. On December 28 they

contained from 9 to 11 chicks; on December 30 the maximum was 12 to 13, and on December 31 they numbered 16 to 20 chicks. From December 27, 1964, to January 15, 1965, during what *Sladen* [1955] calls the creche stage, I observed the activities of the chicks. I selected a group of nests containing 17 chicks about 25 meters away from other penguins. I observed these for 15 minutes at intervals of 3 hours and thus obtained 8 samples for each 24 hours. During the 15-minute periods of observation I noted the momentary activities of the chicks at 3-minute intervals, which gave me $6 \times 5 = 30$ values per hour. The highest possible activity value for the entire group was thus $30 \times 17 = 510$.

As *rest* I counted sleeping, lying down, sitting, standing and panting (the Adélie penguin pants even at $0°C$); as *activity* I counted running, preening, and demanding and consuming food. This activity is re-

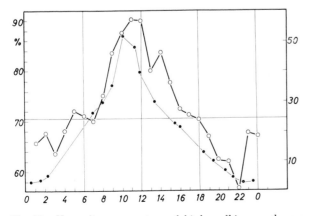

Fig. 17. Heavy line: percentage of birds walking on the sea ice, with scale at left, 100% = walking and resting birds together. Light line: light intensities in thousands of lux, with scale at right.

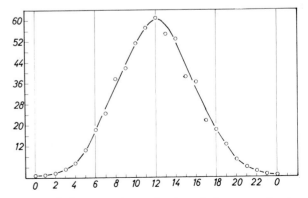

Fig. 19. Light intensities at the time of the count of swimming birds in thousands of lux.

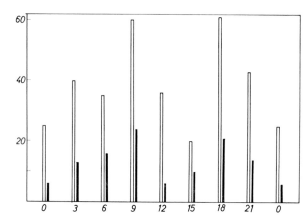

Fig. 20. Activity of chicks. Light bars: general activity. 17 chicks were observed for five 15-minute periods. During each period the activity was noted at three-minute intervals, i.e., six times, so that finally there were for each hour of the day $6 \times 5 \times 17 = 510$ possible units. Ordinate shows units of activity. Black bars: units for preening.

corded in the light bars of Figure 20 for the 8 observation periods. The young birds were most active around 0900 and 1800 hours and least active early in the afternoon ($p = < 0.05$). I recorded the frequency of preening separately (black bars). The preening maxima also occurred around 0900 (highly significant, $p < 0.01$) and 1800 (not significant); the minima, around noon and midnight.

Social relations tendencies. I saw the largest number of chicks in the creches during the rest periods, i.e., at midnight and midday. The groups were smallest at 0600, 0900, and 1800 hours, when the young penguins were most actively feeding. (Most chick activity is begging for and taking food.) The adult birds feed only their own one or two young, and as soon as the young penguins are hungry, move from the creche to their own nests, where often one of the parents is standing while the other is away for food. The number of adult birds present among the young penguins fluctuated between 3 and 16; this was subject to no daily periodicity. During the 8 sampling periods, an average of 10 to 12 adult birds was present in the group.

Future investigations conducted on a larger scale will determine whether the irregularities of these findings mean only slight daily rhythms in the southern summer, or whether they should be ascribed to the sampling method and the relatively small number of observations.

Experiments in Constant Light

Continuous light. Two male penguins were kept to-

gether at two separate times in a partly buried hut 3×3 meters square for 6 and 5 days, respectively. Their movements on the plywood floor and their sounds were recorded by microphone. An electric bulb produced a constant illumination of 100 lux in the center of the room. The recorded activity was analyzed by computer at the Stanford Electronics Laboratory and transferred to film. The results are shown in Figures 21 and 22. Each bird maintained its normal pattern of activity for 2 days; then the activity peak started to shift in a regular fashion. As in all diurnal bird species investigated, the circadian frequency increases in permanent light, i.e., the bird's subjective day remains shorter than 24 hours (Aschoff's rule). In the two experiments the interval between activity peaks was 20.3 ± 1.4 hours and 20.5 ± 0.4 hours.

12–hour light – 12–hour darkness. Figure 23 shows the activity pattern in a 12:12 hour light-dark schedule. The greatest activity peak of the day occurs with the onset of the light.

The first experiment demonstrates the persistence of the activity rhythm (free running rhythm) under constant conditions. The second experiment shows the strong response of the birds to the light that the field observations had already suggested.

DISCUSSION

At Cape Hallett (72°19'S) the polar daylight is continuous from November 10 to February 2. During this time, the maximum difference in the sun's altitude between noon and midnight is 35°22'. Together with the shadow cast over the rookery by a rocky ridge when the sun is low, this causes a regular fluctuation of the light by a factor of 200. *Swade* [1963] found an activity rhythm in arctic rodents at light amplitudes of 4 to 100 foot-candles (factor of 25); the threshold was at 0.7 to 10 foot-candles, where in five of eight experiments a daily rhythm was observed.

A constant light intensity from a clear sky and a constant sun altitude is found only at the pole, where, in the Antarctic, no animal life occurs. Although antarctic vertebrates and land arthropods live in constant daylight during the summer, they are nevertheless exposed to considerable fluctuations of light. That this influences the amount of activity is to be expected, for *Armstrong* [1954], *Hoffmann* [1959], *Karplus* [1952], *Remmert* [1965], and others found that animal activity in the arctic summer depends upon the amount of light present.

At temperate latitudes, the beginning and ending of activity are visible manifestations of basic physiological fluctuations in sine or saw-toothed curves [*Aschoff and Wever*, 1962], the parameters of which are unknown. Instead of the beginning and ending of activity, animal species living in the polar day show complete cycles of increasing and decreasing amounts of activity. This is observable in the population and in individuals as well.

Aside from incidental comments, no systematic investigations are available on the daily rhythms of antarctic animals. During 48 hours of intermittent observations, *Goldsmith and Sladen* [1961] noted no daily rhythm of activity in the emperor penguin colony

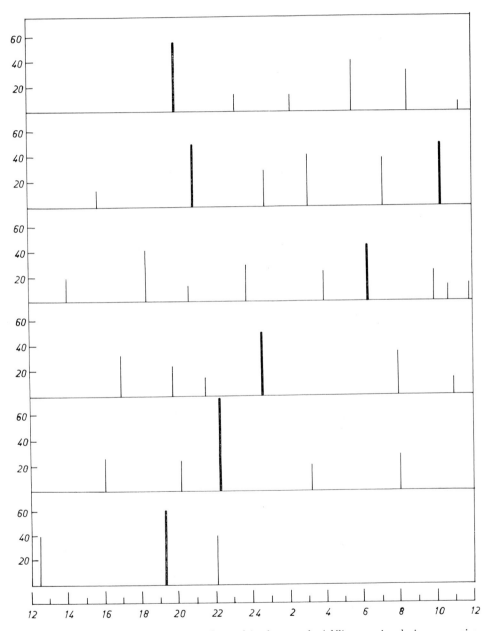

Fig. 21. Activity rhythm (locomotion and sounds) of two male Adélie penguins during successive days in constant light of 100 lux, temperature 0°C, ±1°C. Each bar indicates one bout of activity longer than 10 minutes, placed at the middle of the time of the particular activity bout. Heavy bars are the longest activity peaks for the day. Abcissa: hours of the day; Ordinate: minutes of activity. Period length is 20.3 ±1.4 hrs.

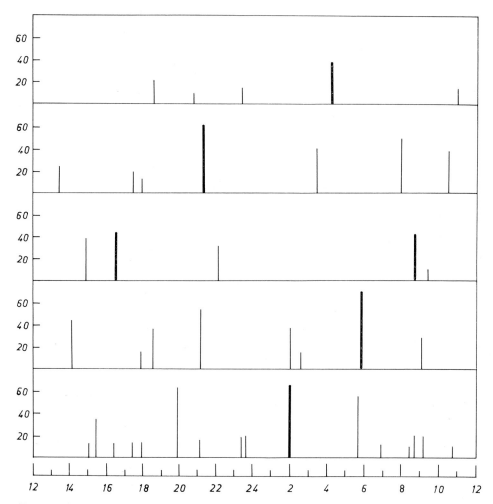

Fig. 22. Activity of a second pair of male Adélie penguins under constant light of 100 lux. Scales as in Fig. 21. Period length is 20.5 ±0.4 hrs.

on Coulman Island (73°S, 169°E). Body temperature measurements taken with a thermistor in the stomach showed no daily rhythms in either emperor or Adélie penguin adults, but their work contains a 24-hour temperature curve (Figure 3) for an emperor penguin chick 6½-months old that shows peaks at 0600 and 1500. These bring to mind the two peaks of activity of the Adélie chicks in midsummer and the two frequency peaks of the ecstatic display of the males. Incidentally, the body temperature of Adélie chicks does not become independent of the ambient temperature until the age of 10 to 15 days.

According to the observations presented here, Adélie penguins show a daily rhythm of activity during the south polar day. The midday minimum observed for all kinds of activities seems possibly connected with reduced dissipation of body heat. When the temperature rose at midday to a maximum of + 3°C at Hallett, even penguins standing still panted heavily. Because of their poor ability to dissipate body heat, when penguins are seized and forced to defend themselves against men, their body temperature rises 2 degrees and does not return to the normal of 37.7°C until 2 hours later [*Goldsmith and Sladen*, 1961].

Comparing the findings to the facts known about arctic birds shows the following similarities and differences.

Similarities. Most arctic birds observe rest for some hours around midnight, generally shortly before midnight, during the period of continuous daylight. *Haviland* [1926] reports midsummer activity of the birds of the Yenisei River was markedly reduced between midnight and 0200 or 0300. *Armstrong* [1954] writes of songbirds in general, "There is fairly general agree-

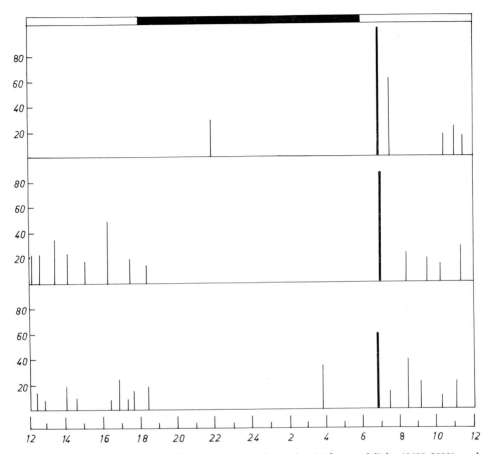

Fig. 23. Activity of two male Adélie penguins in alternating 12 hours of light (0600–1800) and
12 hours of darkness. Scales as in Figure 21.

ment that a lull in passerine song occurs in the Arctic about 2100 hours." The resting periods noted by various writers on the subject differ to some degree. *Palmgren* [1935] observed at the time of the summer solstice a "main resting period" between 1800 and 2300 hours; *Karplus* [1952] between 1700 and 2100 hours; *Armstrong* [1954] between 2000 and 2400 hours; *Dunajeva and Kutscheruk* [1941] between 2300 and 0300 hours. According to *Franz* [1949] in *Musicapa striata* the rest period occurs after midnight, because these birds are still catching insects up to shortly before midnight; 8 other species of birds rest mostly before midnight; the minimum rest period was 3 to 4 hours in *Turdus musicus*, the maximum 6 hours in *Motacilla alba*. According to *Palmgren* [1943], *Chloris chloris* sleeps a minimum of 7 hours. *Armstrong* [1954] estimates the average time at which most arctic song birds go to rest at about 2000 hours.

In conformity with these results, a reduction in walking (Figure 17) and swimming (Figure 18) activity is clearly evident for the Adélie penguin around 2100 hours.

Armstrong [1954] stresses that where the short summer requires quick pairing, birds display and court throughout the night. This is true to a great degree for antarctic penguins.

According to Armstrong, most nest building is done in the morning hours during the polar summer. This corresponds to my findings at Hallett (Figures 6 and 7).

Armstrong [1954:23] states, "Although some of my graphs show that the birds roosted, or reduced their activity, when temperature was high, it is unlikely that it was a direct response to changing temperature." Although strict comparisons cannot be made between flying arctic birds that migrate and are not particularly adapted to extremely low temperatures and the non-flying, cold-adapted antarctic penguins, the influence of temperature upon daily rhythms should nevertheless be reckoned with in both.

The distribution of the ecstatic display activity (Figures 8 to 11) and of the daily rhythms of the chicks (Figure 20) resembles the pattern *Cullen* [1954] found for *Uria lomvia;* most of the birds fly to the colony around 0800 to 0900 hours, and again at 2000 to 2300 hours.

Differences. The pronounced midday activity minimum is, to my knowledge, not known among arctic birds. The Adélie penguins never showed three peaks of activity such as *Klockars* [1941] found in May and June for birds near Helsinki. Nor was it possible to establish an activity maximum around 1400 hours and a minimum at midnight such as *Cullen* [1954] describes for flying by *Fulmarus glacialis.*

The different circadian patterns at different times of the breeding season deserve further attention. During the nestbuilding and egg laying period "nocturnal" activity prevails; the "ecstatic" display of the males shows a morning and an evening maximum; the summer activities such as food gathering reach only one daily peak in the morning, which coincides with rapidly increasing light intensity. Toward fall, however, two activity peaks occur again (chick activity and "reoccupation" displays of males). These changes must be viewed in connection with the life of the Adélie penguins in the polar night.

Acknowledgments. My thanks go to the U.S. National Science Foundation and to the Deutsche Forschungsgemeinschaft for financial assistance that made the research journey possible, and to Richard L. Penney for the suggestion that I undertake the work. George A. Llano, Helen T. Gerasimou, John R. Twiss, Jr., Henry S. Francis, Jr., Edward E. Goodale, and Philip M. Smith contributed largely to the success of my investigations. At McMurdo Sound I had stimulating discussions with John T. Emlen and Richard Penney. At Hallett the twelve mechanics of the U.S. Navy Task Force 43 gave me very valuable aid; the head of the station, Theodore C. Nagel, and the scientific coordinator, Robert C. Judd, prepared for me the best possible working conditions. I thank Gene Levi of the U.S. Weather Bureau for the Hallett weather data. Professor Dr. G. Niethammer is due special thanks for furnishing me with a place at the Alexander Koenig Museum, Bonn, while evaluating my data. Professor Dr. J. Aschoff of the Max-Planck Institut für Verhaltensphysiologie, Professor Dr. Robert I. Bowman, San Francisco State College, and Dr. Oliver L. Austin, Jr., Gainesville, are to be thanked for critical review of the manuscript.

REFERENCES

Armstrong, E. A., The behaviour of birds in continuous daylight, *Ibis, 96,* 1–30, 1954.

Aschoff, J., and R. Wever, Beginn und Ende der täglichen Aktivität freilebender Vögel, *J. Ornithol., 103,* 2–27, 1962.

Cullen, J. M., The diurnal rhythm of birds in the arctic summer, *Ibis, 96,* 31–46, 1954.

Dunajeva, T., and V. Kutscheruk, Material on the ecology of the terrestrial vertebrates of the tundra of South Yamal, *Moscow Soc. Naturalists, N.S. Zool. Sect., 4,* 1–80, 1941.

Emlen, J. T., and R. L. Penney, Distance navigation in the Adélie penguin, *Ibis, 106,* 417–431, 1964.

Franz, J., Über Ernährung und Tagesrhythmus einiger Vögel im arktischen Winter, *J. Ornithol., 91,* 154–165, 1943.

Franz, J., Jahres-und Tagesrhythmus einiger Vögel in Nordfinnland, *Z. Tierpsychol., 6,* 309–329, 1949.

Gain, L., La vie et les moeurs du Pingouin Adélie, *IXe Congr. Intern. Zool., Monaco 1913,* 1914.

Goldsmith, R., and W. J. L. Sladen, Temperature regulation of some antarctic penguins, *J. Physiol., 157,* 251–262, 1961.

v. Haartman, L., Über den Tageshrythmus des Mauerseglers, Apus a. apus (L), *Orn. Fenn., 17,* 7–11, 1940.

Haviland, M. D., *Forest, Steppe and Tundra,* Cambridge University Press, London, 1926.

Hoffmann, K., Über den Tagesrhythmus der Singvögel im arktischen Sommer, *J. Ornithol., 100,* 84–89, 1959.

Karplus, M., Bird activity in the continuous daylight of the arctic summer, *Ecology, 33,* 129–134, 1952.

Levick, G. M., Natural history of the Adélie penguin, *Brit. Antarct. "Terra Nova" Exped. 1910, Zoology 1,* 55–84, 1915.

Mueller-Schwarze, D., Zur Tagesperiodik der allgemeinen Aktivität der Weddell-Robbe (Leptonychotes weddelli) in Hallett, Antarktika, *Z. Morph. Oekol. Tiere, 55,* 796–803, 1965.

Paatela, J. E., Beobachtungen über das Verhalten der Vögel in der Sommernacht, *Orn. Fenn., 15,* 65–69, 1938.

Palmgren, P., Über den Tagesrhythmus der Vögel im arktischen Sommer, *Orn. Fenn., 12,* 65–69, 1935.

Palmgren, P., Zur Tagesperiodik der Finkenvögel, *Orn. Fenn., 20,* 99–103, 1943.

Penney, R. L., The Adélie penguin's faithfulness to territory and mate, *Antarct. Biol., 1,* 401–406, 1964.

Reid, B., *Map of Adélie penguin rookery Seabee Spit,* Dept. Sc. and Industr. Res. New Zealand, 1964.

Remmert, H., Über den Tagesrhythmus arktischer Tiere, *Z. Morph. Oekol. Tiere, 55,* 35–49, 1965.

Roberts, B. B., The breeding behaviour of penguins, British Graham Land Exped. 1934–37, *Sci. Rept. 1,* 195–254, 1940.

Sapin-Jaloustre, J., Ecologie du Manchot Adélie, Actual. Scient. et Industrielles 1270, Publ. Exped. Pol. Franç. Nr. 208, 1960.

Sladen, W. J. L., The Adélie penguin, *Nature, 171,* 952–955, 1953.

Sladen, W. J. L., Some aspects of the behaviour of Adélie and Chinstrap penguins, *Acta XI. Congr. Intern. Orn., Basel 1954,* 241–247, 1955.

Sladen, W. J. L., The Pygoscelid penguins. 1, Method of study, 2, The Adélie penguin, *Falkl. Isl. Dep. Surv. Sc. Rept. Nr. 17,* 1–97, 1958.

Stonehouse, B., The King penguin, *Aptenodytes patagonica,* of South Georgia, 1, Breeding behaviour and development, *Falkl. Isl. Dep. Surv. Sc. Rept. Nr. 23,* 1–83, 1960.

Swade, R. H., Circadian rhythms in the Arctic, unpublished dissertation, Princeton University, 1963.

Warham, J., The Rockhopper penguin, *Eudyptes chrysocome,* at Macquarie Island, *Auk, 80,* 229–256, 1963.

Wilson, E. A., *Natural History,* vol. 2: Zoology, part 2; Aves, Brit. Nat. Antarct. Exped. 1901–04. Sc. Results, London Brit. Mus. Nat. Hist., 1907.

BIOCHEMISTRY OF THE ADÉLIE PENGUIN: STUDIES ON EGG AND BLOOD SERUM PROTEINS

ROBERT E. FEENEY, RICHARD G. ALLISON, DAVID T. OSUGA, JOHN C. BIGLER, AND HERMAN T. MILLER[1]

College of Agriculture, University of California, Davis, California

Abstract. The egg and blood serum proteins of the Adélie penguin (*Pygoscelis adeliae*) were studied and compared with those of other avian species. These studies constituted an introductory program on the comparative biochemistry of the antarctic penguins. The egg white of the Adélie penguin was found to contain most of the homologous proteins found in the egg whites of other avian species. Compared with most other avian egg whites, the Adélie penguin white is unusually high (4–5 per cent) in sialic acid and unusually low in lysozyme (approximately 0.05 per cent). Physical and chemical examinations were made of the purified ovomucoid, ovotransferrin (conalbumin), and ovomacroglobulin. In limited studies of the yolk proteins, the Adélie proteins were found grossly similar to those of the domestic chicken. The blood serum proteins were compared with those of the chicken and also with those of the Humboldt penguin (*Spheniscus humboldti*). The proteins of the two penguin serums differ in several major respects from the proteins of chicken serum and also differ from one another.

Eggs of the Adélie penguin were incubated during the terminal development, and penguin chicks were successfully hatched. Feeding experiments were conducted for a 10-day period.

INTRODUCTION

Blood serum (or blood plasma), milk, and avian eggs are three major fluids of animals that are high in protein, have general biochemical significance, and are relatively easily obtainable in quantities sufficient for biochemical studies. The proteins of these fluids have physiological and embryological importance and are valuable for general fundamental biochemical studies and studies on biochemical taxonomy and genetics. Many of the proteins have unique biochemical properties. These include the gamma globulins of blood serum, which may contain antibodies, and several enzymes or enzyme inhibitors in blood serum and in eggs. The proteins of these three fluids are of further interest because they include homologous and analogous forms of proteins, some of which are found in all three fluids.

Our laboratory has been studying the proteins of blood serum, egg white, and milk, primarily from the following standpoints [*Feeney, 1964a*]:

1. The first involves a fundamental research program on the biochemistry of proteins. The program is directed at understanding the structural basis for bio-

chemical function, i.e. enzymic activity or other unique properties. One of the major approaches in this program is the use of homologous or analogous molecular structures from different species in the same way that the comparative anatomist uses the anatomical structures of different species. The program thus attempts to find the same type of protein in the blood serum, milk, or egg of different species and to relate differences in biochemical function to differences in molecular structure.

2. The second area is that related to biochemical evolution or biochemical genetics [*Anfinsen, 1959*]. Here the primary targets are the differences in protein structures and functions as related to evolution and genetics. It should be obvious that the two programs are closely intertwined with one another.

As a result of preliminary studies, we selected the Adélie penguin (*Pygoscelis adeliae*) as a primary species for investigations of avian blood serum proteins and avian egg proteins [*Feeney et al., 1966*]. Its egg contains sufficient concentrations of certain individual egg-white proteins that are readily amenable to biochemical study and that have properties of particular biochemical interest. Other avian species have similar interesting proteins, but these species lack the attendant aspects of those of the Adélie penguin, which in-

[1] Present address: Department of Chemistry, Lincoln University, Jefferson City, Missouri.

clude: (a) the taxonomic or evolutionary position of the Adélie penguin [*Simpson*, 1946; *Wetmore*, 1960]; (b) the relative isolation of the penguins from other avian species [*Sladen*, 1953, 1965]; and (c) the current availability of large numbers of Adélies for scientific study.

Our laboratory is committed to a definitive study of the blood serum and egg proteins of the Adélie penguin. In addition, certain interrelated physiological, nutritional, and embryological problems will be considered and studied as time and conditions permit and as they relate to the main problems.

GENERAL METHODS AND PROGRAMS

Operations and Programs in Antarctica and at the University of California

Research has been done in three localities: the rookery at Cape Crozier, the BioLaboratory at McMurdo station, and the laboratories at the University of California at Davis. The work at Cape Crozier has so far been restricted to collecting samples. The work at McMurdo station has included preparation of samples, as well as certain preliminary laboratory studies. The laboratory studies at McMurdo are primarily those that must be done on living or unfrozen material. Most all the other studies have been conducted at the laboratories at the University of California, Davis, because of the relative ease of operations there and because many of the studies require sophisticated equipment. The program is now in its third year. As the over-all program has developed into a long range one, it is difficult to classify programs or to divide results on an annual basis, but a general division of objectives can be made.

The 1964–1965 program was a rather short program that embodied a visit of several days to McMurdo station and slightly over a week to the Adélie penguin rookery at Cape Crozier. During the second week of November 55 eggs were obtained from four different colonies. Each egg was less than 24 hours old when removed from the nest. The eggs were cooled and broken open within a few hours of gathering. The contents were separated, and the whites and the yolks were individually bottled and frozen. In this condition they were transported to the laboratory at the University of California at Davis.

In addition to the Adélie eggs, a fresh kiwi (*Apteryx mantelli*) egg was obtained during the return trip through the courtesy of the New Zealand Wildlife Service. The frozen contents of the egg were also transported to the University of California at Davis.

To our knowledge this is the first kiwi egg to be examined in detail biochemically.

Laboratory studies on the penguin and kiwi material were initiated immediately upon our return to Davis. Each of the 55 penguin eggs was examined individually, but the chemistry of the individual constituents was examined on pooled samples. Of course only very limited comparative studies could be done with the single kiwi egg.

Based on results of the 1964–1965 program, an enlarged, long-range program was initiated for 1965–1966. The program on the Adélie penguin was directed at obtaining a large number of eggs for a definitive characterization of the individual egg proteins. Preliminary studies were also to be made on blood-serum proteins. A new program on the blood and muscle proteins of cold-adapted marine species was initiated, which will be described in a later publication.

Permission was obtained from the Office of Antarctic Programs, National Science Foundation, to collect 900 eggs and 6 penguins. A party consisting of Robert E. Feeney (project director), Herman T. Miller (research biochemist), and Richard G. Allison (graduate student) worked at Cape Crozier and at the BioLaboratory at McMurdo, over an approximate 2-month period in October, November, and December of 1965.

Every effort was made to collect eggs and penguins so as to obtain information of value to conservation studies and not to interfere with studies on conservation or with conservation itself. The eggs were taken from 16 different colonies in a part of the rookery outside the area of long-range study by the Johns Hopkins University group.

Only the second egg of the normally two-egg clutch was taken. The first egg was dated (Figure 1), and the second egg was taken fresh when it was laid. All the second eggs were collected from four colonies containing about 400 occupied nests. Table 1 records the numbers of eggs marked and collected from these four colonies through November 16, 1965. The average time between laying of the first and second egg was 3 days for 610 recorded cases. Seven colonies adjacent to the collection area containing a similar number of nests were left undisturbed as a control area. It is planned that this work will provide a basis for future studies on the effects of this removal on these particular colonies.

Of the 735 eggs collected, approximately 100 were used for studies on embryos or membrane stability. Most of the remaining 635 eggs used for egg material were opened at Crozier. After being separated, the

Fig. 1. An Adélie penguin on an egg laid and numbered on November 4, 1965. The egg was marked "4" with indelible ink for identification as the first egg of a normal two-egg clutch and as a reference date to determine the days between the first and second eggs.

egg yolks and egg whites were immediately frozen for transport to the University of California at Davis. Twice when helicopter transport was available within 24 hours of collecting the eggs, they were opened at McMurdo and used for studies that could not be made with frozen material.

The 1966–1967 program was a continuation of the previous programs. Personnel were Robert E. Feeney and graduate students John C. Bigler, Stanley K. Komatsu, and David T. Osuga. (Approximately three-fourths of the field program in 1966–1967 concerned the marine biology program.) A total of 200 Adélie eggs were collected for laboratory studies and for incubation. Collecting and handling were done as described above.

Methods of Protein Analysis

The analytical procedures used in this study have been described in several recent publications from this laboratory [*Feeney et al.*, 1963a, 1966]. In general they included the following, alone or in combination: comparative electrophoretic patterns of the intact egg white and blood serum, direct biochemical assays or analyses of the intact egg white or blood serum, physical and chemical fractionation of the egg white or blood serum to separate and purify the individual components, and determination of the properties of these purified components by physical and biochemical methods. Also used were such procedures as chromatographic fractionations and various types of electrophoretic separations [*Feeney et al.*, 1966; *Rhodes et al.*, 1958] which partially or extensively separate the individual constituents and give profiles or patterns that can be used to describe the constituents qualitatively and quantitatively.

The methods used for assays or analyses of biochemical activities were particularly pertinent to egg white and blood serum in several instances. These included assays of the ovotransferrin and serum transferrin by their chromogenic capacity with ferric iron and estimating the amount of protein present from the salmon-pink color produced with a maximum wavelength at approximately 465 mμ [*Feeney and Komatsu*, 1966]. The transferrins were also determined by radioautograms of patterns of starch gel electrophoresis using Fe^{59} [*Clarke et al.*, 1963]. Lysozyme activities were estimated enzymically with cellular preparations of *Micrococcus lysodeikticus*. Inhibitory activities against the proteolytic enzymes trypsin and chymotrypsin were measured spectrophotometrically with synthetic substrates [*Rhodes et al.*, 1960b]. This procedure is primarily an assay for ovomucoid. Ovoinhibitor was assayed by determining inhibitory activity against fungal proteinase with casein as a substrate as described by *Feeney et al.* [1963b]. Sulfhydryl determinations were done chemically with p-chloromercuribenzoate and with 5,5'-dithiobis(2-nitrobenzoic acid), with and without the addition of sodium dodecyl sulfate as a denaturant [*Fernandez Diez et al.*, 1964]. Sialic acid was determined with thiobarbituric acid [*Feeney et al.*, 1960b]. The presence of riboflavin and the riboflavin-binding capacity (ovoflavoprotein in egg white) were determined by titrations of quenching of fluorescence [*Rhodes et al.*, 1959].

Certain constituents were estimated by their particular physical-chemical properties in the absence of a more definitive biochemical method. These included the ovalbumin and serum albumin which were deter-

TABLE 1. Eggs Marked and Collected

Colony	Egg	Date of Collection (November 1965)													
		3	4	5	6	7	8	9	10	11	12	13	14	15	16
		Number of Eggs Marked or Collected Each Day*													
P9	1st	0	0	3	3	7	8	6	3	2	2	1	2	1	1
	2nd	0	0	0	0	0	2	3	3	3	7	4	2	3	0
P10	1st	0	1	3	1	12	8	8	7	4	0	3	1	0	1
	2nd	0	0	0	1	1	2	4	6	5	5	6	4	0	2
P14	1st	1	0	2	3	3	8	6	2	6	2	0	0	2	0
	2nd	0	0	1	1	0	1	5	3	8	4	2	5	1	1
P15	1st	7	3	32	11	55	25	27	24	10	7	6	7	4	3
	2nd	0	0	1	5	15	12	30	30	30	17	20	13	4	4
Total 1st		8	4	40	18	77	49	47	36	22	11	10	10	7	5
Total 2nd		0	0	2	7	16	17	42	42	46	33	32	24	8	7

* The first egg was marked on the date laid, and the second egg collected on the date laid.

mined by their electrophoretic characteristics, intensities of staining in starch gel electrophoretic patterns, characteristics on chromatography with column of ion-exchange cellulose, and reactions with reagents for sulfhydryls. Other constituents were detected by special procedures, such as the ovomacroglobulin (component 18) in egg white, which was determined immunochemically [Miller and Feeney, 1966], as well as by electrophoretic and chromatographic characteristics.

Procedures for gel electrophoresis included horizontal electrophoresis in slabs of starch gel or polyacrylamide gels [Feeney et al., 1963a]. Two discontinuous buffer systems used were: (a) The pH 8.6 system of Poulik [1957], which consists of 0.076 M Tris, 0.005 M citric acid, and 2.0 M urea in the gel buffer with a bridge buffer of 0.3 M boric acid and 0.06 M NaOH. The electrophoresis was performed at 7 V/cm and 18 MA for 16 hours at room temperature; (b) The pH 5.0 system of Lush [1964] was Tris-citric acid (0.00935 M Tris) and 2 M urea in the gel buffer with succinic acid-NaOH (0.041 M succinic acid) in the bridge buffer. Electrophoresis was done at 17 V/cm and 50 MA for 4 hours at 2°C. The gels were usually stained after electrophoresis by aniline blue-black or nigrosine.

Micro-immunoelectrophoresis experiments were done in a manner similar to that described by Scheidegger [1955]. Electrophoresis in 1% agarose (Marine Colloids, Inc., New York), dissolved in pH 8.2, 0.05 ionic strength barbitol buffer was conducted at room temperature for 4 hours with 18.3 V/cm on 2.5 × 11.5 cm glass slides [Miller and Feeney, 1964]. Immunodiffusion was done according to the method of Ouch-

terlony [1949]. Antisera were produced in rabbits by injection of whole egg white or blood serum in Freund's adjuvant. These were done by Antibodies, Inc., Davis, California.

The more definitive physical methods included those designed to describe molecular size, shape, and charge relationships, such as analytical ultracentrifugation, molecular filtration, and free (moving boundary) electrophoresis. Ultracentrifugation was done in a Beckman Spinco Model E analytical ultracentrifuge. S_{20} values were obtained at a rotor speed of 59,780 rpm. Molecular filtration was done on columns of Sephadex (Pharmacia Laboratories) G-100 and G-200. Free electrophoresis was done in a Perkin-Elmer Model 238 portable electrophoresis apparatus.

Preparation of purified proteins. The egg-white proteins were prepared by appropriate modifications of the methods developed in this laboratory for fractionating and purifying the egg-white proteins of different species in comparative studies. The modifications depended upon the particular properties of the individual constituents, as is the case with the egg-white protein of other avian species [Feeney, 1964a; Feeney et al., 1960a]. The main method employed was sequential chromatography on the ion-exchange agents, diethyl-aminoethyl cellulose and carboxymethyl cellulose. The transferrins and ovomacroglobulin were purified by ion-exchange chromatography and molecular filtration. The ovomucoid was purified by ion-exchange chromatography and fractionating with acetone and trichloroacetic acid. The other constituents were purified by ion-exchange chromatography and precipitating with salt.

Fractionation of egg-yolk proteins was restricted to

Fig. 1. An Adélie penguin on an egg laid and numbered on November 4, 1965. The egg was marked "4" with indelible ink for identification as the first egg of a normal two-egg clutch and as a reference date to determine the days between the first and second eggs.

egg yolks and egg whites were immediately frozen for transport to the University of California at Davis. Twice when helicopter transport was available within 24 hours of collecting the eggs, they were opened at McMurdo and used for studies that could not be made with frozen material.

The 1966–1967 program was a continuation of the previous programs. Personnel were Robert E. Feeney and graduate students John C. Bigler, Stanley K. Komatsu, and David T. Osuga. (Approximately three-fourths of the field program in 1966–1967 concerned the marine biology program.) A total of 200 Adélie eggs were collected for laboratory studies and for incubation. Collecting and handling were done as described above.

Methods of Protein Analysis

The analytical procedures used in this study have been described in several recent publications from this laboratory [*Feeney et al.*, 1963a, 1966]. In general they included the following, alone or in combination: comparative electrophoretic patterns of the intact egg white and blood serum, direct biochemical assays or analyses of the intact egg white or blood serum, physical and chemical fractionation of the egg white or blood serum to separate and purify the individual components, and determination of the properties of these purified components by physical and biochemical methods. Also used were such procedures as chromatographic fractionations and various types of electrophoretic separations [*Feeney et al.*, 1966; *Rhodes et al.*, 1958] which partially or extensively separate the individual constituents and give profiles or patterns that can be used to describe the constituents qualitatively and quantitatively.

The methods used for assays or analyses of biochemical activities were particularly pertinent to egg white and blood serum in several instances. These included assays of the ovotransferrin and serum transferrin by their chromogenic capacity with ferric iron and estimating the amount of protein present from the salmon-pink color produced with a maximum wavelength at approximately 465 mμ [*Feeney and Komatsu*, 1966]. The transferrins were also determined by radioautograms of patterns of starch gel electrophoresis using Fe^{59} [*Clarke et al.*, 1963]. Lysozyme activities were estimated enzymically with cellular preparations of *Micrococcus lysodeikticus*. Inhibitory activities against the proteolytic enzymes trypsin and chymotrypsin were measured spectrophotometrically with synthetic substrates [*Rhodes et al.*, 1960b]. This procedure is primarily an assay for ovomucoid. Ovoinhibitor was assayed by determining inhibitory activity against fungal proteinase with casein as a substrate as described by *Feeney et al.* [1963b]. Sulfhydryl determinations were done chemically with *p*-chloromercuribenzoate and with 5,5'-dithiobis(2-nitrobenzoic acid), with and without the addition of sodium dodecyl sulfate as a denaturant [*Fernandez Diez et al.*, 1964]. Sialic acid was determined with thiobarbituric acid [*Feeney et al.*, 19 The presence of riboflavin and the riboflavin capacity (ovoflavoprotein in egg white) mined by titrations of quenching of [*Rhodes et al.*, 1959].

Certain constituents were estimated b lar physical-chemical properties in th more definitive biochemical method. the ovalbumin and serum albumin

ROBERT E. FEENEY ET AL.

TABLE 1. Eggs Marked and Collected

Colony	Egg	3	4	5	6	7	8	9	10	11	12	13	14	15	16
		— Date of Collection (November 1965) / Number of Eggs Marked or Collected Each Day*													
P9	1st	0	0	3	3	7	8	6	3	2	2	1	2	1	1
	2nd	0	0	0	0	0	2	3	3	2	2	1	2	1	1
P10	1st	0	1	3	1	12	8	8	7	4	0	3	1	0	1
	2nd	0	0	0	1	1	2	4	6	4	0	3	1	0	1
P14	1st	1	0	2	3	3	8	6	2	5	5	6	4	0	2
	2nd	0	0	1	1	0	1	5	3	8	4	2	0	0	2
P15	1st	7	3	32	11	55	25	27	24	10	7	6	7	1	1
	2nd	0	0	1	5	15	12	30	30	30	17	20	13	4	3
Total 1st		8	4	40	18	77	49	47	36	22	11	10	10	7	4
Total 2nd		0	0	2	7	16	17	42	42	46	33	32	24	8	7

* The first egg was marked on the date laid, and the second egg collected on the date laid.

mined by their electrophoretic characteristics, intensities of staining in starch gel electrophoretic patterns, characteristics on chromatography with column of ion-exchange cellulose, and reactions with reagents for sulfhydryls. Other constituents were detected by special procedures, such as the ovomacroglobulin (component 18) in egg white, which was determined immunochemically [Miller and Feeney, 1966], as well as by electrophoretic and chromatographic characteristics.

Procedures for gel electrophoresis included horizontal electrophoresis in slabs of starch gel or polyacrylamide gels [Feeney et al., 1963a]. Two discontinuous buffer systems used were: (a) The pH 8.6 system of Poulik [1957], which consists of 0.076 M Tris, 0.005 M citric acid, and 2.0 M urea in the gel buffer with a bridge buffer of 0.3 M boric acid and 0.06 M NaOH. The electrophoresis was performed at 7 V/cm and 18 MA for 16 hours at room temperature; (b) The pH 5.0 system of Lush [1964] was Tris-citric acid (0.00935 M Tris) and 2 M urea in the gel buffer with succinic acid-NaOH (0.041 M succinic acid) in the bridge buffer. Electrophoresis was done at 17 V/cm and 50 MA for 4 hours at 2°C. The gels were usually ̲ained after electrophoresis by aniline blue-black or ̲sine.

̲-immunoelectrophoresis experiments were done ̲er similar to that described by Scheidegger ̲ctrophoresis in 1% agarose (Marine Col- ̲ew York), dissolved in pH 8.2, 0.05 ionic ̲tol buffer was conducted at room tem- ̲hours with 18.3 V/cm on 2.5 × 11.5 ̲ [Miller and Feeney, 1964]. Immuno- ̲ according to the method of Ouch-

terlony [1949]. Antisera were produced in rabbits by injection of whole egg white or blood serum in Freund's adjuvant. These were done by Antibodies, Inc., Davis, California.

The more definitive physical methods included those designed to describe molecular size, shape, and charge relationships, such as analytical ultracentrifugation, molecular filtration, and free (moving boundary) electrophoresis. Ultracentrifugation was done in a Beckman Spinco Model E analytical ultracentrifuge. S_{20} values were obtained at a rotor speed of 59,780 rpm. Molecular filtration was done on columns of Sephadex (Pharmacia Laboratories) G-100 and G-200. Free electrophoresis was done in a Perkin-Elmer Model 238 portable electrophoresis apparatus.

Preparation of purified proteins. The egg-white proteins were prepared by appropriate modifications of the methods developed in this laboratory for fractionating and purifying the egg-white proteins of different species in comparative studies. The modifications depended upon the particular properties of the individual constituents, as is the case with the egg-white protein of other avian species [Feeney, 1964a; Feeney et al., 1960a]. The main method employed was sequential chromatography on the ion-exchange agents, diethylaminoethyl cellulose and carboxymethyl cellulose. The transferrins and ovomacroglobulin were purified by ion-exchange chromatography and molecular filtration. The ovomucoid was purified by ion-exchange chromatography and fractionating with acetone and trichloroacetic acid. The other constituents were purified by ion-exchange chromatography and precipitating with salt.

Fractionation of egg-yolk proteins was restricted to

partial purification of a "chromoprotein." In one method for preparing the "chromoprotein," the yolk was first diluted with two volumes of 15% glycine, and the suspension was extracted with diethyl ether until no more color was removed. The aqueous phase was dialyzed against 0.5 M Tris buffer at pH 8.5 and then centrifuged for 1 hour at 78,500 G. The clear supernatant fluid was subjected to gel filtration on Sephadex G-200. The yellow fraction obtained was subjected to several further purifications by gel filtration. In another method for preparing the chromoprotein, the procedure of *Bernardi and Cook* [1960] for the fractionation of yolk proteins with $MgSO_4$ was used.

Incubation and examination of eggs. Eggs were incubated in bacteriological incubators in the McMurdo BioLaboratory to study embryological development and biochemical deterioration. Containers of water were placed in the incubator to keep the humidity high. Incubation temperature for biochemical deteriorations was maintained at 42–43.5°C. Incubation temperature for embryological development was maintained at 34–35.5°C, and the eggs were turned daily.

Eggs were opened at specified times and examined for embryonic development. Samples of the egg white were removed for analyses. For the deteriorative studies the characteristics of the thick egg white and physical condition of the egg yolk were determined by measuring the relative heights and widths broken out on a flat surface. These values were expressed as thick white indices or yolk indices [*Rhodes et al.*, 1957, 1960a].

RESULTS AND DISCUSSION

Egg-White Proteins

The gross appearance of whole penguin egg white is similar to that of many other avian species [*Feeney et al.*, 1960a]. It consists of approximately two-thirds thick egg white and one third thin. The white is slightly greenish in color, whereas chicken (*Gallus gallus*) egg white is yellowish. The yellow color in chicken egg white is due to riboflavin. Penguin egg white was found to contain no riboflavin. The egg white of the common duck (*Anas platyrhynchos*) in contrast to chicken egg white contains no riboflavin and is truly white. The duck and the goose (*Anser anser*) do not put riboflavin physiologically in the egg white [*Rhodes et al.*, 1960a]. Apparently the Adélie penguin also puts no riboflavin in the egg white. The source of the greenish color in the Adélie white was

not investigated. One possibility is that the white contains some of the egg-shell pigment.

Table 2 presents a comparison of the egg-white proteins of the Adélie penguin with those of the chicken. The most striking differences between penguin and chicken egg whites are the penguin's comparatively high concentrations of protein-bound sialic acid and comparatively low concentration of the enzyme lysozyme. Other significant differences are in the sulfhydryls and the absence of riboflavin bound to the riboflavin-binding protein in the penguin egg white.

As the absence of riboflavin from the egg white could have been due to a dietary deficiency of riboflavin or to a genetic block of the secretion of riboflavin into the egg white, we tried to differentiate between these two possibilities by administrating large doses of riboflavin orally to laying females. Crystalline riboflavin (0.5 to 1.0 g in 3–4 ml water) was given by esophageal tube, immediately after the first egg was laid and again two days later, to each of the two birds on three nests. (The riboflavin was given to both birds to avoid the possibility of an error in distinguishing sexes.) No yellow color was transmitted to the egg white of the second eggs.

Figures 2a and 2b are starch-gel electropherograms comparing chicken and penguin egg whites. Patterns indicate both quantitative and qualitative differences in the constituents. A pronounced difference is in the area of lysozyme, which is completely absent in the penguin pattern. The presence of small amounts of lysozyme, as found by enzymatic assays, was confirmed by separating lysozyme by ion-exchange chromatography. Solutions of this purified lysozyme, more concentrated than in the original egg white, were examined electrophoretically and showed patterns roughly similar to those of chicken lysozyme. Other differences include the more alkaline characteristics and presence of multiple molecular forms of the ovotransferrin (conalbumin) in the penguin egg white.

Also, the penguin egg white has two discrete components that are absent from chicken egg white. They move rapidly to the negative electrode during electrophoresis and appear more alkaline in isoelectric point than lysozyme, which has an isoelectric point above pH 10. The two components move closely beside one another and are considered to be a "doublet" of the same protein, possibly differing from one another by one or more chemical groupings. The "doublet" proteins are absent from Figures 2a and 2b; they migrate so rapidly that they move off the gel in the times required to obtain separations of the more slowly mov-

TABLE 2. General Properties of Chicken and Adélie Penguin Egg-White Constituents

	Chicken				Adélie Penguin[a]	
	Characteristic Properties	%	Iso-electric Point	M.W.	Characteristic Properties[b]	%
Ovalbumin	Denatures easily; has 4 sulfhydryls.	54	4.6	45,000	More acidic than chicken; has 2 sulfhydryls.	?
Ovotransferrin	Complexes iron; antimicrobial; related to serum transferrin. Has 2 apparent forms.	13	6.0	77,000	More alkaline than chicken; exists in 5 or 6 discrete molecular forms.	4–5
Ovomucoid	Inhibits enzyme trypsin.	11	4.3	28,000	Inhibits enzymes trypsin and α-chymotrypsin (double headed). High in sialic acid.	10
Lysozyme	Enzyme for polysaccharides; antimicrobial.	3.5	10.7	15,000	c	0.05
Ovomucin	Viscous, high in sialic acid. Antihemagglutinin against some species of viruses.	1.5	?	?	c	?
Ovomacroglobulin	Strongly antigenic and extensively cross reacting.	0.5	4.5	800,000	Nearly identical to chicken.	1
Flavoprotein (Riboflavin)	Binds riboflavin.	0.8	4.1	35,000	c	0.3
	Present.	d	d	d	Absent.	0.0
Proteinase Inhibitor	Inhibits enzyme (fungal proteinase).	1.0	5.2	?	c	2.0
Avidin	Binds biotin; antimicrobial.	0.05	9.5	?	e	?
Genetic Globulins	Commonly variable in White Leghorns.	0.5	7	40,000	e	?
Poorly identified proteins						
1. Sulfhydryl containing	Present in trace amount.	<0.02[f]	?	?	Contains highly reactive sulfhydryl.	0.07[f]
2. More alkaline than lysozyme	Possibly absent, or present in very low concentration.	?	?	?	Two rapidly moving bands observed by starch-gel electrophoresis.	?
3. Others	Mainly globulins.	8	4–9	?	e	?
Nonprotein	Primarily half glucose and salts (poorly characterized).	8	d	d	e	?
Sialic acid	All bound to proteins. Approx. half in ovomucin fraction.	0.3	d	d	All bound to protein. Less than 10% bound to ovomucin.	4–5

[a] The molecular weights of the Adélie egg-white proteins are relatively similar to those of chicken, as are all homologous egg-white proteins thus far studied. Isoelectric points vary as is evident in the starch gels, but exact values have not been determined.

[b] Properties were grossly similar to chicken unless otherwise noted.

[c] Proteins identified but properties not determined.

[d] Not applicable.

[e] Presence or absence not investigated.

[f] Expressed as percentages of cysteine as determined with 5,5'-dithiobis (2-nitrobenzoic acid) and a denaturant.

ing constituents. They are readily seen in gels subjected to electrophoresis for about one-third the times usually used.

Preliminary physical-chemical examinations have been done on five of the purified proteins: ovomacroglobulin (C-18), ovotransferrin, ovomucoid, ovomucin, and lysozyme. The ovomacroglobulin characteristics have been studied in more detail than the others and are described elsewhere in comparisons with the ovomacroglobulins of the chicken, duck, and tinamou (*Eudromia elegans*) [*Miller and Feeney*, 1966]. All four ovomacroglobulins are approximately the same size (mol. wt. 600,000–900,000) and show only minor differences in amino acid contents. The ovotransferrin exists in four to six multiple molecular forms, evident as bands on starch-gel electrophoresis. They bind approximately 2 atoms of iron per mole, similar to all other transferrins studied so far [*Clark et al.*, 1963].

The penguin ovomucoid has approximately the same specific inhibitory activity against trypsin as does chicken ovomucoid, but it is the "double headed" or dual inhibitory variety like turkey ovomucoid, inhibit-

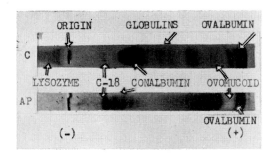

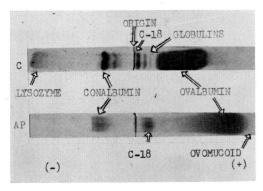

Fig. 2. Starch-gel electrophoretic patterns of chicken (C) and Adélie penguin (AP) egg white. *A (top)* was performed at *p*H 8.6 and *B (bottom)* at *p*H 5.0; both were stained with nigrosine [from *Feeney et al.*, 1966].

ing both trypsin and α-chymotrypsin. The penguin ovomucoid is also similar to turkey ovomucoid in that the activity against trypsin is destroyed by chemical modification with reagents directed at amino groups. Kinetic analyses of the rates of inactivations indicate that the trypsin-inhibitory activity of both the penguin and turkey ovomucoids is inactivated when only one particular amino group in the molecule is modified [*Haynes et al.*, 1967]. Trypsin is inhibited more strongly than α-chymotrypsin, which is apparently bound more weakly.

The contents of carbohydrate and heterogeneity (existence of multiple molecular forms) are related in that different chromatographic fractions contain different amounts of sialic acid. Penguin ovomucoid is generally similar to most other ovomucoids in high content of carbohydrate and absence of the amino acid tryptophan. An exception is the proportionately high content of sialic acid.

The penguin ovalbumin has a molecular size similar to that of chicken and turkey, as determined by sedimentation velocity values, but differs in content and reactivity of sulfhydryl groups. Chicken ovalbumin contains four sulfhydryl groups; none of these readily

react chemically, and all four groups are therefore termed "masked." Three of the four groups are only partially masked and will react with some reagents such as *p*-chloromercuribenzoate; the fourth requires denaturation of the protein to make it reactive. Turkey ovalbumin is similar to chicken ovalbumin in having the three partially masked groups, but it lacks the fourth resistant group. Penguin ovalbumin is still different and has only two sulfhydryl groups, both of them partially masked.

We have discussed the immunological interrelationships of penguin egg-white proteins briefly elsewhere [*Miller and Feeney*, 1966; *Feeney et al.*, 1966]. The penguin egg white gave general immunological cross-reactions with the egg whites of other primitive water birds. When tested against egg whites of the western grebe (*Aechmophorus occidentalis*), Laysan albatross (*Diomedea immutabilis*), and the pink-footed shear-water (*Puffinus creatopus*), antisera to penguin egg white reacted strongly with the ovotransferrins (conalbumins) and to a lesser degree with the other constituents. Comparisons of immunoelectropherograms are given in Figure 3. In contrast, no cross-reaction was detectable between the penguin ovotransferrin and the ovotransferrins of any of the ratites, including the kiwi, cassowary, emu, rhea, and ostrich. A weak but very significant reaction occurred, however, with the egg white of the duck. Strong cross-reactions were obtained with the ovomacroglobulin of the penguin with all other avian species tested. This is in agreement with the general characteristic of this particular constituent in all birds tested.

Egg-Yolk Proteins

Starch-gel electrophoretic patterns of the yolks of the Adélie penguin and the chicken are similar, though both patterns were indistinct. A large part of the materials did not move from the origin, and the patterns had extensive smears and tails, commonly considered to be caused by lipid. The penguin egg yolk formed a gel on freezing and thawing similar to the gel formed with all other species tested [*Feeney et al.*, 1954].

A major difference was the presence of a large amount of yellow color in the ether-extracted penguin yolk, absent from the ether-extracted chicken yolk. The material responsible for this color was partially purified from the ether-extracted yolk and found to be either a protein with a pigment as a prosthetic group, or at least a highly associated complex of a pigment and protein. An attempt was therefore made

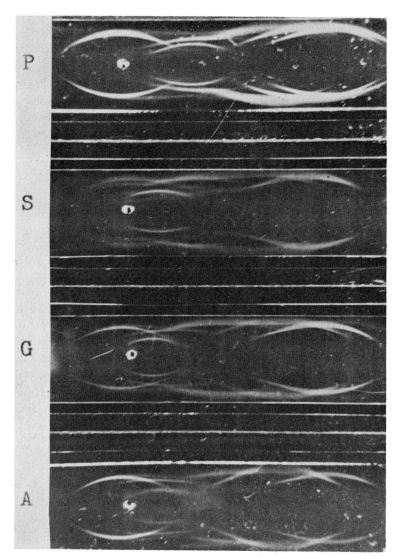

Fig. 3. Immunoelectrophoretic patterns of whole avian egg whites. Antisera in both top and bottom troughs were anti-penguin (Adélie) egg-white sera. Egg whites electrophoresed in center: P, Adélie penguin; S, pink-footed shearwater; G, western grebe; A, Laysan albatross [from *Feeney et al.*, 1966].

to prepare such a yellow fraction by a different method, that of *Bernardi and Cook* [1960] for the separation of the phosvitin, lipovitellin, and livetin fractions by differential solubility in $MgSO_4$. The yellow protein was found in the α and β livetin fraction. The α and β livetin fractions contained many different constituents, as shown by starch-gel electrophoresis at pH 8.6. The yellow constituent was easily seen in the starch gels as a yellow spot just behind the buffer boundary. On further examination of this fraction, the yellow color was characterized as due to the presence of riboflavin. When the yellow material was dissolved in

0.1 *N* HCl and dialyzed against 0.1 *N* HCl, the yellow color passed through the dialysis tubing. The yellow color was now easily extracted by n-butanol. It had the fluorescence and absorption spectrum of riboflavin.

In comparative experiments with chicken egg yolk, a much smaller amount of the yellow protein was found in the α and β livetin fractions.

Blood-Serum Proteins

It was possible to compare the blood serum proteins of the Adélie with those of another penguin, the Humboldt penguin (*Spheniscus humboldti*), as well as with

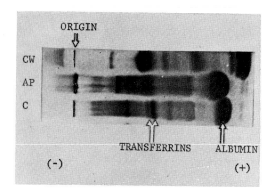

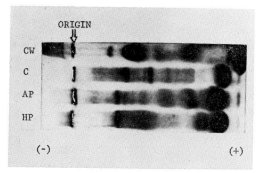

Fig. 4. Starch-gel electrophoretic patterns of chicken egg white and the blood sera of chicken, Adélie penguin, and Humboldt penguin. Fig. 4a (*top*) was performed at *p*H 8.6 and stained with aniline blue-black; Fig. 4b (*bottom*) was performed at *p*H 8.6 and stained with nigrosine. CW refers to chicken egg white. C, AP, and HP refer to chicken serum, Adélie penguin serum, and Humboldt penguin serum, respectively. Sera from different individual birds were used for patterns of Adélie penguins in *a* and *b*. The transferrins of the Adélie penguin serum are present as multiple forms more alkaline than the chicken transferrins. The Adélie penguin serum albumin can be seen to be only slightly more alkaline than that of the chicken serum albumin.

those of the chicken and other avian species (Figures 4a and 4b). The serum albumin and the serum transferrin are indicated for the chicken serum. The location of the transferrins was found in other experiments in which radioautograms were obtained after the addition of radioactive iron (Fe^{59}). The Adélie transferrins were found to be present as 4- to 5-multiple forms[2] and were more alkaline than the chicken transferrins. Other differences between the patterns are easily seen.

The five Adélie penguin sera examined were from

---

[2] Samples of Emperor penguin blood serum became available during the 1966–1967 program and are presently under investigation. Preliminary observations also indicate 4- to 5-multiple molecular forms of the serum transferrins.

two females and three males. The two female sera differed from the three male sera in the same areas of the electropherograms. These differences are related to sex; differences are usually found between sera from laying female chickens and those from nonlaying female or male chickens [*Mok and Common*, 1964].

Incubation and Hatching of Fertile Eggs

These experiments were preliminary in nature and were done to provide possible guidelines for subsequent studies. The relative success of the incubations suggests that incubating and hatching penguin eggs should not present too many difficulties.

In the first trials in 1965, the eggs incubated artificially were incubated previously by the birds for periods varying from 1 to 13 days. Exact records of hatchability were not maintained, because eggs were opened at various intervals to observe viable embryos and to remove samples of egg white and yolk material. Approximately three-fourths of the eggs appeared to have satisfactory embryonic development. Several of these were incubated to hatching (approximately 35 days including the time under the birds). Figures 5a and 5b are photographs of the egg contents after two different periods of incubation (excess yolk material and fluids were removed). The earlier embryo resembles most other avian embryos at the same stage, but the more developed embryo has differentiated sufficiently to show physical features characteristic of the penguin chick.

The second study, done in 1966, included an attempt to feed the penguin chicks. Thirty eggs were dated the day they were laid and left on the nests for periods varying from 26 to 28 days. They were then transferred in a warm incubator to the BioLaboratory at McMurdo station and incubated as described above. Three (10%) were eventually found to be infertile, and 3 were opened for embryo studies after different periods of incubations. Of the remaining 24, 16 (66⅔%) hatched. The other 8 were found to contain dead embryos. The total times of incubations until hatching varied from 34 to 40 days with an average time of 35.6 days. One bird was sacrificed the first day for blood and tissue samples.

The remaining 15 chicks were fed and studied for a period of 10 days, at which time it was necessary to move the experiment to unsatisfactory quarters, and the experiment was ended. The chicks were initially housed in a box in a Jamesway tent heated by an oil stove. The box was equipped with a light bulb for additional heating. Feeding was attempted with tweez-

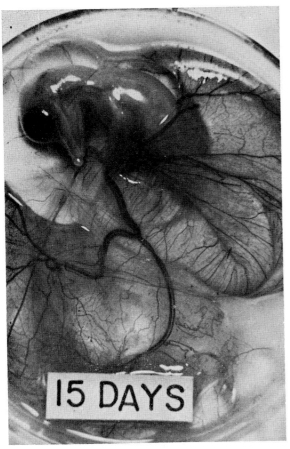

Fig. 5. Adélie penguin embryos incubated under controlled
conditions at two stages of development.

Fig. 5a, a 15-day embryo, was incubated 4 days by the bird
prior to transfer to an incubator for 11 days.

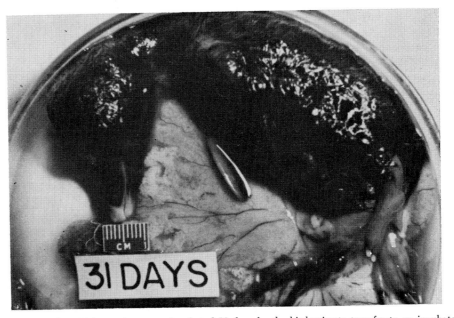

Fig. 5b, a 31-day embryo, was incubated 10 days by the bird prior to transfer to an incubator
for 21 days.

ers and medicine droppers (with an unrestricted opening). Three different diets were used: (1) Dried whole chicken egg, reconstituted with water and warmed to 35°C prior to feeding. (2) Dried shrimp, reconstituted with water and blended in a Waring blender, to which was added dried casein, 10 g/liter. The liquid suspension was warmed prior to feeding. (3) Fresh fish (*Dissostichus mawsoni*) was blended in a Waring blender with a small amount of water, and the suspension was warmed prior to feeding.

Of the above three diets, the fish diet appeared to be much superior to the other two in that the birds consumed it avidly. However, the diets were tested in the order of their numbers, and the fish diet was only tested the last 2½ days of the study.

Of the 15 birds that were continued in the experiment, only 6 lived to its conclusion. The other 9 died at periods varying from 3 to 8 days. All the birds lost weight during the first 3 days; the 6 that lived until the experiment was stopped all showed weight gains the last 2 days.

The following general observations were made during the 9-day experiment.

Parental feeding of the chick in nature is difficult to duplicate. As one parent is always at the nest in the early stages, it is able to feed the chick frequently. In the laboratory a keeper in constant attendance is needed to avoid large time intervals between feedings, or the dosage at each feeding must be increased.

Each chick must be force-fed individually with forceps or a wide-mouthed eyedropper. It is often difficult to keep the chick's mouth open and its tongue out of the way of the instruments. Placing a moist homogenate down the throat with the eyedropper is neater and easier on the chick than simply putting food in the mouth. The moist homogenate is easier to give the chicks than solid food, and more of it can be given at each feeding.

Keeping the chicks dry is a twofold problem. Overmoist homogenate often spills and wets the chicks excessively, and in their tendency to huddle together, they deposit excrement on each other. (The chick's position in the nest, usually "head in" underneath the parent, keeps both the chick and nest relatively free of wet excrement.) This problem might be alleviated by using some type of individual cage with a screen floor.

Incubation of Eggs at High Temperature for Studying Deterioration

The physical and chemical changes that occur in the contents of infertile or fertile eggs held at temperatures higher than favorable for embryonic development are of general scientific interest, mainly because the physical and chemical changes that occur are complicated biophysical reactions. The changes observed are a relaxation and eventual thinning of the thick egg white [*Feeney et al.*, 1952] and a weakening of the yolk membrane, which eventually ruptures. *Feeney et al.* [1956] proposed that the two different phenomena have the same mechanism. The deterioration of the yolk membrane appears to be due to a deterioration of a thick white layer which is enmeshed with the true vitellin membrane of the yolk. Both changes occur independently, however; the egg white thins in the absence of egg yolk, and the yolk membrane weakens when removed from the egg and placed in another environment [*Feeney et al.*, 1951]. The glucose in the egg white also reacts rapidly with several of the egg-white proteins at normal incubation temperatures [*Feeney et al.*, 1963a, 1964]. This deterioration can be seen readily in a number of different types of electrophoretic patterns. The end result is a change in the solubility and electrophoretic properties of the proteins. That this change is the cause of the thick egg-white deterioration has not been proved.

The following are several theories that have been proposed for the deteriorative mechanisms in egg white: (*a*) The ovomucin, a constituent supposedly responsible for the gel nature of the white, is reductively cleaved by reducing agents generated during the incubation [*MacDonnell et al.*, 1951; *Sugihara et al.*, 1955]. All eggs examined have been demonstrated to "thin" rapidly when treated with minute quantities of one of several reducing agents. (*b*) There is a complex between the lysozyme and the ovomucin that dissociates during incubation [*Hawthorne*, 1950]. (*c*) There is no complex between lysozyme and ovomucin initially, but one forms during incubation, changing the gel characteristics and causing thinning [*Cotterill and Winter*, 1955]. (*d*) The solubility of the ovalbumin changes during incubation [*Smith and Back*, 1962]. (*e*) The interaction of the glucose with the protein either directly or indirectly causes thinning. The glucose interaction also could cause thinning indirectly by generating reducing compounds [*Feeney et al.*, 1964].

None of the above theories fits all the available data. Chicken and turkey (*Meleagris gallopavo*) eggs deteriorate rapidly, whereas duck and goose eggs essentially show no deterioration [*Rhodes and Feeney*, 1957]. Duck eggs are relatively low in lysozyme and

TABLE 3.   Deteriorative Changes in Stored Eggs

Species	Storage Conditions			Indices	
	Temperature, °C	Days	No. Eggs	White	Yolk
Penguin	5	3	5	0.06	0.34
	5	8	4	0.06	0.31
	42	3	5	0.02	0.28
	42	8	4	**	**
Chicken*	2	7	12	0.07	0.44
	37	3.5	12	0.04	0.25
	37	7	11	0.03	0.17
Duck*	2	7	12	0.10	0.41
	37	3.5	12	0.09	0.35
	37	7	12	0.09	0.34

* Results for chicken and duck from *Rhodes et al.* [1960].
** Badly deteriorated. Two yolks broke when eggs were opened. Yolks were attached to shells.

very low in sialic acid, and both of these conditions have been implicated as possibly related to deterioration. The possible role of lysozyme is obvious from the above described hypothetical mechanisms. Sialic acid has been considered because the ovomucin fractions contain sialic acid [*Feeney et al.*, 1960b]. As the penguin egg contains even lower amounts of lysozyme (essentially none) and proved the highest in sialic acid ever observed, a study of its deterioration was of particular interest. A biased opinion held that the penguin egg might show high resistance to deterioration, similar to that of the duck and goose eggs. This opinion was wrong; the fact appears to be just the opposite.

Two experiments on deterioration were performed in the 1965–1966 program. One was conducted in a manner similar to that used with chicken or duck eggs. The control and experimental eggs were equilibrated to room temperature. The control eggs then were placed at 2–5°C, and the experimentals were incubated at temperatures slightly above those favorable for embryo development (42–43.5°). Table 3 compares the results obtained in this experiment with those previously obtained with duck and chicken eggs. All the penguin eggs that were incubated showed extensive deterioration. The penguin egg is thus not similar to those of the two waterfowl studied, the duck and goose. A more intensive study is needed to estimate relative rates in comparison with other species and, if possible, employing infertile eggs.

The other studies concerned changes in the egg-white proteins during incubation as seen by starch-gel

electrophoresis. Changes were found similar to those previously reported for the egg whites of chicken and other species [*Feeney et al.*, 1964]. A typical pattern from incubated eggs and control is given in Figure 6. Although the reaction has not yet been definitely identified as the reaction of glucose with the proteins, the most likely conclusion is that the glucose-protein reaction occurs. This demonstration of the occurrence of such changes in gel patterns of the penguin egg white is of importance as a guideline for future studies of these egg whites. Eggs must be refrigerated immediately after collection, and, if possible, the egg white separated and kept frozen until examined.

The Palatability of Penguin Eggs

During the course of three spring periods at the camp at Cape Crozier, the opportunity frequently arose for eating penguin eggs in various culinary forms. During the repeated spring storms, eggs are frequently lost from nests and are broken or unclaimed by the nesting birds. As all eggs in the study areas were marked daily, any eggs found outside the nests were very fresh, less than 24 hours old, and thus suitable for eating. Very fresh eggs were also available as a byproduct of the research program, because egg yolks were broken occasionally during the separation of the yolk and white, and these were not suitable for the biochemical purposes intended.

Eggs were cooked by boiling, frying, or scrambling and were used in omelets or souffles as well as for "binding properties" in meat cookery. The properties

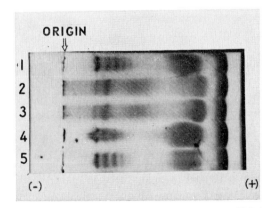

Fig. 6.   Starch-gel electrophoretic patterns at *p*H 8.6 of Adélie penguin egg white stored under different conditions. Sample 1 shows egg white homogenized and frozen while fresh. Samples 2 and 3 are the thick egg white from whole eggs stored at 42°C for 8 days. Samples 4 and 5 are the thick egg white from whole eggs stored at 5°C for 12 days. The stain was aniline blue-black.

of the eggs were sufficiently different from chicken eggs to affect their cookery (functional) properties to a noticeable extent, but they were, by and large, quite acceptable, particularly under the camping conditions at Cape Crozier. The lower temperature of coagulation of the egg white caused a definite rubberiness of the egg white in fried or boiled eggs, somewhat similar to the conditions found with duck eggs [*Rhodes et al.*, 1961]. The greater translucency of the cooked white as compared to cooked chicken egg white also was an unaccustomed factor in the visual acceptability. Individual eggs varied as to the presence or absence of fishy flavor. This may have been due to the differences in the diet of the bird, to the possible presence of genetic differences in the transfer of fishy flavors to the egg white from the body stores, or to differences in the amount of fishy material stored as a result of genetic differences in metabolism and other unrecognized causes. It is probable therefore that the flavor of penguin eggs is primarily a matter of the bird's diet.

Because of these slight but significant differences in properties of the penguin eggs as compared to chicken eggs, the penguin eggs were usually cooked as scrambled eggs mixed with other items such as bacon or sausage, or as souffles with other additives such as onions or mushrooms. Reconstituted dried milk was used in most cases to achieve a smoother cooked product. In the absence of milk the egg product generally was too thick or too fatty. Although the circumstances and the three to six individuals assessing the properties under camping conditions were definitely prejudicial, the penguin eggs were accepted as a frequent constituent of the menu.

Possible Genetic Variations

A limited study was made on the variation in egg-white proteins of individual Adélie penguins. Examinations of 55 egg whites from 50 different individual birds in 4 different colonies and of approximately 20 egg whites from individual birds in various colonies revealed small variations in the sialic acid contents and the electrophoretic mobility of ovomacroglobulin. One egg of the 55 examined showed a clearly different mobility of the ovomacroglobulin. The sample of eggs was too small for anything other than determining the presence or absence of variations in the characteristics of the white. It showed, however, that the incidence of variation in any one rookery is probably small.

GENERAL CONCLUSIONS

Significance of differences between proteins. Extensive differences were found between the Adélie pen-

guin egg-white proteins and the egg-white proteins of the other avian species examined, but the differences between the proteins of blood serum appeared to be much less. Insufficient data were obtained on the egg-yolk proteins to make any conclusions as to differences. However, "differences" are relative and, in the case of both blood-serum proteins and yolk proteins, many important differences may be found when the proteins are examined more extensively. Comparatively large differences among the egg-white proteins of different avian species have been noted by many investigators [*Feeney et al.*, 1960a; *McCabe and Deutsch*, 1952; *Sibley*, 1960]. The teleological and evolutionary significance of such large differences are apparently unknown; at least there is no agreement concerning them at this time. Nevertheless, the relatively large differences in properties and structures of the egg-white proteins of different avian species offer an excellent means of determining taxonomic relationships.

The immunoelectrophoretic and immunodiffusive comparisons indicated strong cross-reactions among the penguin, shearwater, grebe, and albatross. In contrast, with the exception of the particular case of the ovomacroglobulin [*Miller and Feeney*, 1966], the Adélie penguin showed very weak or no cross-reaction with all members of the ratite group. The kiwi, however, reacted strongly with antisera to cassowary proteins. The immunochemical data thus confirmed the close relationship of the penguin to other water birds such as the albatross rather than to the ratites. This relationship is in general agreement with the ideas expressed by *Simpson* [1946]. The degree of the cross-reaction with the duck was interesting and may indicate a closer relationship with the waterfowl than previously suspected.

The apparently more basic isoelectric point of the serum albumin of the two penguins, as compared with human serum albumin, is in agreement with the similar basicity of the serum albumin of the chicken and other avian species [*Krotoski et al.*, 1966; *Mok and Common*, 1964].

Future studies. More studies are planned on unfrozen egg yolk and unfrozen egg white at the BioLaboratory in McMurdo. In addition, we plan further studies on the blood serum of penguins. Work with intact eggs will include continued study of changes in fertile eggs during incubation and a study of the deteriorative effects of incubation under conditions in which embryos do not develop. An important phase of the future studies will be the fundamental and more sophisticated laboratory studies on purified constituents to be done at Davis, California.

Acknowledgments. The assistance of the group from the Johns Hopkins University under William J. L. Sladen in planning and expediting the procurement of penguin eggs is gratefully acknowledged. The individuals from the Johns Hopkins University who contributed to our success in the field were John C. Boyd, William B. Emison, William J. L. Sladen, and Robert C. Wood. Advice and general laboratory assistance was obtained from Susan B. Lind, Gad Feinstein, Royce Haynes, and Judy Miller. The blood serum of the Humboldt penguin and the kiwi egg were provided through the courtesy of the San Diego Zoological Society and the New Zealand Wildlife Service, respectively.

The authors appreciate the financial and physical support of the National Science Foundation.

REFERENCES

Anfinsen, C. B., *The Molecular Basis of Evolution*, John Wiley and Sons, Inc., New York, 1959.

Bernardi, G., and W. H. Cook, An electrophoretic and ultracentrifugal study on the proteins of the high density fraction of egg yolk, *Biochem. Biophys. Acta*, *44*, 86–96, 1960.

Brodkorb, P., Catalogue of fossil birds, *Bull. Fla. St. Mus. Biol. Sci.*, *7*, 179–293, 1963.

Clark, J. R., D. T. Osuga, and R. E. Feeney, Comparison of avian egg white conalbumins, *J. Biol. Chem.*, *238*, 3621–3631, 1963.

Cotterill, O. J., and A. R. Winter, Egg white lysozyme, 3, The effect of *p*H on the lysozyme-ovomucin interaction, *Poultry Sci.*, *34*, 679–686, 1955.

Feeney, R. E., Egg proteins, in *Symposium on Foods: Proteins and Their Reactions*, edited by H. W. Schultz and A. F. Anglemier, pp. 209–224, Avi Publishing Company, Westport, Connecticut, 1964a.

Feeney, R. E., Evolution of proteins, in *Symposium on Foods: Proteins and Their Reactions*, edited by H. W. Schultz and A. F. Anglemier, pp. 345–359, Avi Publishing Company, Westport, Connecticut, 1964b.

Feeney, R. E., H. Abplanalp, J. J. Clary, D. L. Edwards, and J. R. Clark, A genetically varying minor protein constituent of chicken egg white, *J. Biol. Chem.*, *238*, 1732–1736, 1963a.

Feeney, R. E., J. S. Anderson, P. R. Azari, N. Bennett, and M. B. Rhodes, The comparative biochemistry of avian egg-white proteins, *J. Biol. Chem.*, *235*, 2307–2311, 1960a.

Feeney, R. E., J. J. Clary, and J. R. Clark, A reaction between glucose and egg-white proteins in incubated eggs, *Nature*, *201*, 192–193, 1964.

Feeney, R. E., E. D. Ducay, R. B. Silva, and L. R. MacDonnell, Chemistry of shell egg deteriorations: The egg white proteins, *Poultry Sci.*, *31*, 639–647, 1952.

Feeney, R. E., and S. K. Komatsu, The transferrins, in *Structure and Bonding*, Vol. 1, edited by C. K. Jorgensen, J. B. Neilands, R. S. Nyholm, D. Reinen, and R. J. P. Williams, pp. 149–206, Springer-Verlag, Berlin, 1966.

Feeney, R. E., L. R. MacDonnell, and H. Fraenkel-Conrat, Effects of crotoxin (Lecithinase A) on egg yolk and yolk constituents, *Arch. Biochem. Biophys.*, *48*, 130–140, 1954.

Feeney, R. E., D. T. Osuga, S. B. Lind, and H. T. Miller, The

egg-white proteins of the Adélie penguin, *Comp. Biochem. Physiol.*, *18*, 121–130, 1966.

Feeney, R. E., M. B. Rhodes, and J. S. Anderson, The distribution and role of sialic acid in chicken egg white, *J. Biol. Chem.*, *235*, 2633–2637, 1960b.

Feeney, R. E., R. B. Silva, and L. R. MacDonnell, Chemistry of shell egg deterioration: The deterioration of separated components, *Poultry Sci.*, *30*, 645–650, 1951.

Feeney, R. E., F. C. Stevens, and D. T. Osuga, The specificities of chicken ovomucoid and ovoinhibitor, *J. Biol. Chem.*, *238*, 1415–1418, 1963b.

Feeney, R. E., J. M. Weaver, J. B. Jones, and M. B. Rhodes, Study of the kinetics and mechanisms of yolk deterioration in shell eggs, *Poultry Sci.*, *35*, 1061–1066, 1956.

Feinstein, G., and R. E. Feeney, Interaction of inactive derivatives of chymotrypsin and trypsin with protein inhibitors, *J. Biol. Chem.*, *241*, 5180–5183, 1966.

Fernandez Diez, M. J., D. T. Osuga, and R. E. Feeney, The sulfhydryls of avian ovalbumins, bovine b-lactoglobulin, and bovine serum albumin, *Arch. Biochem. Biophys.*, *107*, 449–458, 1964.

Hawthorne, J. R., The action of egg-white lysozyme on ovomucoid and ovomucin, *Biochim. Biophys. Acta*, *6*, 28–35, 1950.

Haynes, R., D. T. Osuga, and R. E. Feeney, Modification of amino groups in inhibitors of proteolytic enzymes, *Biochemistry*, *6*, 541–547, 1967.

Krotoski, W. A., D. C. Benjamin, and H. E. Weimer, Effects of starch concentration on the resolution of serum proteins by gel electrophoresis: A six-species comparison, *Can. J. Biochem.*, *44*, 545–555, 1966.

Lush, I. E., Egg albumen polymorphisms in the fowl: loci II and III, *Genet. Res.*, *5*, 39–49, 1964.

MacDonnell, L. R., H. Lineweaver, and R. E. Feeney, Chemistry of shell egg deteriorations: The effect of reducing agents, *Poultry Sci.*, *30*, 856–863, 1951.

McCabe, R. A., and H. F. Deutsch, The relationships of certain birds as indicated by their egg-white proteins, *Auk*, *69*, 1–18, 1952.

Miller, H. T., and R. E. Feeney, Immunochemical relationships of proteins of avian egg whites, *Arch. Biochem. Biophys.*, *108*, 117–124, 1964.

Miller, H. T., and R. E. Feeney, The physical and chemical properties of an immunologically cross-reacting protein from avian egg whites, *Biochemistry*, *5*, 952–958, 1966.

Mok, C. C., and R. H. Common, Studies on the livetins of hen's egg yolk, 2, Immunoelectrophoretic identification of livetins with serum proteins, *Can. J. Biochem.*, *42*, 1119–1131, 1964.

Ouchterlony, O., Antigen-antibody reactions in gels, *Acta Pathol. Microbiol. Scand.*, *26*, 507–515, 1949.

Poulik, M. D., Starch gel electrophoresis in a discontinuous system of buffers, *Nature*, *180*, 1477–1479, 1957.

Rhodes, M. B., J. L. Adams, N. Bennett, and R. E. Feeney, Properties and food uses of duck eggs, *Poultry Sci.*, *39*, 1473–1478, 1960a.

Rhodes, M. B., P. R. Azari, and R. E. Feeney, Analysis, fractionation, and purification of egg-white protein with cellulose-cation-exchanger, *J. Biol. Chem.*, *230*, 399–408, 1958.

Rhodes, M. B., N. S. Bennett, and R. E. Feeney, The flavoprotein-apoprotein system of egg white, *J. Biol. Chem.*, *234*, 2054–2060, 1959.

Rhodes, M. B., N. Bennett, and R. E. Feeney, The trypsin and chymotrypsin inhibitors from avian egg whites, *J. Biol. Chem.*, *235*, 1686–1693, 1960*b*.

Rhodes, M. B., and R. E. Feeney, Mechanisms of shell egg deterioration: Comparisons of chicken and duck eggs, *Poultry Sci.*, *36*, 891–897, 1957.

Scheidegger, J. J., Une micro-méthode de l'immunoélectrophorèse, *Int. Archs. Allergy appl. Immun.*, *7*, 103–110, 1955.

Sibley, C. G., The electrophoretic patterns of avian egg-white proteins as taxonomic characters, *Ibis*, *102*, 215–284, 1960.

Simpson, G. G., Fossil penguins, *Bull. Am. Mus. Nat. Hist.*, *87*, 1–99, 1946.

Sladen, W. J. L., Ornithological research in Antarctica, *BioScience*, *15*, 264–268, 1965.

Sladen, W. J. L., Adélie penguins, *Nature*, *171*, 952–955, 1953.

Smith, M. B., and J. F. Back, Modification of ovalbumin in stored eggs detected by heat denaturation, *Nature*, *193*, 878–879, 1962.

Sugihara, T. F., L. R. MacDonnell, C. A. Knight, and R. E. Feeney, Virus antihemaggultinin activities of avian egg components, *Biochim. Biophys. Acta*, *16*, 404–409, 1955.

Wetmore, A., A classification for the birds of the world, *Smithsonian Misc. Coll.*, *139*, 1–37, 1960.

Williams, J., Serum proteins and the livetins of hen's egg yolk, *Biochem. J.*, *83*, 346–355, 1962.

SALT AND WATER METABOLISM OF THE ADÉLIE PENGUIN

DONALD S. DOUGLAS

Department of Biological Sciences
George Washington University, Washington, D.C.

Abstract. Studies made of the salt and water metabolism of the Adélie penguin, *Pygoscelis adeliae*, in the rookery at Cape Hallett, Antarctica, included examination of: (1) the extrarenal excretion of salt and water and certain other aspects of nasal gland function, (2) the renal excretion of salt and water, and (3) changes in the thiocyanate space. Techniques of handling, restraining, and experimenting were developed that permitted all experiments to be carried out by a single person.

The relationship between both body and nasal gland size and the nasal gland secretory function was examined in the Adélie chick. The gland was found functional at hatching. During growth linear correlations were found to exist among (1) the weight of the glands and the net weight of the chick, (2) the maximum rate at which the glands can eliminate salt and the net weight of the chick, and (3) the maximum rate of secretion of salt and the weight of the glands. The relative growth rate of the nasal gland is about one-half that of the net body weight. The rate of maximum chloride excretion increases at about the same rate as the growth of the glands. No correlation was found between the maximum concentration the glands could produce and their size.

The chloride concentration and the rate of flow of the cloacal discharge were studied in the adult during the incubation period when its fresh water intake is nil. The flow rate was found to be about one-seventh of that reported for chickens deprived of water. Chloride concentration in the discharges was uniformly low. These results indicate that the bird conserves both water and electrolytes during the incubation period.

The chloride concentration of the cloacal discharge from the feeding chicks was quite high, ranging from 200 to 400 meq/l. This contrasts with that of feeding adults where the maximum concentration was about 130 meq/l, and most were less than 60 meq/l. Urinary flows in the chicks were difficult to estimate owing to the presence of fecal matter. In some cases there was evidence of oliguria, but in others there appeared to be a polyuria.

Thiocyanate space in the incubating adult male appeared to decrease during his fast, but it also showed an increase in the range of values during this time. The mean of all the values for thiocyanate space was 29.1 ± 2.5 per cent of the total body weight.

An increase in thiocyanate space accompanied the infusion of hypertonic salt solutions. This increase, presumably at the expense of intracellular water, usually was not sufficient to dilute the administered load to the equivalent of normal plasma levels.

Infused solutions of hyperosmotic urea did not stimulate the nasal glands to secrete. Slight secretory activity was seen in two birds after a few hours, but the rate of flow was not enough to permit collection. Marked osmotic diuresis followed the infusion of urea.

INTRODUCTION

The Adélie penguin (*Pygoscelis adeliae*) is a member of one of the most completely marine-adapted families of birds. It is also a bird that undergoes a lengthy fast while it is on land and may have to go without water for almost as long as it fasts. Thus it is necessary for the Adélie to be able to meet environmental conditions of high salinity for extended periods of time, and then to withstand dehydration concomitant with no water intake for a somewhat shorter, but perhaps metabolically more arduous, period.

For six to eight months of the year, the Adélie presumably lives at the fringes of the pack ice where its only source of fresh water is the snow overlayer, which is subject to the salt sprays accompanying the high seas of the antarctic fall and winter. The life history of this bird from about the last of February through about the middle of October, the antarctic winter, is not well known. It is not present then either at its breeding sites along the coast of Antarctica or at any of the more northern subantarctic islands on which man has wintered.

Frederick Cook's [1900] account of the *Belgica*'s besetment in the pack ice of the Bellingshausen Sea in 1898 describes penguins about the ship during the winter months and includes a photograph captioned "The Small Pack Penguin," which is clearly an Adélie. The account of this 13-month experience, as well as the accounts of the conditions under which Shackleton made his escape from the Weddell Sea pack ice after losing the *Endurance* emphasize how well the Adélie adapted to a rigorous environment.

Monthly temperature minimums in the latitudes of its range are from about $-7°C$ during midsummer to about $-40°C$ in winter. Few winter days are without winds, most of which are southerly, from the polar ice cap. Storms are a regular feature of the antarctic winter, and the Adélie may then be faced by a choice between heavy seas and the churning ice pack.

The summers of this penguin are hardly less remarkable than the winters. Its breeding sites are along the coasts of the continent and the islands within the antarctic convergence. Treks of up to 50 miles over solidly frozen sea may be necessary to reach the proper rookery in early spring, though when the ice breaks up later in the summer the rookery may be only yards from the open sea. Snow is on the ground when the Adélie returns to nest in late October; however, this snow cover is usually gone, except for the large drifts, by the latter part of November. With it goes the most readily accessible source of fresh water to the breeding and incubating adult. Moreover, the Adélie does not often come off its nest and eggs, except in aggressive action against an intruding penguin or a worrysome skua, or in retreat from a fastmoving mammal—dog, man, or seal. Should the bird leave its nest untended for even a few minutes to eat snow, its eggs would almost certainly be lost to the skua (*Catharacta skua*) on the continental shores, or to the sheathbill (*Chionis alba*) and the skua on the Antarctic Peninsula and the subantarctic islands in the American quadrant of the continent.

During this breeding and incubation period the adult male fasts for as long as six weeks and may lose up to 40% of his body weight. The female's fast is not quite so spectacular. She departs for the sea as soon as she lays the clutch of two eggs and returns in two or three weeks to relieve her mate.

The Adélie chick also faces interesting problems in salt and water metabolism. The most evident problem arises when the chick must be fed within about 72 hours of hatching. The material the parents regurgitate to the young is that on which they themselves feed during the summer—crustacean plankton (euphausids).

To correlate experimental physiological information on the salt and water metabolism with the natural history of the breeding Adélie, a field program supported by the National Science Foundation was undertaken. During the austral summers 1959–1960 and 1960–1961, I was stationed on the site of the Adélie penguin rookery at Cape Hallett in the Ross Sea area of the Antarctic.

The first season was largely devoted to developing handling and experimental techniques. This took considerable time, as the only previous ecological and physiological investigations on penguins were concerned with the species' temperature problems [*Sapin-Jaloustre*, 1953 and 1960; *Eklund*, 1957; *Goldsmith and Sladen*, 1961]. No convenient or reliable means were known for one person to handle, restrain, and experiment with unanesthetized penguins. It was also necessary to develop ways of collecting blood, urine, and nasal secretion samples.

The second season's program was partially thwarted by a delay in the arrival of equipment. Also, fine weather throughout this summer enabled the chicks to mature and leave the rookery earlier than anticipated from the previous season's experience and published descriptions of their growth and development [*Sladen*, 1958]. This curtailment of the effective working season diminished the completeness of the program.

TECHNIQUES

Animal material. Aside from its own unique problems in salt and water metabolism the Adélie penguin is a particularly favorable bird for physiological studies. Adults are relatively large and compact, weighing between 3.5 and 5.5 kg during the summer. Their size permits serial blood sampling without side effects from proportionately heavy blood loss. Both their compactness and their natural erect posture make them easier to handle than flying birds of comparable weight.

The animals used in this study were apparently healthy adult and young Adélies taken from one of four marked and banded colonies in the Hallett rookery. When incubating adults were brought into the laboratory for experimentation, the eggs were also brought in and kept in an incubator. At the end of the experiment, the eggs and the adult were returned to the nest. In about 75 per cent of the cases these birds succeeded in hatching at least one chick.

Capture and handling techniques. A landing net about 24 inches in diameter with a 4-foot handle was helpful for capturing birds with a minimum of chasing and disturbance. The net also helped prevent ruinous, headlong dashes through nesting areas which might result in scattered and broken eggs.

Birds were marked with aluminum flipper bands designed by W. J. L. Sladen and provided by the U. S. Antarctic Bird Banding Program. Red plastic tape (Scotch brand) applied to the surface of the band lasted well through the winter at sea and provided a useful and easy way of distinguishing the birds I had banded from those banded by others. The Adélie holds to its nest very well when incubating, thereby permitting regular checking of banded colonies.

The birds were weighed on a 10-kg-capacity Chatillon dynamometer. They were held by the feet, head down, and calmed as much as possible. The bird's neck was flexed until its chin rested on its chest. In this position it was placed in a 5-gallon battery can attached to the dynamometer. Lying thus on its neck in a confined space usually kept the bird quiet long enough for the pointer to stabilize so that the dial could be read to the nearest 20 grams, but especially spirited birds needed up to four or five attempts.

Restraining methods. The first major technical problem was that of restraining the bird for experiments of several hours' duration. Initially the penguin was placed on its belly in a trough with its extended flippers secured laterally and its feet secured posteriorly. Subsequently the trough was placed on end, and the bird was supported in a natural standing position on a platform, its back in contact with the board. This arrangement permitted collection of cloacal discharge, nasal gland secretion, and blood without change in the bird's position.

Adélie chicks were more difficult to handle, especially as they approached molting size. It was not possible to restrain them without inflicting damage. Therefore, they were put into metabolism cages of various sizes made of tin cans and polyethylene funnels. A wire mesh across the bottom of the can supported the chick and strained out most of the feces.

It was found during the first season's work that air temperatures of 18°–27°C in the closed space of the laboratory contributed noticeably to the discomfort and agitation of restrained birds. Therefore the doors

of the laboratory hut were opened during experiments, and the temperature was taken down to that of an average summer day outdoors, between 2° and 7°C.

It was also found that a brief period on the board with minimum of noise and movement in the laboratory helped quiet the birds, and they frequently appeared to doze. Visual stimuli seemed to produce greater agitation than sharp, short noises or dull, constant sounds. The human voice at normal talking levels usually produced appreciable agitation in a restrained bird, but talking or crooning to the birds in a low-pitched monotone seemed to calm them. Most of the experiments were conducted between 10 p.m. and 8 a.m. to minimize disturbance by conversation with interested visitors from the station.

Blood sampling. Blood sampling and fluid infusion techniques underwent continual improvement. The Adélie has only a small skin surface that is not covered with a dense feather coat and no readily discernible superficial blood vessels. When this project started the only known method for obtaining blood samples from penguins was by cardiac puncture (Sladen, personal communication, 1959). This method precluded any sort of infusion or serial blood sampling, and considerable effort had to be devoted to developing a satisfactory technique for carrying out the planned experiments.

With the evolution of the restraining board the first season, I was able to obtain good, repeated blood samples from a small vein on the surface of the outermost toe. For tapping this vein, 23-gage needles were the most generally useful. It was hoped that one foot could be used for sampling and the other for infusions, but the pain and/or pressure receptors stimulated during injection produced strong leg withdrawal efforts and violent general responses. These movements could not be prevented without damage to the limb or further increasing the animal's agitation. The necessary rapidity of the injection frequently caused edema.

A more suitable infusion site was found later in the brachial vein. This is the humeral vein that *Watson* [1883] describes in the Spheniscidae from dissections on the king penguin, *Aptenodytes patagonicus.* The vein is relatively small, and I first saw it in the living state on an Adélie that was returning from feeding. Dissections earlier had revealed nothing large or convenient enough for tapping. This bird, a male, had been fasting for about 40 days and consequently was quite thin. He had been at sea about 48 hours. Upon emerging from the water his flippers were thoroughly wet, and a blue line was discernible on their under surface. Slight pressure applied to the axillary region acted as a tourniquet, and the line became the sought-for usable vein. The stubby feathers and the relatively thick skin of the flipper usually prevent its being seen clearly. The vein can be catheterized with an 18-gage thin-walled needle and polyethylene tubing (Clay-Adams PE 50; 0.58 mm i.d. × 0.65 mm, o.d.). The flippers were then used for infusion and the feet for blood sampling.

Collection of fluid discharge. The problem of cloacal discharge collection was readily settled, once the restraining board was in the vertical position, by placing a polyethylene funnel beneath the bird's vent. The funnel drained through a hole in the platform on which the bird stood into preweighted collecting vials. These were supported beneath the platform by a sling that permitted easy changing with one hand. In most cases evacuation of the cloaca was of the bird's own volition, but occasional sample releases were induced by pressure on the abdomen near the cloaca.

To collect the nasal gland secretion the bird's head had to be held constantly with a vial over the upper beak. In no other way could quantitative collection and continuous flow of the secretion be insured. Vials taped to the bird's head soon aroused it to violence, and secretion ceased. Anything held to or over the beak for a length of time required intermittent readjustment, or even removal, to minimize temper displays.

Treatment of samples. Cloacal discharge and nasal gland secretion samples were collected in preweighed vials, weighed, and then refrigerated with ice in a box made from a packing crate outside the laboratory.

Blood samples were drawn into heparinized syringes. They were centrifuged in 1-ml Wintrobe tubes in an International clinical centrifuge at top speed for 25 minutes. The hematocrit was read, and the plasma decanted into a vial. The plasma samples were stored in the station's frozen meat locker prior to analysis.

Analytical procedures. Chemical analyses were routinely carried out for the chloride and thiocyanate ions in plasma in the nasal gland secretion, and in cloacal discharge samples as the experiment demanded and the size of the sample permitted. Total osmotic pressure was determined for some of the plasma samples using a melting point method.

Chlorides were titrated using the Cotlove chloridometer [*Cotlove et al.*, 1958]. Standard curves were constructed in the range of the samples being titrated. Three points sufficed to bracket a series and provided

consistent straight lines from which the unknowns were read.

Nasal gland secretions in the range of 600–800 meq/l of chloride required samples of about 12 μl, while plasma (90–130 meq/l) required 100 μl, and urine (2–415 meq/l) required samples of 50 to 200 μl. The sample was diluted with approximately 4 ml of a nitric acid-acetic acid solution. The precision of this dilution was found to be unimportant in the range of 3.0 to 4.5 ml.

Thiocyanate ion determinations were made by a modification of the colorimetric method described by *Eder* [1951]. About 350 μl of plasma were required for each analysis. A standard curve was constructed and unknowns read directly from it. Error on replicates was almost always less than 2 per cent and regularly was less than 1 per cent.

In determining the thiocyanate space, 2–3 ml of a 2.50 g/100 ml of NaSCN solution were infused into the bird so as to obtain an approximate plasma concentration of 5 to 10 mg/100 ml for the SCN ion. The actual amount infused was determined by weighing the syringe before and after delivery.

To estimate the thiocyanate space a semilog plot of ion concentration versus time was drawn, and the curve describing the time course of the disappearance of the ion was extrapolated to zero time. The space was then taken as the volume that would have been required to contain the known amount of sodium thiocyanate at a concentration equal to that extrapolated at time zero.

Total osmotic pressure was estimated from the freezing point depression of the sample as indicated by the disappearance of ice crystals in a sealed capillary. The sample, about 10 μl, was drawn by capillarity into the tube and jarred into the center. The ends were sealed with sealing wax. Without dry ice the instantaneous freezing necessary for small crystal formation and thus for precise determination of the melting point was a problem. After supercooling in a brine and snow mixture at $-17°C$, samples were sprayed with ethyl chloride to induce rapid crystallization. The spray alone was not sufficient to freeze the samples which had to be thoroughly supercooled first to obtain small crystals.

The point of disappearance of the last crystal was observed in a 2-liter brine bath made of plexiglass and insulated with 1″-thick styrofoam. A Beckman down-reading differential thermometer, calibrated for the $0°C$ point on deionized samples of glacier water, was used to measure the temperature of melting of the last crystal.

The optical system consisted of crossed polaroids mounted on a commercial thermometer-reader of 4 to 5× magnification. The sample was held by a spring clamp and immersed at least 1¼ inches beneath the surface of the bath. In this arrangement it was not possible to have the sample tube in contact with the bulb of the thermometer, but holding the thermometer about 2.5 cm from the center of the sample tube minimized any temperature gradient that might exist despite the vigorous stirring.

The bath warmed by absorbing heat from the air. The rate of warming was kept as close to 5 to 6 centigrade degrees per hour as possible. Attempts to warm more slowly resulted in an erratic rate of warming. Thus a larger bath, better insulation, or both, would have been required to attain constant lower rates. Faster warming produced such rapid melting of the sample that a distinct endpoint could not be observed.

The best precision regularly obtainable under these conditions was a spread of 0.02°C between the extreme values of replicate determinations for any sample. Spreads of 0.03°C to 0.04°C were considered satisfactory, although in such cases more determinations were made to increase the reliability of the average value calculated.

RESULTS

Function and Growth of the Nasal Gland

It is now known that the nasal gland of marine birds can secrete a sodium chloride solution that is more concentrated than sea water. This makes it possible for them to ingest quantities of marine organisms whose body fluids range in osmotic pressure up to that of sea water, and to excrete the excess salt. Since the first food the Adélie chick receives is regurgitated crustacean plankton, it was of interest to learn how soon after hatching the nasal gland can produce its strongly hypertonic solutions and to determine the relative rate of growth of the gland and of the functional changes, if any, that accompany its growth.

The experiments here reported were performed on 25 birds ranging in size from hatchlings of 83 grams up to adults weighing 4460 grams. Chicks of known ages were taken from nests that contained one other apparently healthy sibling and were given an intraperitoneal salt load of 1 ml of 10 per cent NaCl per 100 g body weight. The nasal gland secretion was collected for at least two hours to ensure that the samples would include a period of maximum average rate of secretion for each bird. The actual duration of collection of each sample was determined by the rate of

TABLE 1. Nasal Secretion from Hatching Chicks

Chick No.	Weight, g	Secretion Chloride Concentration, meq/1	Approximate Amount, µl	Approximate Collection Time, min
1	98.0	651	400	25
		578	300	25
		429	100	25
2	89.8	596	250	25
		518	100	20
3	93.7	672	200	20
		660	200	20
		549	200	20
		404	50	20

fluid secretion. After several samples had been collected and when the rate of secretion was obviously slowing, the bird was killed by decapitation. The salt glands were removed and weighed. The stomach contents were also removed so that the net body weight of the chick could be determined.

We found repeatedly in these and in other experiments that whenever the bird became agitated the flow of nasal gland secretion and the concentration of the secretion fell rapidly, often to nil. Several experiments in this series had to be abandoned or the results discarded because disturbances to the bird produced erratic or discontinuous secretion.

Nasal gland function in the hatching chick. The Adélie chick hatches with a functional, if not actually functioning, salt secreting gland. Several chicks were seen to be sneezing a clear fluid from the beak shortly after they had emerged from the shell. As it was not possible to collect any of this fluid, three chicks were given salt loads shortly after they had cracked out of the egg. The extraembryonic membranes were moist and the blood vessels in them were still red. Drops of moisture were noted on the beaks of all three chicks. Table 1 shows the results obtained when 1 ml of 10% NaCl was injected intraperitoneally into each chick. While the volume of secretion in all cases was very small, the concentrating ability of the gland is manifestly adequate to cope with anything an adult bird might encounter in the sea.

Functional growth of the nasal gland. Figure 1 plots the weights of the nasal glands from the experimental chicks against their net body weights. The correlation coefficient was calculated to be 0.987. A regression line calculated by the method of least squares has been

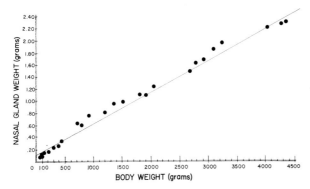

Fig. 1. Relationship of nasal gland weight to total body weight. The regression line calculated to fit the data and shown in the figure has the equation: $y = (0.52 \times 10^{-3})x + 0.14$.

drawn through the points. The slope of the line is 0.52×10^{-3} and the intercept is $+0.14$.

Figure 2 plots the relation between maximum secretory rate and nasal gland weight. The slope of the regression line is 1.12×10^{-1} and the intercept occurs at -0.012. Although there appears to be an appreciable dispersion of the points about this line, the correlation coefficient was again very high, 0.984. With a correlation this close to 1.0, the value of the standard error of correlation would be so small that only a very few points could fall inside a two standard error distance from the line. Therefore no standard error has been indicated on either graph.

One final note of interest is that plotting nasal gland

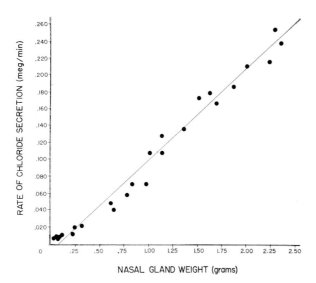

Fig. 2. Relationship of maximum secretory rate of the nasal glands to nasal gland weight. The regression line calculated to fit the data and shown in the figure has the equation: $y = (1.12 \times 10^{-1}) x - 0.012$.

weight against total body weight shows the rate of gland growth to be 52 per cent the rate of bird growth. Thus the nasal gland grows in tissue mass at about one-half the rate that the whole bird grows in body mass.

Maximum concentration of nasal gland secretion. Figure 3 shows the maximum chloride concentration of the nasal gland secretion versus the weight of the nasal glands. There is no correlation between the concentrating ability of the glands and their size. The average of the values is 788 meq/l. A single standard deviation of 43.4 to each side of the mean includes 20 of the 25 plotted points, further emphasizing the lack of correlation between the maximum concentration produced and the weight of the glands. It is noteworthy that four of the five points outside the single standard deviation occurred in very young chicks where the low rate of fluid secretion necessitated longer collection periods.

Average secretory rate profiles. The average rates of secretion for each of the collection periods in two of the chick experiments are plotted in Figure 4A and B. Part A graphs an experiment in which there were no outside disturbances and the chick did not become visibly upset during the collection. The average rate of secretion increased for about an hour, attained a maximum, and then began to decline, all in a relatively smooth manner. In contrast, the effect of agitation on secretion is clearly seen in B. The arrows indicate the times at which the chick became noticeably restless. In the first instance all secretion ceased within 5 minutes of the first sign of unrest. Although the actual rate of drop formation was not timed, it was noted that the interval between successive drops increased until collection was stopped. Following the second disturbance the chick calmed relatively quickly, and secretion did not cease, although it fell to almost half the rate of the immediately preceding period. The recovery from each of these periods of depressed secretion is striking.

All the experiments in this series on the chick's nasal gland and nearly all the salt-loading experiments on adults in which the nasal gland secretion was collected showed profiles similar to these two.

Chloride concentration of food in the alimentary tract. Several samples of food were taken from chicks that had just been fed. This was done by sitting quietly near the nests of adult birds that shortly before had returned from feeding at sea. The food-begging pattern of the chick signaled the beginning of

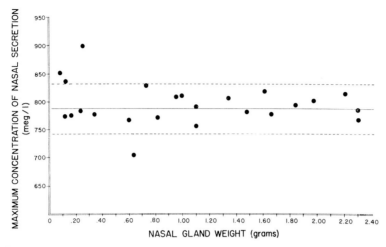

Fig. 3. Relationship of maximum concentration of nasal gland secretion to
the weight of the nasal glands. The solid line is drawn through the mean value
of the maximum concentrations, 788 meq/l. The broken lines to each side are
a single standard deviation distance from the mean, ±43.3.

feeding by regurgitation. Immediately after feeding it was possible to grasp the chick and milk most of the meal back up its esophagus into a vial. The chick was then returned to the nest, thoroughly frightened, but otherwise none the worse for its experience. To determine the chloride concentration of the contents of the various regions of the upper alimentary canal, samples were taken of the contents of the pyloric and cardiac regions of the stomachs of several chicks killed by decapitation.

The food and gut content samples were diluted with water, boiled, and the resulting mixture diluted with water to a weight twice that of the original sample. The solids were allowed to settle, and the fluid was analyzed for chloride. Replicate determinations on all samples corresponded to well within 3 per cent. The concentrations are expressed in milliequivalents of chloride per gram of food material.

The results obtained are shown in Table 2. Not only the relative uniformity of the concentrations, but also their lowness were surprising. A 3.5 per cent solution of NaCl, roughly the concentration of sea water, contains about 600 meq of chloride per liter. Unfortunately no water-content determinations were made of these food samples. More complete salt and water analyses were planned on material to have been obtained during the 1961–1962 season. As these were not obtained, the question of the actual ionic composition and the osmotic pressure of the food given the chick remains open, as does the question whether or not the krill given the chick has been significantly altered in the stomach of the adult with respect to its ionic composition and water content.

Cloacal discharge. The lack of a complete partition between the urodaeum and the coprodaeum of the cloaca and the concomitant lack of separate external openings for the alimentary and urogenital systems contribute to the difficulty of understanding the physiology of the avian kidney. The presence of solid uric acid in the urine of birds further complicates matters. This both threatens obstruction of cannulae introduced to act as artificial urethrae and prevents the total weights and volumes of urine from being used as accurate measurements of the rate of fluid output.

The data obtained were from titration of the fluid of the cloacal discharge after the solid matter, including uric acid, had settled to the bottom of the collection vial. No attempt was made to separate the urodaeum from the coprodaeum, to create an artificial anus, to exteriorize the ureters, or to cannulate the cloaca. It is therefore necessary to bear in mind two points. In the first place, the chloride titrated cannot be considered as definitely of renal origin. An unknown amount may have traveled through the gut when salt-loading was by stomach tube. When the bird was loaded by infusion or by intraperitoneal injection, the assurance of the chloride being from the kidneys is greater. Secondly, no data are available on the action, if any, of the cloaca upon the material before its discharge. Thus it cannot be known whether the concentrations obtained were purely those of ureteral urine.

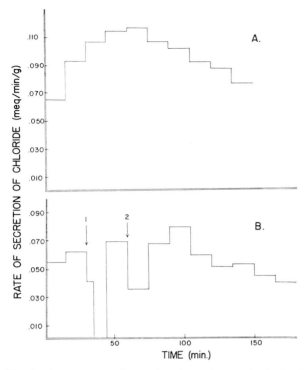

Fig. 4. Average rate of secretion from the nasal gland of Adélie penguin chicks. (A) An experiment in which there was no disturbance of the chick. Secretion continued until decapitation. (B) The effect of disturbance of the chick on the rate of secretion. The first decrease followed the entrance of two persons into the laboratory (arrow 1). The second followed their moving around prior to leaving (arrow 2). Note also that the maximum rate of secretion of the undisturbed chick on a meq/min/g basis is about 1.5 × the maximum rate of secretion of this chick.

The weight of each sample was taken as an estimate of the amount of fluid discharged. In some cases the volume of the fluid after the solid matter had settled was determined with a syringe. Without information on the movement of solids, salts in solution, and water through the lumen of the alimentary canal, it was felt that more precise determinations of the fluid volume of the cloacal discharge from feeding and non-fasting birds would be of little value. On the other hand, the fluid volume of the discharge from fasting birds was found to bear a sufficiently linear relationship to the weight of the entire sample to permit an estimate of the relative state of polyuria or of oliguria through weighing.

Cloacal discharges in feeding chicks. Table 3 is a summary of the cloacal chloride concentrations obtained from each chick, both before and after salt loads of 1 ml of 10 per cent NaCl per 100 grams body

weight. Although the weight of each sample is known, the discharge was usually so heavily loaded with intestinal matter that estimating the cloacal flow could shed little light on renal flow. Consequently amounts and flow rates have not been included.

Two points in this table are significant. First, high pre-loading cloacal chloride concentrations occur in any size chick and thus are not unique to any specific age. Moreover, these chloride concentrations are more than adequate to handle the concentrations of chloride previously shown to be present in the food. Second, no clear pattern of chloride concentrations follows the administration of the salt load. Some of the highest values recorded bracket a moderately high value (chick #8) while two other very high values bracket two clearly moderate values (chick #6). The periods of apparent anuria also merit notice. The longest period between discharges was about 7 hours (chick #3), and lapses of 2 to 4 hours were not uncommon in other chicks.

Cloacal discharges in feeding adults. The low chloride concentrations obtained from feeding adult birds is in marked contrast to those from feeding chicks. Table 4 summarizes the results obtained from some of these adults. That all had been feeding recently was manifest by the odor of digesting krill on their breath and the recognizable body forms in their excreta. Moreover, the color of the fecal matter was a bright orange-pink, characteristic of the living euphausid, and

TABLE 2. The Chlorine Concentration of Food Regurgitated to and Stomach Contents of Adélie Chicks

Source of Sample	Chloride Concentration, meq/kg
Esophagus	186
Esophagus	158
Esophagus	184
Esophagus	162
Esophagus	183
Esophagus	157*
Esophagus	181
Stomach, Cardiac (1)	146
Stomach, Cardiac (2)	134
Stomach, Cardiac (3)	180
Stomach, Cardiac (4)	215
Stomach, Cardiac (5)	155
Stomach, Pyloric (1)	144
Stomach, Pyloric (2)	122
Stomach, Pyloric (3)	171
Stomach, Pyloric (4)	193
Stomach, Pyloric (5)	146

* This sample largely fish—gray color and vertebrae. It was also more completely digested than the other samples.

TABLE 3. The Chloride Concentration (meq/l) in the Cloacal Discharge of Feeding Chicks before and after Salt Loading

Bird No.	Weight, g	1st hr	2nd hr	3rd hr	4th hr	5th hr	6th hr	7th hr	8th hr	9th hr	10th hr	11th hr	12th hr	13th hr	14th hr	15th hr	16th hr	17th hr	18th hr	et seq
1	85	—	207	—	405	—	296	—	—	—	—	349	—	345	—	—	—	—	—	397
2	94	—	229	—	—	—	—	—	210	—	—	—	—	—	—	—	275	—	⋮	193
3	126	Lost	Lost	—	Lost	—	—	—	—	333*	—	—	304	—	—	—	—	360	—	—
4	133	338	293, 332	356	401	318	—	342	—	—	—	—	401	—	—	—	425	—	—	401, 355, 250
5	182	269*	79*	320	332	323	252, 285	—	—	—	—	—	—	—	—	—	—	—	—	—
6	236	312–771	—	311	332	330	379	—	—	—	—	—	—	—	—	—	—	—	—	—
7	256	267	—	—	Lost	Lost	—	—	—	—	—	—	375	—	—	—	—	—	—	—
8	268	320	321	—	—	—	—	—	—	—	—	—	—	—	—	—	—	—	—	—
9	283	275*	288	242	—	313	—	338	—	—	281	230*	—	—	—	—	—	—	—	228, 199, >405, 188, >405, 202
10	349	221	—	242, 262*	257	—	—	—	—	—	—	—	—	—	—	—	—	—	—	—
11	413	—	—	296	—	315	—	308	—	—	—	—	—	—	—	—	—	—	—	—
12	454	—	—	293	—	—	—	—	—	—	—	—	—	—	—	—	—	—	—	—
13	707	317	—	*	292	—	309	—	—	364	—	—	—	—	—	—	—	—	—	—
14	745	259	169	—	—	203	—	—	—	—	322	—	—	—	—	—	—	—	—	—
15	850	261, 207	—	—	—	180	—	—	180	—	91*	—	—	158	—	—	—	—	—	—
16	902	232	210*	293*	—	310	—	—	—	—	228	237	—	—	—	—	—	—	—	(et seq) 242, 181, 99, 75
17	1243	211, 232	209*	—	—	297	—	—	166, 207	—	—	—	—	—	174	—	—	—	—	242
18	1250	219	—	286	171, 261	265	—	—	—	—	—	—	—	—	—	—	—	—	—	—
19	1250	—	—	227	233, 237	241	—	—	—	—	—	—	—	—	—	—	—	—	—	—
20	1260	—	33	—	—	—	10	—	—	—	—	—	—	—	32	—	—	—	—	15, 23, 16
21	1350	—	202	191*	301, 286	290	252	—	165	—	—	—	—	—	—	—	—	—	—	—
22	1600	—	310	—	—	—	290*	210	146	200	—	—	—	—	—	—	—	—	—	—
23	1990	—	290*	—	—	335	—	—	—	—	—	—	—	—	—	—	—	—	—	—
24	2025	—	235,* 173	—	—	—	—	—	—	—	—	—	—	—	—	—	—	—	—	—
25	2825	—	—	—	200	144	—	—	—	—	—	—	—	—	—	—	—	—	—	—
26	3060	104*	70	183, 191	149, 27	—	—	—	—	—	—	—	—	—	—	—	—	—	—	—
27	3250	—	—	235	139	—	—	—	—	—	—	—	—	—	—	—	—	—	—	—

Concentrations in italic denote times before salt loading. Asterisks indicate point at which chicks were salt-loaded.

TABLE 4. The Cloacal Discharge from Feeding Adult Adélies

Bird No.	Weight, kg	Time since Last Discharge, hr	Chloride Concentration, meq/l	Amount, g	Rate, g/hr/kg
A	5.43	0.03	29	0.57	3.14
		0.06	10	10.55	29.20
		0.12	6	6.16	9.45
		0.33	19	18.38	10.11
		1.00	107	34.13	6.28
B	5.19	0.50	117	30.53	11.83
		0.25	127	21.18	16.82
C	5.72	0.08	36	1.14	2.40
		0.05	22	6.07	21.60
		0.20	32	11.03	9.65
		0.08	33	7.26	15.20
D	6.86	0.33	31	largely krill	
		0.50	24	15.00	4.00
		0.42	54	3.73	1.26
E	5.23	0.03	80	2.71	15.60
		0.08	47	—	—
		0.12	32	1.83	2.92
		0.27	37	3.50	2.48
		0.16	35	2.59	3.04
		0.08	37	0.73	1.67
		0.20	34	3.07	1.77
		0.25	59	2.93	2.24
F	5.35	0.20	47	5.39	5.04
		0.15	36	2.83	3.53
		0.08	32	1.84	4.12
		0.10	29	2.26	4.22
		0.18	28	—	—
G	7.31	0.03	103	1.03	4.22
		0.53	112	—	—

filled with black flecks—the eyes of the more completely digested krill. The color of the feces after several hours of digestion is a deeper orange-brown. Body forms are then not recognizable, and the eyes are fewer or are obscured in the more compact, darker mass. In the feeding birds discharges often consist of a clear colorless liquid and white uric acid. The duration of collection of these samples was comparatively short, none longer than 2 hours.

All samples show a chloride concentration of considerably less than 200 meq/l and in most cases less than 50 meq/l. Moreover, four of the five values greater than 90 meq/l came from two birds, each of which showed a tendency to increase in concentration with time. One point which was quite apparent at the time of collection was the mild diuretic state of all these birds. This was first suggested by the greater frequency of evacuation as compared with that of the fasting adult birds, and contrasts with the relative in-

frequency already noted in the feeding chicks. That some of the fluid appearing in the cloacal discharge may have resulted from the movement of magnesium and sulphate ions through the intestine cannot be excluded. The chicks, fed by regurgitation from the adult bird's stomach, may not receive as large amounts of these ions.

Whenever the samples were not so heavily loaded with partially digested krill that the result would have been meaningless as an index of renal flow, the rate of cloacal discharge has been calculated. It was noted that the fluid flow rates from samples devoid of krill or from those in which the krill was readily separable were rather higher than those shown by fasting adults.

Finally, it is noteworthy that all these birds were secreting to some degree from the nasal gland at the time of capture and continued to do so while on the restraining board. The results of the collection of this secretion are presented later. None of these birds was given a salt load.

Cloacal discharges in fasting adults. In Tables A and B of the Appendix are presented selected results from several adult birds on which both preloading and salt-loaded cloacal discharge data are available. The experiments have been arranged chronologically, and each bird has been given a number to facilitate comparison between values obtained before and after loading. These are all non-feeding birds. The times shown in the tables are the fractions of an hour since the preceding cloacal evacuation. Preloading cloacal discharges from 10 birds that were given loads other than NaCl follow the matched birds in Table A.

Of particular interest here is a consideration of the rate of flow. To provide a reference for comparisons, the range and frequency of flow rates in birds before salt loading have been determined. Of the 131 samples obtained prior to salt loading, 119 permitted a calculation of the flow that would be representative of what the ureteral flow might be. The range of values was found to be from 0.1 g/hr/kg to 3.7 g/hr/kg. Figure 5 is a histogram of the distribution of the rates. For convenience in plotting, 0.2 g/hr/kg was arbitrarily selected as the unit range. On the basis of the results shown in Figure 5, rates of 3 g/hr/kg and above are clearly indicative of polyuria in fasting birds. It is likely that flows of 2 g/hr/kg also represent polyuria, as 88 per cent of the values are less than this.

As no direct measurements of renal function were attempted, these flows cannot be interpreted as clear

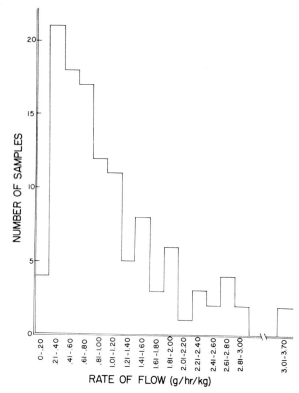

Fig. 5. Distribution of cloacal flow rates in fasting adult Adélie penguins. The arithmetic mean is 1.04 g/hr/kg. The modal range is 0.21–0.40 g/hr/kg. The median falls in the range 0.71–0.90 g/hr/kg.

reflections of kidney activity, except perhaps in the more extreme cases of oliguria and polyuria. It is noteworthy that all of the preloading flows greater than 2 g/hr/kg came during the latter part of the season, when the birds have access to open seas and to drift snow and are relatively unrestricted either by the environment or by breeding responsibilities. Thus, even though they may have been without food or water for the 48 hours before the experiment, they are likely to have been in a more favorable water balance than the birds that had fasted for many days.

Effect of salt loads on the cloacal discharge. Several types of salt loads were given to the adult birds. Solutions of 3.5, 10, and 20 per cent NaCl (w/v) were infused, and 3.5 and 20 per cent solutions were administered by stomach tube. The actual amount of salt given any bird varied with the principal aim of the experiment. That the chloride concentrations of the discharge continued for the most part to be lower than 110 meq/l is therefore quite striking.

Of 34 adult fasting birds that were salt-loaded, 18 showed cloacal discharge chloride concentrations less

than 110 meq/l for all samples. However, 8 had at least one value greater than 200 meq/l. Perhaps even more significant was the tendency in all birds for the concentration to fall in the succeeding samples. In only two instances was there any suggestion of an increase, birds 2 and 4. Bird 4 died without ever secreting from the nasal gland, and the plasma chloride at the time of its third cloacal discharge was 137 meq/l. This is 10 meq/l higher than found in any salt-loaded bird that lived.

If, for the sake of discussion, the value of 2 g/hr/kg is accepted as the point for considering the bird in polyuria, then 8 of the 10 cases with concentrations higher than 110 meq/l showed evidence of polyuria at some time following loading. However, no other correlation is apparent between concentration and flow rate, nor is any pattern discernible in the time relations of the more voluminous flows. Furthermore, the high concentrations and flow rates obtained from bird 1 may be attributed in part to the loading solution running through the alimentary canal. No suitable marker such as powdered charcoal was available at the time to eliminate this possibility.

Trends or correlations among salt loads, cloacal discharge chloride concentration, and cloacal flow rates are obscure. For every experiment that may be cited in support of a type of response, a comparable experiment can be shown that either fails to support or actually refutes it.

Chloride concentration in cloacal discharge. The results presented here have been brought together from a variety of experimental conditions. The only criteria for grouping have been the ages of the birds, their state of feeding, and whether the birds were salt-loaded. No attention has been given to the amount or method of administration of the salt loads. Also, no consideration has been given to the time sequences of the samples.

Figure 6 is a bar graph of the chloride concentrations obtained from five groups of birds. The chloride concentrations of the non-feeding and fasting adult birds before salt loading are shown in section 1. More than 90 per cent of the samples have plasma chloride concentrations lower than the average. (The average preloading plasma chloride concentration for the adult birds in these experiments was 109.7 meq/l. No values are available for the chicks.) The discharges of adults taken into the laboratory after they had been fasting were marked by heavy, bile stained, mucous laden masses. A total of 131 samples from 57 different birds

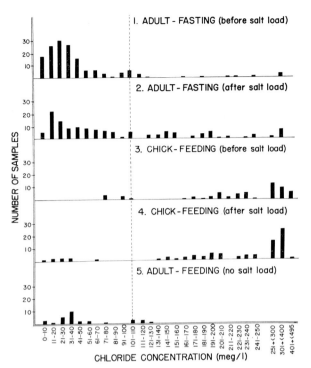

Fig. 6. Distribution of the chloride concentrations in the cloacal discharge of five groups of Adélie penguins.

are included. About 72 per cent of the values are in the distinctly low range of 0 to 50 meq/l. Another 20 per cent fall between 50 meq/l and the average plasma chloride concentration of 110 meq/l.

A similar picture is seen in section 5 showing data from adults captured just as they were coming out of the sea from feeding. A total of 31 samples was obtained from 9 different birds. About 80 per cent of the values are in the low range of 0 to 50 meq/l, and the three above 110 meq/l are not greatly so.

In marked contrast to these results from feeding adults are the concentrations obtained from the feeding chicks before they were salt-loaded (Figure 6, section 3). Nearly 90 per cent of the concentrations are well above the average adult plasma concentration of 110 meq/l. All the chicks had been fed in the 24 hours prior to their being taken for the experiment, and none was without an appreciable amount of food in the gut. Of the 54 values tallied, representing 28 birds, only 11 per cent are less than 110 meq/l, and none of these is less than 70 meq/l. The remaining 89 per cent are in a high range of 160 meq/l to 415 meq/l.

Although the samples obtained from adults (Figure 6, section 2) tend to have higher chloride concentration after salt loading than before, about 68 per cent

still remained less than 110 meq/l. In the chicks, the response to salt loading (Figure 6, section 4) appears to be one of continuing high cloacal chloride concentrations with about 90 per cent of the values greater than 110 meq/l. However, the values after salt loading tend toward a downward shift from the extremely high ones found before loading.

Changes in the thiocyanate space of adult Adélie penguins. It was of interest to have information on the behavior of the extracellular fluid space in two different situations. The first involved changes in the extracellular volume of fasting adult male Adélies. The second concerned the extracellular volume changes that were expected in association with salt loading and the onset of the nasal gland secretion. Sodium thiocyanate was selected as an index for estimating changes in the extracellular space. The relative simplicity of both the experimental technique and of the chemical analysis for the ion justified its use even though it is subject to criticism as a substance for determining extracellular volume.

Thiocyanate spaces in fasting adults. Male birds that had been incubating and therefore fasting were taken from the banded colonies and infused with known amounts of NaSCN. Regularly spaced blood samples were drawn to determine the disappearance curve. At least one hour between infusion and the first blood sample was necessary to avoid inexplicable increases in plasma thiocyanate ion concentration. This sample was followed by at least two, and usually three, hourly samples. Then the bird was either released or used for experiments on the response of the thiocyanate space to osmotic and salt loading.

Of the experiments performed, 16 were technical successes in that both infusion and sampling were satisfactorily quantitative. The known fasting time of these birds ranged from 0 to 36 days. The average thiocyanate space for all 16 birds, expressed as per cent of the total body weight, was 29 per cent. The range of values was 23.1–32.2 per cent, and the standard deviation was ±2.5 per cent.

Figure 7 graphs the thiocyanate space against the duration of the fast. The solid line is drawn through the mean value, and the dashed lines are a distance of one standard deviation from the mean. The range of the thiocyanate space values increases as the duration of the fast increases. Whereas the seven birds fasting 10 days or less showed a range of 30.0–32.2 per cent, the nine that fasted 10 to 36 days showed a range of 23.1–30.8 per cent.

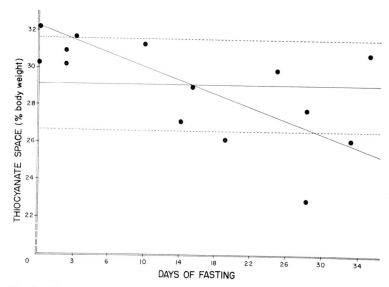

Fig. 7. The relationship of the thiocyanate space to the duration of the fasting period in the adult male Adélie penguin. The thiocyanate space is expressed as a per cent of the total body weight of the bird. The solid line is drawn through the mean value of the thiocyanate spaces, 29.1 per cent. The broken lines to each side are a single standard deviation distance from the mean, 2.5 per cent. The dotted line is a regression line fitted by the method of least squares. The equation for this line is $Y = -0.21 \times +32.3$.

The calculated correlation coefficient between duration of fast and the thiocyanate space is -0.554. Except for the single case of 30.8 per cent at 36 days, all birds fasting longer than 10 days had thiocyanate spaces no greater than 30.0 per cent of their body weights. The slope of a line fitted by the method of least squares has a value of -0.21. These values suggest a negative correlation between extracellular fluid volume and length of fasting. On the other hand, this apparent decrease may have resulted from the thiocyanate ion entering the cellular spaces more readily when the bird had fasted longer.

Thiocyanate space shifts after salt loading. Following the completion of the blood sampling required for determining the thiocyanate space, salt loads were administered to some of the birds to permit an estimation of the fluid volume shifts that were expected to follow the infusion of hyperosmotic solutions.

Table 5 shows the results of this series of experiments. The only constant feature that emerges from the figures is the fact that the thiocyanate space expands when NaCl is infused into the bird as 1.7 eq/l and 3.4 eq/l (10 and 20 per cent) solutions.

Of unknown significance is the observation that in seven of the eight birds the expansion was between 2 and 4 per cent of the body weight. In six of these

seven birds the water drawn from the intracellular spaces was not sufficient to dilute the chloride administered to preloading plasma levels, if it is assumed that all the chloride remains outside the cells. In the other two birds the extracellular expansion was sufficient to have diluted the infused chloride to less than the normal plasma level. In only three cases was enough plasma left after thiocyanate analysis to permit chloride determinations. Two of these showed a slight rise in plasma chloride, even though the extracellular expansion was great enough to have diluted the entire load to levels well below normal plasma levels.

Other Aspects of Nasal Gland Function

Initiation of nasal gland secretion. The salt gland appears to operate in an "on-off" fashion. The first stimuli found to elicit secretion from the nasal gland were hyperosmotic solutions of sodium chloride and of sucrose. These experiments indicated that the nasal gland receptor mechanism was responding to an osmotic load and was not exclusively sensitive to plasma concentrations of sodium or chloride. Moreover, it seemed likely that infusing materials such as sucrose in hyperosmotic solutions would actually lead to a decrease in the plasma sodium and chloride concentrations through plasma dilution by osmotic withdrawal of water from the intracellular fluid.

TABLE 5.   Changes in Thiocyanate Space with Salt Loading

Bird Weight, kg	SCN-Space ml	Salt Load, meq	Extrapolated Volume after Load, ml	Estimated Volume after Load, ml	SCN-Space Change, ml	Calculated Load Dilution, meq/l
3.69	1030	27.40	1050	1130	+80	342.5
3.60	1110	27.40	1295	1380	+85	322.5
3.95	1078	27.40	1232	1406	+174	157.5
3.45	1000	25.65	1183	1285	+102	250.9
3.58	1072	17.12	1072	1352	+280	61.2
4.20	1225	35.95	1285	1386	+101	356.0
4.00	1295	34.20	1295	1445	+150	228.0
3.51	1130	9.00	1212	1332	+120	75.0

It was thus of interest to determine what response, if any, would be obtained from the nasal gland when hyperosmotic solutions of a substance that readily penetrates cell membranes, such as urea, were infused. To determine the presence and value of any osmotic threshold, the melting point of plasma samples was determined for birds that had received urea loads and for birds that had received salt loads. All loads were infused into fasting adult birds at a rate of 1 ml/min through cannulated right flipper veins. The results of six of these experiments are presented in Table 6.

In comparing the effects of the salt and urea loads it should first be noted that all the urea infusions constituted greater osmotic loads than any of the salt infusions. The least urea load was 132 mM of urea, and the greatest salt load was 35 mM of NaCl or approximately 70 mM of osmotically active particles. In two of the birds loaded with urea a very slight splattering from the beak began about 3 hours after completion of the infusion, but the flow rate was not great enough to yield a titratable sample. All of the salt-loaded birds, on the other hand, began secreting at collectable rates within 1–10 minutes from the time loading was completed.

The second point to note is that the urea-loaded birds showed cloacal flow rates markedly above those of salt-loaded birds. All the birds showed rates above 3 g/hr/kg, with some ranging up to 35 g/hr/kg. By the time nasal gland secretion began, on even the slightest scale for the urea-loaded birds, 100–250 ml of fluid had been lost through the cloaca.

No osmotic threshold for the initiation of secretion was discernible. The values of the plasma osmotic pressure show that increase of this, per se, does not trigger the response. Rises of at least 50 milliosmoles in plasma osmotic pressure were found in the urea-loaded birds. None of these birds secreted as a direct result of the rise. On the other hand, the plasma osmotic pressure of salt-loaded birds rose no more than 10 milliosmoles before secretion began. In two cases a decrease of about 10 milliosmoles occurred at the onset of secretion, followed by a slight increase.

Spontaneous secretion from the nasal gland. A sight common in the rookery during the late spring and throughout the summer months was that of a clear, colorless fluid dripping from the tip of the beaks of both adults and older chicks. Some of the birds brought into the laboratory during this time dripped from the beak after being placed on the restraining board. Many other birds, although not actually permitting drops to form on their beaks, splattered a salty fluid on the board prior to the beginning of an experiment. Attempts to collect this discharge were not always successful, especially when the discharge rate was low and the only evidence of discharge was the accumulation of salt on the board. Furthermore, whenever the bird's head was held to collect the discharge, the flow ceased shortly thereafter. A few minutes after releasing the head and removing the vial, splattering usually began again. It was thus impossible to obtain an accurate estimate of the rate of flow.

The concentrations of the secretions that were collected are shown in Table 7. The most striking feature of these 48 samples from 27 birds is that the concentrations are low. The range is from 188 to 744 meq/l, and most samples lie between 200 and 400 meq/l. Moreover, 7 of the 9 values above 500 meq/l were obtained from two birds that secreted more copiously and for longer periods than any of the others. Both had been feeding and were only a few minutes out of the sea. As with all the others, however, these stopped secreting after a short interval, and subsequent samples were obtained only by waiting several minutes after removing the vial.

Attempts were made in the rookery to collect the nasal discharge from birds seen to be secreting steadily. Rates of 4 to 6 drops a minute for periods of 5 to 15 minutes were not uncommon, and rates of 1 to 2 drops a minute were seen regularly. Whenever these birds were caught and held to collect the secretion, none could be obtained. Most of the birds with high rates of dripping (4–6 per minute) were adults that had krill or traces of krill in their cloacal discharges. Few young chicks were seen dripping; they seemed to discharge their nasal secretions with a clearly audible sneeze or a quick head shake. In a few this occurred as often as every 4 to 6 minutes, provided they remained both undisturbed and awake.

The birds that had the less frequent discharges (1–2 per minute and less) often made a rasping sound by rubbing the tongue against the roof of the mouth. This tongue rubbing seemed to be a regular phenomenon, even when no secretion could be seen. All the birds producing it in the laboratory splattered a salty solution on the restraining board. It is also noteworthy that while nasal dripping and the sound correlated with a functioning salt gland were clearly detectable on warm summer days, neither was noted so frequently during the colder days of spring.

Secretion of the thiocyanate ion by the nasal gland. Analysis of the secretion from the nasal gland indicates that the secretory mechanism is highly specific for sodium and/or chloride. Only very small amounts of potassium, magnesium, or sulfate normally appear in the secretion [*Schmidt-Nielsen*, 1957]. It was therefore of interest to determine how the secretory mechanism would respond to an anion with roughly the same physical-chemical properties as chloride. The thiocyanate space experiments provided an opportunity to determine to what extent this ion, which does not normally occur in appreciable quantities in the plasma, would be excreted by the nasal gland. Consequently the secretions from birds that were given a thiocyanate injection and then salt-loaded were analyzed for both thiocyanate and chloride. The results of these experiments, recently reported elsewhere [*Douglas*, 1966], may be summarized as follows: (1) Thiocyanate appears in the nasal secretion, and (2) the secretion/plasma ratio of thiocyanate is about 5.5, similar to that of chloride.

DISCUSSION

The Adélie chick. The Adélie chick hatches weighing between 85 and 95 grams. Approximately 10 grams of this is yolk and yolk sac enclosed within the body cavity. If we assume that Brody's equation for the basal metabolic rate in the thermoneutral zone applies to the relatively quiet, brooded chick, then its daily energy expenditure is expressed by: kcal/day = 89 × (wt. in kg) 0.64 [*Brody*, 1945: 371]. Thus a chick with a net body weight of about 85 grams at hatching would have a basal energy expenditure of about 19 kcal/day. If the caloric value of chicken yolk [693 kcal/100 g, *Nutritional Data*, 1957] is taken as sufficiently close to that of Adélie penguin yolk, then 10 grams of residual yolk provides some 69 kcal. At a consumption rate of 19 kcal per day, this should last about 3.6 days. A few direct measurements have shown that within about 3 days the yolk sac of the unfed chick weighs about 2 g, which agrees rather well with the calculation of energy reserves in the yolk at hatching (Douglas, unpublished data). From the standpoint of the energy supply for survival of the hatchling this is quite reasonable. The feeding grounds of the adults may be some distance from the rookery; the adult may not return from sea precisely when the eggs are hatching. A variety of factors, including foul weather, may cause a time lapse between hatching and first feeding.

In any case, the chick must be able to handle hyperosmotic food at its first feeding. A few chicks were seen to receive a regurgitated meal from one of the parents before their down was completely dry. While these may have been rather exceptional cases, it was not unusual for chicks to begin receiving food within the first 24 hours of life outside the shell.

The regurgitated food yielded chloride concentrations that, while somewhat above normal plasma levels for both chicks and adult birds, were not greater than those the kidney appears able to produce.

To estimate the approximate chloride concentration of the living euphausid ingested by the adult bird, we can assume a water content equivalent to 75 per cent of the total body weight [*Vinogradov*, 1953: 376]. *Flemister* [1958] determined that the extracellular fluid volume of the marine Mangrove crab, *Goniopsis cruentatus*, is between 30 and 46 per cent of the total body water. *Krogh* [1939] analyzed the chloride concentrations of blood and tissue in the two decapod crustacea, *Eriocheir* and *Carcinus*, as follows: blood, 450 meq/l and 485 meq/l, respectively; and tissue, 139 meq/l and 57 meq/l. These figures indicate that the actual chloride concentrations of the euphausid are probably in the range 130 meq/kg–230 meq/kg. As this range brackets the values obtained for the chloride concentration of krill fed to the chicks by their

TABLE 6. Initiation of Nasal Gland Secretion

| Weight, kg | Load | Elapsed Time, hr | Cloacal Discharge | | | Plasma Osmotic Pressure, mOs |
			Chloride Concentration, meq/l	Amount, g	Rate, g/hr/kg	
4.00	20% urea, 40 ml (132 mM)	0	25	4.5	1.1	392
		0.3	11	5.9	4.4	
		0.6	9	4.3	4.3	414
		1.7	8	28.1	6.1	
		2.7	3	16.1	4.0	
		4.1	6	18.6	3.5	
		5.6	13	18.3	3.1	
		7.1	11	12.9	2.2	441
		8.3	9	12.1	2.4	
		9.8	15	9.9	1.7	

No nasal gland secretion until about 6½ hours after infusion of load. Slight splattering only.

Weight, kg	Load	Elapsed Time, hr	Chloride Concentration, meq/l	Amount, g	Rate, g/hr/kg	Plasma Osmotic Pressure, mOs
3.84	20% urea, 60 ml (198 mM)	0	8	9.4	1.2	392
		0.5	5	5.7	2.7	
		0.8	3	11.0	11.5	
		0.9	8	21.5	35.0	
		1.3	7	11.4	7.1	
		1.9	29	32.4	12.8	
		2.4	27	24.4	11.0	
		3.3	15	29.1	10.1	
		4.6	6	32.1	6.7	
		5.5	2	15.6	4.4	
		10.7	13	25.6	1.4	
		12.7	25	9.4	1.0	
		14.7	13	8.1	1.1	450

No evidence of there having been any nasal gland secretion.

Weight, kg	Load	Elapsed Time, hr	Chloride Concentration, meq/l	Amount, g	Rate, g/hr/kg	Plasma Osmotic Pressure, mOs
3.49	20% urea, 65 ml (223 mM)	0	8	2.0	?	410
		1.3	12	9.2	2.1	
		1.4	18	7.2	15.9	468
		1.6	16	26.7	30.8	
		1.9	11	32.7	33.4	
		2.2	6	17.7	15.2	
		2.6	5	21.9	18.8	
		3.1	3	29.5	15.4	
		3.9	1	26.4	10.1	
		5.0	1	18.5	4.6	
		6.4	1	25.8	5.7	
		8.4	2	15.2	1.5	
		14.9	8	14.9	0.8	505

Slight nasal gland splattering about 3 hours after load.

Weight, kg	Load	Elapsed Time, hr	Chloride Concentration, meq/l	Amount, g	Rate, g/hr/kg	Plasma Osmotic Pressure, mOs
3.95	3.5% NaCl, 56 ml (33.7 mM)	0	10	6.8	1.5	398
		1.7	87	6.8	1.1	392
		4.1	48	2.7	0.4	403
		6.1	32	5.7	0.7	
		13.9	27	17.7	0.6	
		16.5	26	7.6	0.7	
		17.9	25	4.9	0.8	
		21.3	28	1.3	0.1	412

Nasal gland began secreting at a collectable rate after infusion of about 45 ml (27 mM), and began to slacken after 1½ hours

Weight, kg	Load	Elapsed Time, hr	Chloride Concentration, meq/l	Amount, g	Rate, g/hr/kg	Plasma Osmotic Pressure, mOs
4.11	3.5% NaCl, 30 ml (18.9 mM)	0	—	—	—	371
		7.9	17	18.3	0.6	382
		9.5	24	3.5	0.6	403

Nasal gland began secreting at a collectable rate after infusion of about 28 ml (17 mM), and began to slacken after 1¼ hours
First cloacal discharge came 34 minutes after beginning of infusion, or 7.9 hours after beginning thiocyanate space determination

TABLE 6. Initiation of Nasal Gland Secretion (*continued*)

Weight, kg	Load	Elapsed Time, hr	Cloacal Discharge			Plasma Osmotic Pressure, mOs
			Chloride Concentration, meq/l	Amount, g	Rate, g/hr/kg	
4.20	10% NaCl, 21 ml (36 mM)	0	78	2.6	1.9	334
		1.0	8	2.4	0.6	322
		2.0	19	5.3	1.3	371
		5.0	20	6.1	0.5	
		7.0	38	5.0	0.6	
		9.0	41	5.2	0.6	
		12.0	53	6.9	0.5	
		15.0	40	12.5	1.0	
		18.1	40	4.0	0.3	
		19.5	32	5.3	0.9	368

Infusion fairly rapid, about 7 ml/min; copious nasal secretion within 2 minutes of end of loading.

parents, nothing suggests that the krill is significantly altered in its chloride composition between the time the adult ingests it and regurgitates it to the chick.

In the absence of direct experimental data, it is reasonable to assume that the euphausids have a body fluid osmotic pressure similar to that of other marine crustacea. They therefore would be expected to be isosmotic with sea water and consequently hyperosmotic to vertebrate body fluids. Thus the chick at hatching must cope with the same osmotic problems as the feeding adult.

The ability of the nasal glands to produce sodium chloride solutions more concentrated than sea water [*Schmidt-Nielsen* et al., 1957] is the principal means by which a marine bird such as the Adélie handles ingested hyperosmotic fluids. The nasal glands in the adult Adélie are well developed. The observation that chicks splatter a salty fluid while still partially surrounded by extraembryonic membranes indicates that the gland is functional at hatching.

A correlation coefficient of 0.98 was found between the weight of the salt glands and the maximum rate at which they can excrete chloride (Figure 4), which is excellent. (*Maroney* [1958] considers a correlation coefficient of 1.0 a perfect functional correlation.) As the weight of the glands is linearly related to body weight (Figure 1), one would expect the excellent correlation shown between net weight of the bird and the maximum rate of secretion of the gland (Figure 2).

The fasting adult Adélie. The Adélie penguin loses up to 40 per cent of its body weight in an annual period of starvation during breeding and incubation. Some of the more subtle physiological implications of the long fasting period of the male Adélie may be

overshadowed by the emphasis placed on the metabolic cost of the fast. Not only does the male go without food during the 32–42 days that he is ashore, but he also must be able to get along with almost no water during this period.

The source of fresh water for the Adélie is snow. In the rookery at Hallett one commonly sees birds lying in a snow bank pecking and eating snow. In the laboratory after a salt-loading experiment several pen-

TABLE 7. Spontaneous Nasal Discharge of Adult Adélies

Bird No.	Chloride Concentration, meq/l	Bird No.	Chloride Concentration, meq/l
1	522	13	188
2	240	14	256
	339		620
	371	15	328
	333	16	328
3	326	17	413
	258		311
4	192	18	222
5	336	19	316
6	202	20	495
	224	21	272
7	284	22	235
	245	23	249
8	245	24	238
9	244	25	317
10	566		444
	322	26	277
	658	27	293
	656		369
	516	28	351
11	234		296
	217	29	390
12	688		331
	730		
	744		

guins appeared not to recognize liquid as a source of fresh water. They ignored a bowl of water, but would peck at a small pile of cotton or shredded paper, presumably because these resembled snow. When finally released, many birds headed for the nearest snow drift and quickly began eating.

Snow is not always readily available to the incubating bird. Most of the Cape Hallett rookery is on a low, relatively level, terminal moraine which is normally cleared of deep snow by the prevailing winds off the polar plateau. The layer of snow on the colony mounds when the birds arrive in mid-October is gone within a few days. This is due partly to warming temperatures, but even more to the traffic of birds during the nest-building and courtship periods. At Cape Hallett and at Cape Adare some snow drifts may persist throughout the summer, but these are several hundred feet from most of the colonies.

Even when drifts are present near a colony, birds that leave their nests to eat snow risk losing their eggs to the cold or to predators. The skua and the sheathbill are vigilant and quick to take advantage of such opportunities. Although snow storms do occur during the summer, heavy moist falls occurred in 1959–1960 only after the chicks began hatching, and little snow fell during 1960–1961. Not all penguin rookeries are as windswept as those at Hallett and Adare, but as *Sladen* [1958] and *Sapin-Jaloustre* [1960] clearly show for the Antarctic Peninsula and the Adélie Coast rookeries, snow-covered colonies suffer a high loss of eggs during incubation. *Sladen* [1958: 69] reports mortalities as high as 100 per cent for colonies at Hope Bay, and "the greatest losses occurred where snow lay thickest." The colonies for which I have records were all free of snow within days of the return of the breeding birds. Thus, the incubating Adélie cannot rely on freshly fallen snow to meet its water requirements, nor can it leave the nest unguarded to go to a snow drift. Therefore, it is faced with a problem of conserving not only its energy reserves but also its water and electrolytes.

The values for the cloacal flow are of about the same magnitude as the more recent values in the literature for urine flow in chickens. Early estimates of urine output [*Sharpe*, 1912; *Mayrs*, 1924; *Davis*, 1927], based on urine collected from cannulated ureters in adult chickens over short periods, yielded average calculated flows of 20–40 ml per hour. Assuming an average weight of 2.2 kg, these were flows of about 9.1–18.2 ml/hr/kg. More recently *Hester et al.* [1940] and *Hart and Essex* [1942] have shown such esti-

mates to be erroneously high. Whereas urine output from cannulated ureters is quite high for the first 30 minutes, later flows are one-fifth to one-sixth as high. A daily estimate based on the first collection gave an output of about 670 ml per day, but that based on the second collection was only 120 ml per day. Handling of the bird and the operation of cannulation were suggested as reasons for the initial high flows. Also, these birds were all anesthetized, and *Davis* [1927] showed that ether anesthesia tends to increase urine flow. I have observed similar disturbance diuresis in unanesthetized chickens and to a lesser extent in unanesthetized herring gulls, as has *Skadhauge* (personal communication, 1965) in unanesthetized chickens.

Korr [1939] reports experiments with unanesthetized chickens that had been given no water for 24–72 hours. When these birds were normally hydrated, urine flows from cannulated cloacae were about 20 ml per hour. Following dehydration the flows fell to about 5.4 ml per hour. For a 2.2 kg bird this would mean a flow rate of about 2.45 ml/hr/kg, which is in the upper range of the cloacal flow obtained from unanesthetized, non-cannulated, incubating Adélies.

Weight data for the penguin provide a basis for calculations showing that the discharge rates found are of a meaningful magnitude. Let us assume that an incubating male Adélie that weighs about 5 kilograms when he comes ashore will spend about 35 days without food and 25 days without food or water. Let us further assume that he loses 1.7 kg during this period. *Sladen*'s graphs [1958] indicate that weight loss during incubation is relatively linear, so that this bird will have an average weight of 4.2 kg for the period. In the 25 days without water about 25/35 of 1.7 kg, or 1.4 kg, will be lost.

The mean cloacal flow rate, about 1.0 g/hr/kg, is used to estimate a likely cloacal water loss during this time. At this rate, in an average period of 25 days without water, a bird with an average weight of 4.2 kg would lose 2.6 kg by cloacal flow. This is 1.5 times the actual total weight loss and obviously too high. If, on the other hand, we use the upper end of the modal range for cloacal flows, 0.4 g/hr/kg, the total cloacal flow over the same period for the same bird is about 1.05 kg, leaving about 350 g or less that may be lost as respiratory water.

To compare this value with one calculated for the respiratory water loss, let us assume that *Brody*'s [1945] equation for the basal metabolic rate in the thermoneutral zone applies to the relatively quiet, incubating Adélie. If the average weight over the period is 4.2 kg, then the net energy expenditure for a 25-day fast would be about 5100 kcal. Assuming that all this energy comes from fat, which has a caloric value of about 9.5 kcal/g, the fast entails the metabolism of about 550 g of fat. About 1 l of oxygen is consumed in the metabolism of 2.02 g of fat (recalculated from *Kleiber*, 1961: 125) so that during this period some 275 l of oxygen are consumed. Finally, if we assume that *Dawson*'s [1958] value for the amount of respiratory water lost in the cardinal is a reasonable approximation of that lost by the Adélie, then the respiratory water loss during the fast would be 245 g. This value is sufficiently similar to that allotted for water loss through the lungs to lend support to the assumption that the cloacal flow for the incubating Adélie is in the neighborhood of less than 0.4 g/hr/kg. Thus the cloacal flow in the fasting penguin is about one-seventh the ureteral urine flow reported by *Korr* [1939] for chickens deprived of water.

It appears consonant with the data presented and with what is known of avian renal physiology to regard the urinary system of the Adélie penguin as well suited to water conservation and to adopt as a working hypothesis the view that its excretory functions are primarily the removal of nitrogenous wastes, predominantly uric acid [*Davis*, 1927], and the maintenance of acid-base balance [*Wolbach*, 1955; *Orloff*, 1961].

The nasal glands regulate osmotic balance through the excretion of sodium and chloride.

The results obtained from the spontaneous nasal discharges of the fasting birds are compatible with this interpretation of the division of labor. The wide range of chloride concentrations and of dripping rates may indicate the ability of the nasal gland to respond differentially to the intensity of the stimulus up to a point of maximum effort. Thus, regular water loss from the lungs and the cloaca necessitates a regular osmotic compensation effected by the removal of sodium chloride. On the other hand, it would seem more efficient for the gland to put out less fluid at near maximum concentrations whenever it is secreting. The low concentrations reported here would then probably be a consequence of cessation of action by the gland.

The decrease in thiocyanate space during fasting suggests a decrease in the extracellular body water in the fasting Adélie. The increased dispersion of points reflected in the correlation coefficient of −0.54 probably reflects the unpredictable behavior of the thiocyanate ion. *Levitt and Gaudino* [1950] question the reliability of thiocyanate as an indicator of extracellular space. They found that the ion not only penetrates

cell membranes, it also seems to penetrate in varying degrees under apparently similar conditions. As they were unable to find a consistent relationship between the inulin and thiocyanate spaces, they doubt the latter's value for measuring changes in extracellular volume. Figure 7 shows some change during this period; whether this represents an increased permeability of cell membranes to the thiocyanate ion or a loss of extracellular water is not known.

If data were available that would permit us to relate thiocyanate spaces to lean body weight, a more definite trend might emerge. Presumably thiocyanate enters adipose tissue in negligible amounts and at slow rates compared to its penetration into cells. Thus the expression of thiocyanate space as per cent lean body weight would raise the values in the early fast and would alter those later in the fast only slightly.

The renal response to salt loading. The significance of the response of the excretory system to salt infusions in the Adélie is not clear. In most cases the cloacal flow remained about the same after salt loading as before. Some birds had a clear, though transient, diuresis; in others the cloacal flow rate increased only slightly. The increased flows were accompanied by marked and similarly transient increases in the chloride concentrations. The feature common to all the salt-loading experiments was the eventual low chloride concentration of the discharges, even when the plasma chloride concentration increased by 10 to 15 meq/l.

Bressler [1960] reports similar results in dogs. These animals responded to infused solutions of 10 per cent sodium chloride with a fall in urine sodium and chloride concentrations, accompanied by a rise in the plasma concentration of these ions. He hypothesized a passive reabsorption in the proximal tubule to account for these seemingly paradoxical results. *Holmes et al.* [1961] report a two-phase response to the ingestion of hypertonic saline solutions in hydrated domestic ducks, *Anas platyrhynchus*. Their birds, loaded by a stomach tube with 20 ml of 20 per cent sodium chloride, responded with an initial period of diuresis and high sodium concentrations. At varying times during this phase the nasal gland began secreting. If a weight of 2.6 kg is assumed for these birds, recalculation of their data shows flow rates of about 26, 20, and 15 ml/hr/kg for three 20-minute collection periods following salt loading. Sodium concentrations of 630, 455, and 250 meq/l were recorded during these periods. Unfortunately there is no way to determine how much of the fluid and sodium that appeared in the cloacal discharge came directly through the alimentary canal. Nearly one-half the salt load and all the fluid load are accounted for in the first sample of 23 ml. There appears to be a trend toward decrease in cloacal flow and sodium concentration, but no information was given on renal activity after the onset of nasal gland secretion except that no urine was collected.

Korr's [1939] results with chickens shed some light on those obtained with the Adélie. Following the infusion of 10 per cent sodium chloride into hydrated birds, he obtained a "maximal increase in urine flow of 0.89 ml/min/kg," which is equivalent to an increase of about 54 ml/hr/kg. The preloading flow is not given. During this time the urine chloride concentration rose, although the total osmotic pressure fell. Furthermore, this diuresis could be sustained with a constant intravenous infusion. Since only one fasting Adélie responded with a diuresis of 50 g/hr/kg and only a few showed flow rates above 5 g/hr/kg in response to salt loads, these birds are probably in a state of water depletion. This is in agreement with the earlier discussion on the fasting birds and consonant with an apparent decrease in the extracellular fluid volume.

Elucidation of the problem of the renal response to salt loads in the Adélie and in birds in general is complicated by the open question as to the role of the cloaca. Davis [1927] states, although he gives no supporting figures, that urodaeal chloride concentrations from chickens are very much less than ureteral chloride concentrations. Korr [1939] suggests that the site of water reabsorption is the rectum. Both Davis and Korr believe that water reabsorption in chickens is not against an osmotic gradient, but that it is accomplished with the simultaneous reabsorption of chloride. Weyrauch and Roland [1958] disagree on both points; they dismiss the possibility of appreciable amounts of cloacal fluid re-entering the rectum on the basis of an apparent 90 per cent efficiency of Houston's valve. Their experiments with Na^{24} and I^{131} led them to conclude that no significant reabsorption of electrolyte or water from the cloaca occurs. More recently Skadhauge and Schmidt-Nielsen [1965] report data indicating negligible salt or water reabsorption from the chicken's cloaca. These findings were consistent whether the birds had been hydrated, subjected to a salt load of 12 meq/kg, or deprived of water for 40 hours.

Schmidt-Nielsen et al. [1963] suggest that in marine birds the cloaca acts in concert with the kidneys to permit the excretion of a semi-solid urine through water reabsorption allowed by the active transport of sodium. This suggestion was based, in part, upon an extrapolation of the earlier results reported in chickens and, more firmly, upon their observation that under salt loading the cloacal effluent of gulls showed declining sodium concentrations. Recent experiments using a cloacal cannula in herring gulls have shown that this decline in salt concentration occurs before urine comes into contact with the cloacal mucosa [Douglas, 1966].

The initiation of nasal gland secretion. One of the first reports to appear on the salt-secreting function of the nasal gland [Fange et al., 1958] notes that hypertonic solutions of sodium chloride are not the only stimuli that will elicit gland secretion. Hypertonic solutions of sucrose infused into cormorants, *Phalacrocorax auritus*, elicited a secretion which, while having a lower flow rate than that found after salt loads, had a similar concentration. A subsequent report on the herring gull indicated that the gland was under nervous control and that its innervation was of parasympathetic origin [Fange et al., 1958]. These findings suggest a receptor in the central nervous system but still leave open the question as to what the receptor was responding.

The results of the sucrose-loading experiment eliminate specific plasma levels of sodium or chloride as thresholds for stimulation. The results of the urea-loading experiments in the Adélie appear to eliminate a specific plasma osmotic pressure as the stimulus. Whereas infusions of hypertonic saline in amounts of 4.3 to 8.6 meq/kg initiated secretion, infusions of urea 8 times as great (33 to 64 meq urea/kg) did not cause secretion. These birds had been fasted only long enough to clear their alimentary canals. Rather frequent discharges after the urea load, some at rates as high as 30 g/hr/kg, rapidly brought them to a dehydrated condition. Although slight splattering from the beak was then observed, no nasal secretion could be collected.

The plasma osmotic pressure determinations confirm what the results of the experiments on the birds suggest. Specific levels of or specific changes in plasma osmotic pressure are not sufficient to elicit nasal gland secretion. Such levels or changes are possibly necessary conditions for the normal initiation of secretion. McFarland [1964] reports that a minimum sodium load of 8.4 meq/kg was required to initiate secretion in his apparently well-hydrated gulls. This load led to increased plasma sodium concentrations of about 20 meq/l, which compares well with the data of *Nechay et al.* [1960] indicating an increase of about 27 meq/l in the serum of the ring-billed gull, *Larus delawarensis,* necessary for secretion. A simple conversion of figures reveals that increases in serum sodium chloride of 27 meq/l correspond to increases in osmolality of about 50 mos/kg (milliosmoles per kilogram) from the salt. Probably dilution of other solutes by the fluid shifting from the intracellular to the extracellular compartment made the actual plasma osmotic pressure increase lower than this.

These results are in accord with *Gilman's* [1938] measurements of plasma osmotic pressure changes following the infusion of hypertonic salt and urea solutions into dogs. His experiments were designed to shed light on the question of the source of the urge to drink. Solutions of 20 per cent NaCl produced strong drinking responses, while an equivalent osmotic load given as 40 per cent urea did not. Dogs given urea drank much less than those given salt and consequently took a longer time to return to normal. The increases in plasma osmotic pressure were nearly equal in all the cases, but the serum specific gravity, a reflection of plasma protein concentration, fell after salt loading, whereas it showed no change following urea loading. This reduction after saline injection indicates a shift in fluid out of the cells and into the extracellular spaces. More recent studies using the same approach are in general agreement with Gilman's conclusions. *Wolf* [1950] found that cellular shrinkage calculated to be 1–2 per cent of the initial cell size led to drinking in man, dogs, and rats. *Fitzsimmons* [1964] prefers to express the drinking threshold in terms of effective osmotic pressure. He presents data indicating that increases as small as 1.6 per cent are sufficient to initiate drinking. Cellular shrinkages of 1 or 2 per cent are not difficult to envision in the salt-loaded bird. *McFarland* [1964] reports blood volume increases by as much as 80 per cent following injection of 8.4 meq sodium chloride in gulls. Most of this presumably came from the intracellular fluid compartment. Increases in effective osmotic pressure of only 1 to 2 per cent are readily understood, but such increases are so slight their direct measurement is very difficult.

Andersson [1953] reports experiments that localized the site of the receptor governing water intake in goats. Using 2 per cent sodium chloride solutions, he was able to elicit strong drinking responses with injections of about 100 μl into regions of the hypothalamus near the third ventricle and near the paraventricular nucleus. Injections into more lateral portions of the hypothalamus produced no effect on drinking.

The parallels between the stimulus to drink and the stimulus to secrete from the nasal gland suggest a receptor in the marine bird analogous to the thirst receptor in dogs and goats. The location of the receptor is unknown, although the results of *Fange et al.* [1958] indicate that it is in the central nervous system. The hypothalamus would not be an unlikely site. This receptor appears to be sensitive primarily to an osmotic gradient across its cell membranes. A clear understanding of what actually occurs when the receptor stimulates the nasal glands to secrete and when the glands cease secreting, even though the administered load has not been completely eliminated, awaits experimental clarification.

Acknowledgments. This research was carried out in the Department of Zoology, Duke University, Durham, North Carolina, under the direction of Dr. Knut Schmidt-Nielsen, and was supported under grants G-8953 and G-13593 from the National Science Foundation.

APPENDIX

TABLE A. Representative Values for the Rate and Composition of the Cloacal Discharge of Fasting and Non-Feeding Adult Adélie Penguins before Salt Loading

The first 11 birds in this table appear in Table B. The remaining 10 birds were subjects of other than salt-loading experiments.

Bird No.	Weight, kg	Time since Last Discharge, hr	Chloride Concentration, meq/l	Amount, g	Rate, g/hr/kg
1	3.56	—	102	5.5	—
2	3.29	1.0	35	5.1	1.6
		2.0	40	3.0	0.9
		1.0	38	5.2	1.6
3	3.43	2.0	23	3.0	0.4
4	3.77	0.5	10	0.5	0.3
		0.7	22	2.0	0.8
5	3.68	0.6	26	0.8	0.4
		0.3	29	1.6	1.5
6	4.83	0.3	95	0.8	0.5
7	6.05	0.5	11	1.3	0.4
		0.6	37	2.3	0.6
8	6.12	0.6	24	1.2	0.3
9	5.08	1.3	104	2.56	0.5
10	4.20	1.3	16	2.1	0.4
		0.3	78	2.6	1.9
		1.8	109	1.9	0.2
11	3.95	1.2	10	6.8	1.5

TABLE A. Representative Values for the Rate and Composition of the Cloacal Discharge of Fasting and Non-Feeding Adult Adélie Penguins before Salt Loading (*continued*)

The first 11 birds in this table appear in Table B. The remaining 10 birds were subjects of other than salt-loading experiments.

Bird No.	Weight, kg	Time since Last Discharge, hr	Chloride Concentration, meq/l	Amount, g	Rate, g/hr/kg
12	3.89	1.0	12	2.9	0.7
		1.0	16	3.6	0.9
		1.2	42	4.3	0.9
		0.5	40	2.8	1.6
13	4.17	1.0	26	5.1	1.2
		1.8	18	17.2	2.3
		1.0	15	6.1	1.5
		0.6	9	4.1	1.5
		1.0	25	7.9	1.9
14	3.45	1.0	41	0.4	0.1
		3.3	12	5.6	0.5
		2.0	10	1.6	0.2
		0.5	8	1.3	0.7
15	4.20	0.2	56	0.6	0.8
		0.5	93	0.9	0.5
		0.6	71	0.7	0.3
16	4.18	0.3	27	2.1	1.5
		1.0	28	4.7	1.1
		0.5	45	2.1	1.0
		0.5	45	1.7	0.8
		0.6	18	1.0	0.1
		0.3	87	1.8	1.3
17	3.60	0.6	94	2.6	1.1
		2.2	60	7.6	1.1
		3.1	50	11.1	1.1
18	3.79	—	13	6.0	—
		0.6	40	4.2	1.7
		0.8	41	9.4	3.3
		0.5	23	5.1	2.3
		1.4	12	4.1	0.8
		1.3	7	7.4	1.5
19	4.23	1.0	46	4.5	1.1
		1.0	66	8.1	1.0
		0.5	65	7.8	3.7
		0.5	37	5.4	2.5
		1.0	34	11.1	2.6
20	3.84	1.0	16	1.3	0.3
		1.0	9	3.9	1.0
		2.2	8	9.4	1.1
21	3.99	1.8	245	13.7	1.9
		0.5	264	9.2	4.3
		2.0	163	3.4	0.4
		2.0	119	6.4	0.8
		3.9	64	13.5	0.9
		2.5	106	4.7	0.4
		3.0	66	5.9	0.5
		1.0	64	2.9	0.7

TABLE B. Representative Values for the Rate and Composition of the Cloacal Discharge of Fasting and Non-Feeding Adult Adélie Penguins after Salt Loading

Bird No.	Weight, kg	Load	Time since Last Discharge, hrs	Chloride Concentration, meq/l	Amount, g	Rate, g/hr/kg
1	3.56	50 ml	1.2	160	27.3	6.5
		of	1.1	313	19.9	5.2
		3.5%	2.5	219	6.0	0.6
		(stom)	4.6	88	24.7	1.5
			11.0	44	17.6	
			1.0	53	12.6	0.7
2	3.29	7.5 ml	0.7	61	3.1	1.2
		of	2.0	66	5.6	0.8
		10%	1.3	61	2.8	0.6
		(iv)	11.0	98	25.3	0.6
3	3.43	7.5 ml	1.5	76	2.4	1.0
		of	1.5	78	2.7	0.5
		10%	1.0	48	3.6	1.0
		(iv)	0.8	43	5.7	2.0
			7.8	18	24.0	0.9
4	3.77	95 ml	0.6	11	2.7	1.0
		of	0.6	7	3.1	1.3
		3.5%	0.7	10	2.1	0.8
		(stom)	0.6	45	4.4	2.0
			0.8	54	3.7	1.2
			0.3	110	0.5	0.4
5	3.68	8.4 ml	0.4	153	1.1	0.7
		of	0.2	202	4.4	5.9
		20%	0.6	193	5.8	2.4
		(iv)	0.4	149	5.2	3.4
			0.8	110	6.5	2.4
			1.2	170	9.7	2.2
			1.0	44	5.6	1.5
6	4.83	20 ml	0.6	148	28.1	8.7
		of	0.1	145	11.2	28.9
		10%	0.3	309	6.1	5.1
		(iv)	0.4	315	3.9	1.9
			0.8	248	1.6	0.4
			2.0	56	4.8	0.5
			8.0	14	20.8	0.5
			2.0	18	7.6	0.8
7	6.05	13 ml	0.1	175	12.5	20.6
		of	0.1	148	44.5	56.3
		20%	0.2	135	44.0	33.1
		(iv)	0.1	129	21.4	29.5
			0.1	129	14.4	23.8
			0.3	128	11.9	6.8
			3.2	79	22.5	1.2
8	6.12	13 ml	0.4	127	10.9	4.8
		of	0.5	196	5.8	1.9
		20%	0.5	188	0.7	0.2
		(iv)	1.0	107	5.6	9.1
			1.0	78	10.4	1.7
			1.2	63	11.4	1.6
			1.1	28	9.3	1.4

TABLE B. Representative Values for the Rate and Composition of the Cloacal Discharge of Fasting and Non-Feeding Adult Adélie Penguins after Salt Loading (*continued*)

Bird No.	Weight, kg	Load	Time since Last Discharge, hrs	Chloride Concentration, meq/l	Amount, g	Rate, g/hr/kg
9	5.08	10 ml	0.5	155	15.4	6.1
		of	1.1	244	10.6	1.8
		20%	0.5	52	2.3	0.9
		(iv)	0.3	246	1.6	1.0
10	4.20	21 ml	1.0	8	2.4	0.6
		of	1.0	19	5.3	1.2
		10%	3.1	20	6.1	0.5
		(iv)	2.0	38	5.0	0.6
			2.0	41	5.2	0.6
			3.0	53	6.9	0.5
			2.9	40	12.5	1.0
			3.1	40	4.0	0.3
			1.4	32	5.3	0.9
11	3.95	56 ml	1.6	87	6.8	1.0
		of	2.5	48	2.7	0.4
		3.5%	1.9	32	5.7	0.7
		(iv)	7.8	27	17.7	0.6
			2.6	26	7.6	0.7
			1.5	25	4.9	0.8
			3.3	28	1.3	0.1

REFERENCES

Andersson, B., The effect of injection of hypertonic NaCl solutions into different parts of the hypothalamus of goats, *Acta Physiol. Scand., 28*, 188–201, 1953.

Bressler, E. H., Reabsorptive response of renal tubules to elevated sodium and chloride concentrations in plasma, *Am. J. Physiol. 199*, 517–521, 1960.

Brody, S., *Bioenergetics and Growth*, Reinhold Publishing Company, New York, 1945.

Cook, F. A., *Through the First Antarctic Night*, Doubleday and McClure, New York, 1900.

Cotlove, E., H. V. Trantham, and R. L. Bowman, An instrument and method for automatic, rapid, accurate, and sensitive titration of chloride in biologic samples, *J. Lab. Clin. Med., 51*, 461–468, 1958.

Davis, R. E., The nitrogenous constituents of hen urine, *J. Biol. Chem., 74*, 509–513, 1927.

Dawson, W. R., Relation of oxygen consumption and evaporative water loss to temperature in the cardinal, *Physiol. Zool., 31*, 37–48, 1958.

Douglas, D. S., Secretion of the thiocyanate ion by the nasal gland of the Adélie penguin, *Nature, 209*, 1150–1151, 1966.

Douglas, D. S., Low urine chloride concentrations in salt-loaded gulls (abstract), *The Physiologist, 9*, 171, 1966.

Eder, H. A., Determination of thiocyanate space, in *Methods in Medical Research*, Vol. 4, edited by M. B. Visscher, pp. 48–53, Yearbook Publishers, Chicago, 1951.

Eklund, C. R., and F. E. Charlton, Measuring the temperatures of incubating penguin eggs, *Am. Scientist, 47*, 80–86, 1959.

Fange, R., K. Schmidt-Nielsen, and H. Osaki, The salt gland of the herring gull, *Biol. Bull., 115*, 162–171, 1958.

Fange, R., K. Schmidt-Nielsen, and M. Robinson, Control of secretion from the avian salt gland, *Am. J. Physiol., 195*, 321–326, 1958.

Fitzsimmons, J. T., The effects of slow infusions of hypertonic solutions on drinking and drinking threshold in rats, *J. Physiol., 167*, 344–354, 1964.

Flemister, L. J., Salt and water anatomy, constancy and regulation in related crabs from marine and terrestrial habitats, *Biol. Bull., 115*, 180–200, 1958.

Gilman, A., The relation between blood osmotic pressure, fluid distribution and voluntary water intake, *Am. J. Physiol., 120*, 323–328, 1937.

Goldsmith, R., W. J. L. Sladen, Temperature regulation of some antarctic penguins, *J. Physiol., 157*, 251–262, 1961.

Hart, W. M., and H. E. Essex, Water metabolism of the chicken with special reference to the role of the cloaca, *Am. J. Physiol., 136*, 657–668, 1942.

Hester, H. R., H. E. Essex, and F. C. Mann, Secretion of urine in the chicken (*Gallus domesticus*), *Am. J. Physiol., 128*, 592–602, 1939.

Hokin, L. E., and M. R. Hokin, Evidence for phosphatidic acid as the sodium carrier, *Nature, 184*, 1068–1069, 1959.

Holmes, W. N., J. G. Phillips, and D. G. Butler, The effect of adrenocortical steroids on the renal and extrarenal responses of the domestic duck (*Anas platyrhynchus*) after hypertonic saline loading, *Endocrinol., 69*, 483–495, 1961.

Kleiber, M., *The Fire of Life: An Introduction to Animal Energetics*, 454 pp., John Wiley and Sons, New York, 1961.

Korr, I. M., The osmotic function of the chicken kidney, *J. Cell. Comp. Physiol., 13*, 175–194, 1939.

Krogh, A., *Osmotic Regulation in Aquatic Animals*, Cambridge University Press, London, 1939.

Levitt, M. F., and M. Gaudino, Measurement of body water compartments, *Am. J. Med., 9*, 208–215, 1950.

Maroney, M. J., *Facts from Figures*, Penguin Books, London, 1956.

Mayrs, E. B., Secretion as a factor in elimination by the bird kidney, *J. Physiol., 58*, 276–287, 1924.

McFarland, L. Z., Minimal salt load required to induce secretion from the nasal salt glands of sea gulls, *Nature, 204*, 1202–1203, 1964.

Nechay, B. R., J. L. Larimer, and T. H. Maren, Effects of drugs and physiologic alterations on salt gland excretion in sea gulls, *J. Pharmacol. Expmtl. Therap., 130*, 401–410, 1960.

Nutritional Data, fifth edition, Heinz Nutritional Research Division and Mellon Institute, H. J. Heinz Co., Pittsburgh, 1963.

Orloff, J., M. Kahan, and L. Brenes, Renal tubular effects of ammonium salts on electrolyte transport, *Am. J. Physiol., 201*, 747–753, 1961.

Sapin-Jaloustre, J., L'établissement de la thermorégulation chez le Manchot Adélie, *C. R. Acad. Sci., 237*, 1443–1444, 1953.

Sapin-Jaloustre, J., *Ecologie du Manchot Adélie*, 211 pp., Hermann, Paris, 1960.

Schmidt-Nielsen, K., Extrarenal excretion of salt in birds, *J. Elisha Mitchell Scientific Soc., 73*, 235, 1957.

Schmidt-Nielsen, K., and R. Fange, The function of the salt gland in the brown pelican, *Auk, 75*, 282–289, 1958.

Schmidt-Nielsen, K., C. B. Jorgensen, and H. Osaki, Extrarenal salt excretion in birds, *Am. J. Physiol., 193*, 101–107, 1958.

Schmidt-Nielsen, K., A. Borut, P. Lee, and E. Crawford, Jr., Nasal salt excretion and the possible function of the cloaca in water conservation, *Science*, *142*, 1300–1301, 1963.

Sharpe, N. C., On absorption from the cloaca in birds, *Am. J. Physiol.*, *66*, 209–213, 1923.

Skadhauge, E., and B. Schmidt-Nielsen, Cloacal storage and modification of urine in the fowl (abstract), *Fed. Proc.*, *24*, 643, 1965.

Sladen, W. J. L., *The Pygoscelid Penguins*, Falkland Islands Dependencies Survey, Report 17, 97 pp., Her Majesty's Stationary Office, London, 1958.

Vinogradov, A. P., *The Elementary Chemical Composition of Marine Organisms*, translated by J. Efron and J. K. Setlow, 647 pp., Sears Foundation for Marine Research, Yale University, Memoir 2, Copenhagen, 1953.

Watson, M., Report on the Spheniscidae, in *Report on the Scientific Results of the Exploring Voyage of H. M. S. Challenger, 1873–1876*, Vol. 7, Her Majesty's Stationary Office, London, 1883.

Weyrauch, H. M., and S. I. Roland, Electrolyte absorption from the fowl's cloaca: resistance to hyperchloremic acidosis, *J. Urology*, *79*, 225–263, 1958.

Wolbach, R. A., Renal regulation of acid-base balance in the chicken, *Am. J. Physiol.*, *181*, 149–156, 1955.

Wolf, A. V., Osmometric analysis of thirst in man and dog, *Am. J. Physiol.*, *161*, 75–86, 1950.

FEEDING PREFERENCES OF THE ADÉLIE PENGUIN AT CAPE CROZIER, ROSS ISLAND

WILLIAM B. EMISON

Department of Pathobiology, Johns Hopkins University, Baltimore, Maryland

Abstract. Field studies on the feeding preferences of the Adélie penguin in the Ross Sea region of the Antarctic were carried out during the austral summers of 1964–1965 and 1965–1966. A technique was devised whereby food could be removed from the stomachs without harming the birds, thus making possible the collection of a large number of samples. Stomach contents collected from 201 parents feeding chicks at the Cape Crozier rookery and from 6 parents at the Beaufort Island and Franklin Island rookeries were preserved and brought to the United States for analysis.

The two common food organisms were *Euphausia crystallorophias* and a small shoaling fish, *Pleuragramma antarcticum*. Also identified from the stomach contents, but incidental in the diets, were *Euphausia superba*, two fish species of the family Chaenichthyidae, and 14 species of amphipods.

By number of organisms, the diets of the Cape Crozier Adélies were composed of 91–95% euphausiids, 4–8% fishes, and less than 2% amphipods; by volume they were about 60% euphausiids, 39% fishes, and 1% amphipods. The stomach contents collected at Beaufort Island and Franklin Island had about the same species composition, but the per cent compositions of the common food groups differed.

The Adélies nesting in the Ross Sea region fed on small marine organisms that swarm in the upper layers of the water. The birds selected organisms over 15 mm in length.

INTRODUCTION

Adélie penguins (*Pygoscelis adeliae*) are an important trophic component of the Ross Sea marine community. During the chick feeding period, Adélies make constant trips to sea and return to their land rookeries with semidigested food to feed their chicks.

Although considerable literature relates to the ecology of the Adélie penguin in its terrestrial environment, very few studies have as yet been made of its relation to its marine environment. Yet a minimum of 75% of its adult life is spent in this environment, either directly in the water or on the pack ice that rims the antarctic continent. In fact Adélie penguins should be considered essentially aquatic animals that return to land only to breed.

The southern oceans comprise an extremely productive area, and some investigators consider it the most fertile oceanic region of its size in the world. Studies have shown that the standing crop of zooplankton is at least 4 times as great as that in the tropics [*Foxton*, 1956]. Within this rich marine environment the Adélie penguins are important consumers, and a main portion of this study examines the relationship between the Adélie and the marine organisms upon which it preys. This was done by collecting and analyzing food being brought from the sea by parents at Cape Crozier, Antarctica, from December 1, 1964, to January 7, 1965, and from December 10, 1965, to February 5, 1966.

Most of the chicks at Cape Crozier hatched during the second week of December, and most left the rookery during the first week of February. Most of the collecting was done during this 8-week period. This portion of their breeding cycle is summarized as follows from *Sladen* [1958]. A successful pair of Adélies normally feed only their own young and may rear either one or two chicks. During the guard stage, which spans the first three or four weeks after hatching, one parent remains feeding and guarding the chick(s) while the other is at sea collecting food. They exchange roles when the parent returns from the sea. At the end of the guard stage the chicks are left alone while both parents feed independently at sea. This, termed the creche stage, is characterized by small groups (creches) of chicks awaiting the return of their parents with food. At the end of this stage the creches disperse (dispersal stage), but the chicks remain on or near their nest sites for several days before they leave the rookery for sea, independent of their parents.

The Study Area

Two Adélie rookeries lie next to each other about 8 kilometers northwest of Cape Crozier (77°29′S, 169° 34′E), which forms the eastern end of Ross Island (Figure 1). The western rookery, which covers about 1.6 kilometers of coastline and extends inland almost 1.2 kilometers, is situated on windswept lava slopes and reaches an elevation of 135 meters at its inland extremities. About 150,000 pairs of Adélies (Watson, personal communication) breed in this rookery, which they reach from three beaches separated by lava cliffs jutting abruptly out of the sea. The eastern end is bordered by an overhanging ice cliff about 30 meters high, from which a region of snow- and ice-covered cliffs stretches another 500 meters. At the eastern end of these cliffs lies the eastern rookery, extending 500 meters along the shore and inland less than 400 meters. This rookery, which contains some 25,000 pairs of breeding Adélies (Watson, personal communication), is situated on flat, lava-strewn ground less than 30 meters above sea level, and it has only a single beach. Both rookeries are composed of many distinct colonies, ranging in size from fewer than ten to several thousand nests. These occupy well-drained, snow-free sites and are easily defined by the accumulations of guano built up over the years.

In addition to the major study conducted entirely at the western Cape Crozier rookery, a few stomach samples were collected at the Beaufort Island (76°55′S, 167°05′E) and Franklin Island (76°07′S, 168°20′E) rookeries (Figure 1).

Definitions

Stomach samples. Collective term for complete and partial stomach contents collected from individual birds.

Complete stomach sample. Bird killed and all contents of stomach removed and preserved.

Partial stomach sample. A portion of the contents removed and the bird released alive.

Peak of chick hatching. First date on which the nests examined (50 nests selected at random per day during chick hatching) contained more chicks than eggs.

Peak of chick departure. First date on which 50% or more of the chicks had departed the rookery (estimated).

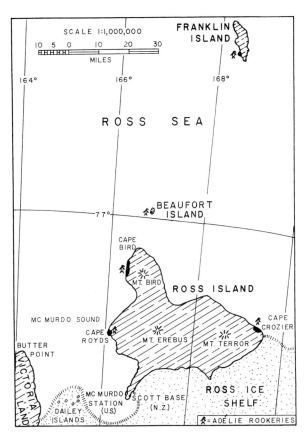

Fig. 1. Ross Island, Beaufort Island, and Franklin Island Adélie penguin rookeries.

Chick-rearing period. Period when adults are feeding chicks, i.e., guard and creche stages.

Feeding trip. The journey of a parent from the rookery to sea, where it catches and retains food in its stomach, and back again to the rookery, where it regurgitates the gathered food to its chicks.

METHODS

Field Methods

Stomach samples were collected from 207 adult Adélie penguins during the 1964–1965 and 1965–1966 chick-rearing periods.

Complete stomach samples were removed from 37 Adélies. Figure 2 shows that 31 of these were collected at the Crozier rookery; of the remaining 6 (not shown in Figure 2), 5 were collected at the Beaufort Island rookery on January 19, 1966, and 1 in the pack ice near the Franklin Island rookery on January 12, 1966.

The birds were collected either at their nest sites or on the rookery beaches. Most of the Adélies collected on the beaches were injured either by leopard

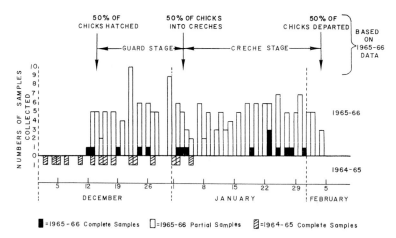

Fig. 2. Distribution of stomach samples collected at Cape Crozier.

seals (*Hydrurga leptonyx*) or by being caught between pieces of wave-tossed pack ice while attempting to land.

Partial stomach samples were collected from 170 Crozier Adélies feeding chicks during the 1965–1966 chick-rearing period (Figure 2). The sampling technique by which small amounts of food were collected without harming the bird is described below.

The material collected was preserved in 8% formalin and later placed in 70% alcohol in the laboratory.

All complete stomach samples taken were weighed to the nearest gram before being placed in preservative.

All partial stomach samples were placed in preweighed plastic bottles of preservative and were reweighed at the end of a collecting day. The difference between the two weights was taken as the weight of the sample.

During the guard stage, data on weight of food brought into the rookery by parents feeding chicks were obtained by capturing an adult fresh from the sea on its nest before it had begun to feed the chick(s). The parent was weighed in a canvas restraining device suspended from a spring scale hung in a large crate that protected it from the wind on three sides; accuracy was to the nearest 15 grams. The bird was then marked, released, and watched until it returned to its nest. It was then reweighed at varying intervals (6–12 hours), before being relieved by its mate. The difference between the first and second weighings was usually large, indicating most of the food had been regurgitated to the chicks. The final one or two weighings before nest relief (usually 1–2 days after it had taken over the nest), usually showed a leveling off of weight loss. The difference between the first and last weighings was taken as the amount of food fed to the chicks.

During the creche stage, chicks (and not adults) were caught and weighed just before being fed and then immediately released. When the chick(s) were captured, the parent was marked to prevent confusion with other adults in the colony. After feeding was observed, the chick(s) were weighed again as soon as the adult left the colony. Observation was continued, because the adult often did not leave the rookery but returned shortly to feed the chicks again. The chicks were then weighed again. The gain in weight between the first and last weighing of the chick(s) was taken as the amount of food brought in.

Analysis of Stomach Contents

Laboratory analysis of the preserved contents was both qualitative and quantitative. The identifications of all organisms reported in this paper were either made or confirmed by a specialist in the particular group concerned (see acknowledgments). Usually these identified only a few specimens of each species and, unless stated otherwise, all subsequent work and conclusions drawn are entirely the responsibility of the author.

The number of individuals in each major group of

TABLE 1. Diagnostic Parts Used to Determine Number of Individuals in Each Major Group of Organisms

Organisms	Diagnostic Parts
Euphausiacea	Eyes
Fishes	Eye lenses
Amphipods	Exoskeletons
Copepods	Intact organisms
Cephalopods	Beaks
Marine algae	Fragments
Hydrozoa	Fragments
Ectoparasites (of fishes)	Exoskeletons
Endoparasites (of fishes)	Intact organisms

organisms in the samples was determined by counts of diagnostic parts (Table 1). Complete counts of all organisms were made of the 170 partial stomach samples and of 10 of the smallest complete stomach samples collected. The remaining 27 complete stomach samples were rough sorted, i.e., all except the extremely abundant organisms (euphausiids and small fishes) were removed and counted. Then 1/3 or 1/2 of the remaining contents by weight (preservative poured off) were separated. Complete counts were made of this portion and the final results corrected for the entire sample.

Volumetric measurements of organisms in the 15 complete samples collected during 1964–1965 was done by dividing the remains of the organisms into major groups, allowing them to dry partially for 3 to 4 hours, and measuring their displacement of water.

Partial Stomach Sampling Technique

The laboratory sorting of the 15 complete stomach samples collected at Cape Crozier during 1964–1965 revealed some individual variation in the percentages of food items in each stomach. Obviously smaller samples taken from a large number of Adélies feeding chicks provide more precise data for the rookery as a whole than large complete samples from a few adults and, if taken regularly throughout the chick-rearing period, will also show whether the percentages of the common food items change appreciably during the season. With these points in mind a sampling technique was devised in 1965–1966 for removing small amounts of food from the stomachs quickly without harming the birds. The technique was modified from a method used by D. H. Thompson during 1964–1965 to remove adult Adélie stomach contents to feed a chick he was hand-rearing (personal communication).

Apparatus. The sampling device (Figure 3) consists of a hollow plexiglass tube 60 cm long, 2.3 cm in inside diameter, and 2.5 cm in outside diameter, a rubber stopper 2.3 cm in diameter, and a piece of wire more than 60 cm long. The plexiglass tube is marked at measured intervals so that its depth in the bird can be determined when suction is applied, and its ends are filed to remove any rough edges. A wire loop in the stopper is easily engaged or disengaged by a hook in one end of the long piece of wire.

Operating procedure. When an adult fresh from sea is noted entering the colony and either going into a nest relief ceremony with its mate or a feeding chase with its chicks (showing it to be a parent just return-

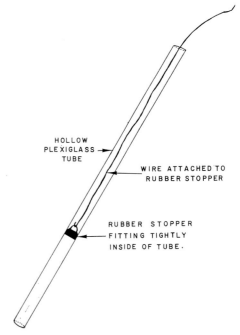

Fig. 3.   Adélie penguin stomach sampling device.
(For dimensions see text.)

ing to feed its chicks), it is caught with a hand net and taken a short distance away from the nest to avoid further colony disturbance.

The person sampling sits on the ground with the Adélie held firmly between his knees, opens its beak with the fingers of the left hand, and inserts the sampling tube into the oral cavity with the right hand. Care must be taken at this point that the tongue is under the tube and not sticking into it. Sliding the left hand down the back of the head and grasping the neck at the base of the skull, he maneuvers the tube farther into the oral cavity with the right hand (Figure 4A). When the end of the tube reaches the entrance to the esophagus, resistance from the bird is usually apparent, and at this stage it is critical that no force be used. Usually within a few seconds the bird swallows and the tube starts down the esophagus. It is essential that the left hand maintain upward pressure on the head to keep the neck stretched and the journey of the tube down the esophagus as straight as possible (Figure 4B). Usually with a little pressure the tube slides quickly down the esophagus and into the contents of the stomach. If resistance is met, the bird generally relaxes within a few seconds if no force is used. When the desired depth is attained, as determined by the mark on the tube at the bird's gape, the left hand slides up the bird's head and grasps the tube

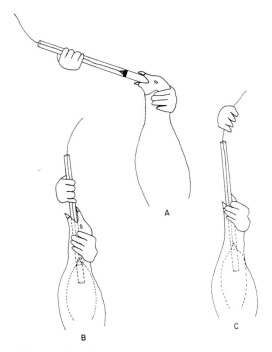

Fig. 4. Steps in obtaining partial stomach samples from Adélie penguins (see text).

between the thumb and forefinger at the gape, the rest of the hand still applying gentle pressure to keep the head and neck upstretched. With the fingers of the left hand holding the tube down, the right hand grasps the wire at the free end of the tube and pulls the rubber stopper slowly up the tube (Figure 4C), sucking stomach contents into it. This accomplished, the left hand again grasps the neck region and the right quickly removes the tube from the bird. The stomach contents sucked into the tube are mixed with mucous and in a semi-solid state; therefore they remain in place within the tube.

The wire and stopper are then pulled out of the tube and the stomach contents allowed to drop into a pre-weighed plastic bottle containing 8% formalin. The tube is cleaned by wiping the outside with a dry paper towel and running another paper towel through the inside with the long wire disconnected from the stopper. The bird is then marked on the breast with flomaster ink and released.

Each Adélie sampled was watched and notes taken on its behavior after its release. Usually the bird returned to its nest site and within a few minutes began feeding its chicks. The ink on the breast remained for the entire 8 weeks; thus surveillance of the rookery where the samples were being collected would have revealed any unnatural behavior by the sampled birds.

The marked birds were observed constantly coming and going throughout the 8 weeks with no indication of any change in behavior.

Blood was observed inside the tube after its removal from the stomachs of two parents; both were individually marked before being released. Both birds were seen feeding chicks at their nest sites 3 and 6 weeks later. The chicks were of normal size and apparently had been fed regularly during this time. No blood was observed inside the tube of the other 168 Adélies sampled, but occasionally streaks of blood observed on the outside of the tube came from mouth lesions caused by the bird biting the tube during insertion and pinching papillae in the oral cavity between the mandibles and the tube.

Quantitative Results

Weights. Weights and dates of the 37 complete stomach samples and 46 weighings of adults and chicks obtained are shown in Figure 5. The 10 weights obtained before the peaks of chick hatching were all complete samples removed from parents with only eggs in their nests. All the other complete content weights obtained were either from parents with chicks in their nests or from injured birds collected on the beaches. The stomach content weights of the 10 parents incubating eggs were lower than those taken later, but no significant changes are apparent in weights of contents taken throughout the chick-rearing period.

Several factors were probably involved in the extremely wide range of weights obtained after hatching; the most important seemed to be the degree of digestion of the contents. Other factors were: (1) individual variation, probably to some extent related to age and sex of the parent; (2) differences in the weighing and sampling methods; (3) species composition of the contents (stomachs with a large proportion of fishes tended to be heavier than those containing mostly crustaceans); and (4) number of chicks being fed (parents feeding two chicks brought in 100–200 grams more food on the average than those feeding one).

The Beaufort and Franklin Island weights correspond closely to the Crozier weights taken at about the same time.

The relatively low weights of the 1964–1965 Crozier stomach samples after chick hatching were possibly a result of the pack ice moving in and out of the Crozier vicinity until January 3, forcing the parents to travel over several miles of broken-up ice to reach the rookery on some days.

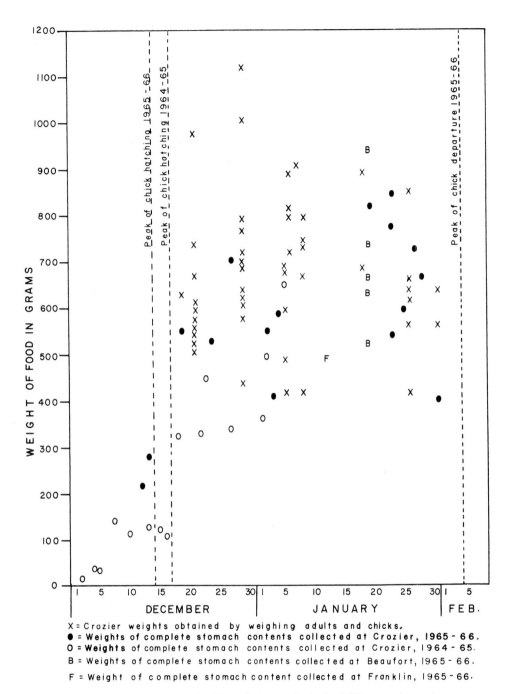

Fig. 5. Weight of food brought into rookeries by Adélie penguins.

The partial samples collected from the 170 Crozier adults feeding chicks during 1965–1966 average 45 grams per sample.

Numbers. The average numbers of food organisms in relation to weights of stomach contents are shown in Table 2. These increase as the stomach content weight increases up through the 301–400 grams weight range. Variations in the heaviest stomachs reflect the degree of digestion of the contents. The counts of food organisms were, for the most part, based on hard parts that are little altered by digestion in the stomach (Table 1) ; therefore a relatively accurate count of individuals is possible even though the bulk of the organisms has been partly digested and absorbed. The

TABLE 2. Numbers of Food Organisms in Relation to Weights of 37 Complete Stomach Contents Collected

Range of Stomach Content Weights, g	Number of Stomachs Examined	Average Number of Food Organisms per Stomach
0–50	3	343
51–100	–	–
101–150	5	1528
151–200	–	–
201–300	2	2609
301–400	4	3656
401–500	5	5737
501–600	7	4716
601–700	4	5540
701–800	4	6659
801–900	2	9845
901–1000	1	5632

farther advanced the digestion, the higher the number of individuals present is likely to be in a sample of a given weight. The large average number of food organisms in the 801–900 grams weight range is a result of a single stomach from an injured Crozier male that contained 12,500 food organisms, most well digested euphausiids. This is nearly 4000 more organisms than were found in any of the other stomachs examined.

The partial stomach contents collected from the 170 Crozier Adélies averaged 382 organisms per sample.

Table 3 shows the abundance and relative importance in the diet of the items identified from the stomach contents. Only the organisms in group 1 are manifestly of importance in the Adélies' diet. Group 2 organisms constitute by number less than 0.1% of all food items taken, and those in group 3 are not considered food.

The numbers of Euphausiacea a pair of successful Adélie parents removed from the Ross Sea to feed their young at Cape Crozier was estimated by multiplying the average number of euphausiids in each complete sample (5,300) by the average number of feeding trips the parents made. Observations of banded and marked parents at marked nests showed that in rearing either one or two chicks successfully each pair of parents made between 35 and 40 feeding trips during the chick-rearing period. Thus each successful pair of Crozier parents removed somewhere between 180,000 and 200,000 euphausiids from the Ross Sea during the 1965–1966 season.

Per cent compositions. Figure 6 shows the mean per cent compositions by number (computed by finding per cent composition of each sample and then determining means for total samples collected) of Euphausiacea, fishes, and Amphipoda in the different groups of samples collected. The partial and complete stomach samples collected at Crozier during the 1965–1966 season were combined after the mean per cent differences between each of the three principal food organisms proved not significant statistically at the 95% level (Table 4).

Table 4 also shows that while the mean per cent differences of the Euphausiacea and fishes between the 1964–1965 and 1965–1966 Crozier samples were statistically significant at the 95% level, the difference between the amphipods was not. Statistical comparison of the 1965–1966 Crozier and 1965–1966 Beau-

TABLE 3. Abundance and Relative Importance of Items Found in Adélie Stomach Contents

	Organisms	Number Counted from All Samples
Group 1, abundant and considered principal food organisms in the Adélie diet	Euphausiacea*	>200,000
	Fishes*	>17,000
	Amphipoda	>3,000
Group 2, rare and not considered important in the Adélie diet	Copepoda	24
	Cephalopoda	2
	Hydrozoan fragment	1
	Marine alga fragment	1
Group 3, found in stomachs, but not considered food items in the Adélie diet	Nematoda†	Common
	Cestoda†	Common
	Isopoda†	2
	Stones‡	Common
	Egg shell fragments‡	3
	Bones‡	1

* Estimate based on sub-sample counts.
† Parasites of fishes.
‡ Probably consumed in rookery.

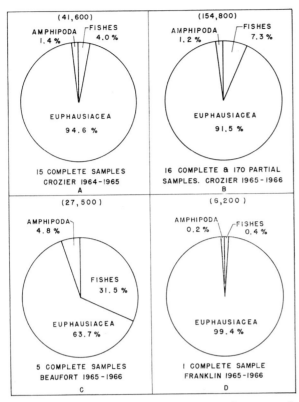

Fig. 6. Composition of Adélie penguin stomach contents by numbers of organisms. Figures in parentheses indicate the total number of organisms.

the season the Euphausiacea preponderate in the stomach contents with little variation from week to week.

The per cent composition by volume was determined for the 15 complete stomach contents collected at Crozier during the 1964–1965 season (Figure 8). These measurements, although rough because of the rapid digestion of the fishes, show the importance of the less numerous, but larger fishes in the diets. The fishes comprise a minimum of 39% of the contents by volume, as compared with only 4.0% by number. Conversely, the Euphausiacea comprise only 60% by volume and 94.6% numerically.

Qualitative Results

Euphausiacea. The two identified species, *Euphausia crystallorophias* Holt and Tattersall, and *E. superba* Dana were separated by dissecting-scope examination of the antennular peduncles. *E. superba* has a dorsal lobe at the distal end of the first segment of the peduncle, which *crystallorophias* lacks [*John*, 1936].

It was possible to identify between 100 and 150 euphausiids selected at random from each of 28 of the 36 complete stomach samples collected at the Crozier and Beaufort rookeries. The remaining 8 samples were well digested and contained only from 10 to 78 intact and identifiable euphausiids. From the single complete sample collected at Franklin Island 196 euphausiids were identified. Between 5 and 30 euphausiids selected at random from each of 166 (out of 170) of the partial stomach samples were identified; the remaining 4 partial samples contained no intact specimens. These findings are tabulated in Table 5.

In addition to the identifications of the random samples, the remaining euphausiids in the 37 complete stomach contents were sorted grossly by picking out and identifying all very large individuals (usually *E.*

fort samples shows the differences not significant at the 95% level. The sample size (5) from Beaufort was inadequate, and the large, but not statistically significant, differences show the need for more comparison of the foods at different Adélie rookeries.

Figure 7 shows the mean per cent compositions by number of the samples collected each week during the 1965–1966 Crozier chick-rearing period. Throughout

TABLE 4. Comparative Compositions of Principal Food Organisms in Different Groups of Samples Collected

Samples Compared	Number of Samples	Euphausiacea		Fishes		Amphipoda	
		Mean %	P*	Mean %	P*	Mean %	P*
Cape Crozier partial 1965–1966	170	91.5		7.4		1.1	
versus			>.05		>.05		>.05
Cape Crozier complete 1965–1966	16	91.7		6.6		1.7	
All Cape Crozier samples 1965–1966	186	91.5		7.3		1.2	
versus			<.05>.01		<.05>.01		>.05
All Cape Crozier samples 1964–1965	15	94.6		4.0		1.4	
All Cape Crozier samples 1965–1966	186	91.5		7.3		1.2	
versus			>.05		>.05		>.05
All Beaufort Island samples 1965–1966	5	63.7		31.5		4.8	

* Probability of the observed difference between samples occurring by chance (*t* test).

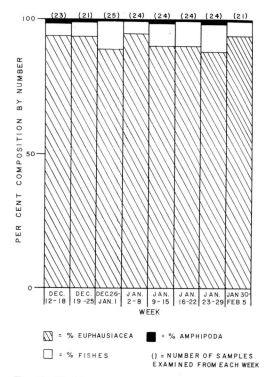

Fig. 7. Weekly composition of stomach contents by number of major food organisms during the 1965–1966 chick-rearing period at Cape Crozier.

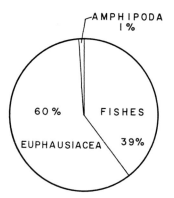

Fig. 8. Composition by volume of 15 complete stomach samples collected at Cape Crozier 1964–1965.

superba). The two methods yielded a minimum per cent frequency of occurrence for both species shown in Table 6.

Measurements were made in conjunction with the identifications. As some identified specimens were broken in half and could not be measured, the number of measurements is less than the number of individuals identified. Because of frayed and missing telsons on many individuals, all euphausiid measurements were made from the anterior tip of the rostrum to the distal tip of the pleon. Most workers use the distal tip of

the telson rather than the distal tip of the pleon, so my measurements for *E. crystallorophias* are 2 to 4 mm shorter than those of other publications.

Figure 9A shows the sizes of all *E. crystallorophias* and *E. superba* measured in this study. Figure 9B compares sizes of *E. crystallorophias* consumed during the first 4 weeks of the 1965–1966 chick-rearing period with those consumed during the last 4 weeks at Cape Crozier. Figure 9C compares sizes taken during comparable time periods in 1964–1965 and 1965–1966 at Crozier. Figure 9D compares lengths of *E. crystallorophias* consumed by the Crozier Adélies during the last 4 weeks of the 1965–1966 chick-rearing period with those consumed by the Beaufort Adélies on January 19, 1966.

Figure 9A shows that the size range of *E. superba* taken by the Adélies during December, January, and the first part of February was 24–36 mm. To compare these sizes with *E. superba* growth rates given by *Bargmann* [1945] and *Marr* [1962], they should be increased by 4–5 mm to allow for the missing telson in

TABLE 5. Relative Abundance of Euphausiacea in Adélie Penguin Stomach Contents

	Cape Crozier 1964–1965, 15 Complete Samples	Cape Crozier 1965–1966, 166 Partial Samples	Cape Crozier 1965–1966, 16 Complete Samples	Beaufort Island 1965–1966, 5 Complete Samples	Franklin Island 1965–1966, 1 Complete Sample
Number of euphausiids examined	1326	2164	1725	409	196
Average number per sample	88	13	108	82	196
Number and per cent of *E. crystallorophias*	1322, 99.7	2147, 99.2	1723, 99.9	407, 99.5	193, 98.5
Number and per cent of *E. superba*	4, 0.3	17, 0.8	2, 0.1	2, 0.5	3, 1.5

TABLE 6.   Frequency of Occurrence of Euphausiacea in 37 Complete Stomach Contents

	Cape Crozier 1964–1965, 15 Complete Samples	Cape Crozier 1965–1966, 16 Complete Samples	Beaufort Island 1965–1966, 5 Complete Samples	Franklin Island 1965–1966, 1 Complete Sample	Totals
Number and per cent of samples with *E. crystallorophias*	15, 100	16, 100	5, 100	1, 100	37, 100
Number and per cent of samples with *E. superba*	3, 20	5, 31	2, 40	1, 100	11, 30

my measurements. From the growth rates given by these other workers, the individuals consumed by the Adélies are between 1 and 2 years of age, which *Bargmann* [1945] considers to be adolescents that will not spawn until about 2 years old. The slight increase in lengths of *E. crystallorophias* consumed during the last 4 weeks of the 1965–1966 Crozier chick-rearing period over those consumed during the first 4 weeks as shown in Figure 9B probably reflects the normal monthly growth increment for this time of year. Figure 9C shows that the majority of *E. crystallorophias* taken during 1964–1965 measured between 14 and 19 mm in length, whereas the majority of those taken during a comparable period in 1965–1966 were slightly larger, measuring between 14 and 23 mm in length. This variation was also shown when the 1965–1966 lengths of *E. crystallorophias* taken by the Crozier Adélies were compared with those taken by the Beaufort Adélies, which were also consuming an obviously smaller size group measuring 9–12 mm.

Fishes.   The three species present, *Pleuragramma antarcticum,* Boulenger (Family Nototheniidae), and *Chaenodraco wilsoni,* Regan, and *Chionodraco kathleenae,* Regan (Family Chaenichthyidae), when more than 75 mm in length could usually be identified by diagnostic skeletal parts in the stomach contents. Fishes less than 75 mm in length were usually well digested, with only the eye lenses remaining unaltered. Occasionally the rostrums, skulls, and body segments were also recovered in the top portion of some samples. H. H. DeWitt (personal communication) examined a number of the best specimens and felt they were mostly *P. antarcticum* and an occasional chaenichthyid. When rostrums were found, the differences between the nototheniid *P. antarcticum* and the chaenichthyids were very obvious, even in the extremely small individuals, but as no rostral spines could be distinguished on the small chaenichthyid specimens, further identification was impossible.

Because the small fishes (less than 75 mm in length)

digested rapidly, it was difficult to obtain large numbers of them for examination. A few relatively intact specimens were found in each of 5 of the Crozier samples collected in 1964–1965, 41 of the Crozier samples collected in 1965–1966, and in all 5 of the Beaufort Island samples collected in 1965–1966; the single Franklin Island sample did not contain any intact small fishes. Table 7 shows that 90% of the small fishes identified were *P. antarcticum;* the remaining 10% were Chaenichthyidae.

Table 8 shows the relative abundance of the fishes more than 75 mm in length. The Chaenichthyidae could not be separated without the rostrums, which were missing in about ⅔ of the cases; these appear under "Chaenichthyidae spp." in the table. The "unknown" column lists an excess of large eye lenses over corresponding skeletal parts.

Of the 37 complete samples, all but 2 contained fish remains; the 2 exceptions were from Crozier adults collected before the 1964–1965 chick-hatching peak, and both contained relatively small amounts of food (15 and 32 grams). Because of the well digested condition of most of the small fishes, no attempt has been made to determine per cent frequency of occurrence by species, although the preponderance of *P. antarcticum* in the few samples examined (Tables 7 and 8) indicates that this is the species Adélies usually feed on.

To obtain a rough estimation of lengths of fishes taken by the Adélies, the most intact specimens were measured and their lengths correlated with the diameters of their eye lenses. Measurements of the lenses found in the contents have been divided into two size groups (Table 9). About 98% of the fishes taken by the Adélies were less than 75 mm in length (eye lense diameter less than 2 mm). The remaining 2% over 75 mm in length (eye lense diameter over 2 mm) ranged up to about 200 mm in length (diameter 5–6 mm).

Amphipoda.   Table 10 lists the 14 species of 5 families found in the stomach contents. A few specimens of each species were identified by specialists (see

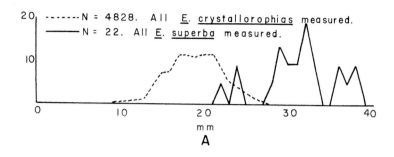

A

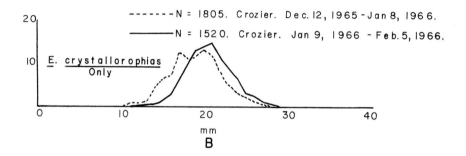

B

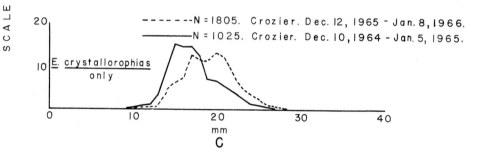

C

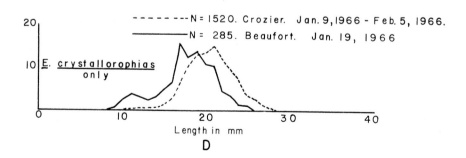

D

N = Number of euphausiids measured in obtaining length frequency distribution in each group of samples.

Fig. 9. Frequency distribution by lengths of euphausiacea consumed by Adélie penguins (see text).

TABLE 7.   Relative Abundance of Small* Fishes Found in Adélie Penguin Stomach Contents

	Cape Crozier 1964–1965	Cape Crozier 1965–1966†	Beaufort Island 1965–1966	Totals
Number and per cent of *Pleuragramma antarcticum*	16, 100	127, 91	10, 67	153, 90
Number and per cent of Chaenichthyidae	–	13, 9	5, 33	18, 10
Totals for each group of samples taken	16	140	15	171

* Less than 75 mm in length.
† Complete and partial samples combined.

acknowledgments). Subsequent identifications and counts were made by comparisons with the identified specimens and from distinctive features given in the literature.

Total counts made of all amphipods present in the Adélie stomach contents (based on heads when the animals were fragmented) are shown in Table 11 (1965–1966 partial and complete samples are combined in column 2). *Orchomenella plebs* and *O. rossi* are very similar, and the individuals lacking the distinguishing features are lumped as *Orchomenella spp.* *Uristes murrayi* were usually found broken into two pieces (only two individuals were found intact). Both segments could usually be found in the contents, but in some cases only the front portion was recovered. I have placed these under *Uristes sp.*

The genus *Orchomenella* was the most important group of amphipods the Adélie penguins took at the three Ross Sea rookeries. Nearly all of these were either *O. plebs* or *O. rossi*. The only other orchomenelid present was *O. pinguides*, of which a single individual was in each of 2 partial samples. As both these samples also contained specimens of *Paramoera walkeri*, a littoral species that was present in only 3 partial samples, it suggests that *O. pinguides*, at least in the 6–7 mm size range, may also be littoral in habit. (See Figure 10.)

The genera *Uristes*, *Cheirimedon*, and *Cyphocaris* were taken in small numbers in an occasional sample collected at Cape Crozier.

Hyperia macrocephala made up 7% of all the amphipods taken, although usually fewer than five individuals occurred in any one sample. The one exception, an adult collected at Beaufort Island, contained 121 individuals, nearly half of all the *H. macrocephala* found in this study. Possibly this may be a species that occasionally swarms, but it is also possible that the bird had been feeding in a different area or in an atypical manner, as it contained the largest percentage of fishes (61%) of all the complete samples collected.

Both *Hyperia macronyx* and *Hyperiella dilatata* are relatively small amphipods (Figure 11) and were often found in the intact stomachs of *Pleuragramma antarcticum* the Adélies consumed. Undoubtedly a number of these amphipods were released into the Adélie stomach contents by the digestion of the fishes' stomachs.

Only four *Hyperoche medusarum* were found. Barnard [1930] points out that members of this genus appear to be distinctly rare in comparison with other hyperiids. Of the two species of the genus *Eusirus* found, *E. microps* constituted by number about 5% of the total amphipods taken, while *E. perdentatus* made up less than 0.5%.

Epimeriella macronyx made up 8% of the total amphipods examined, making it the most abundant amphipod not of the genus *Orchomenella*. A few small individuals were recovered from the stomachs of some of the larger fishes; however, most *E. macronyx* were in the length range of 15–30 mm (Figure 12), and were ingested directly by the Adélies.

TABLE 8.   Relative Abundance of Large* Fishes Found in Adélie Penguin Stomach Contents

	Cape Crozier 1964–1965	Cape Crozier 1965–1966†	Beaufort Island 1965–1966	Franklin Island 1965–1966	Totals and Per Cent
Pleuragramma antarcticum	31	124	8	1	164, 63.3
Chaenodraco wilsoni	3	6	1	–	10, 3.9
Chionodraco kathleenae	–	–	3	–	3, 1.1
Chaenichthyidae spp.	2	24	3	–	29, 11.2
Unknown	12	36	5	–	53, 20.5
Totals for each group	48	190	20	1	259

* 75–200 mm in length.
† Complete and partial samples combined.

TABLE 9. Relative Abundance in Relation to Size of All Fishes Found in Adélie Penguin Stomach Contents

	Cape Crozier 1964–1965	Cape Crozier 1965–1966*	Beaufort Island 1965–1966	Franklin Island 1965–1966	Totals
Number and per cent of fishes less than 75 mm in length (eye lenses 2 mm or less in diameter) †	1800, 97	7825, 98	7250, 99	25, 96	16900, 98
Number and per cent of fishes 75–200 mm in length (eye lenses over 2 mm in diameter) ‡	48, 3	190, 2	20, 1	1, 4	259, 2
Totals for each group of samples	1848	8015	7270	26	17159

* Complete and partial samples combined.

† Numbers based on sub-samples counts, corrected for entire sample and rounded to nearest 25 fishes.

‡ Numbers based on complete counts (p. 202).

Table 12 shows the frequency of occurrence of amphipod species in the 37 complete stomach samples collected. The 170 partial samples collected during the 1964–1965 season at Cape Crozier have been omitted as all percentages were too low to be of significance.

An attempt was made to obtain total lengths (anterior margin of head to tip of uropods) of 20 individuals selected at random from each species in each sample group (Figures 10, 11, and 12). Individuals of a number of species were either rare or broken up in the contents and these are represented by only a few measurements. Except for the large female *Orchomenella rossi* shown in Figure 10, no attempts were made to sex or age the individuals measured.

Marine algae. A single fragment of *Iridaea obovata*, a sublittoral species growing to a depth of about 10 meters(Zaneveld, 1966), was found in the stomach of a 1965–1966 Crozier Adélie.

Parasites. Cestode larvae were found in most stomachs containing fish remains. All appeared to have been released by the digestion of the fishes. A few were examined and identified as plerocercoids of pseudo-phyllidean tapeworms. Nematodes, although less numerous than cestodes, were common in most stomachs containing fish remains. Six were identified as juveniles of *Contracaecum* sp. These also were probably released by the digestion of the fishes. Two isopods (praniza larvae of the family Gnathiidea) were found in the stomach of a 1965–1966 Crozier Adélie. In the larval stages these animals are ectoparasitic on fishes.

Stones. A number of stones were found, usually on the top of the stomach contents indicating the birds had ingested them after reaching the rookery. Only rarely were stones found at the bottom of the contents, and those few were probably consumed at the rookery after the chick was last fed. Nothing indicated that the Adélies ate these at sea.

Egg shells. Penguin egg shell fragments found on the top of the contents of two 1965–1966 Crozier parents were undoubtedly ingested at the rookery.

Bones. A single bone fragment (probably penguin) was found near the top of the contents of a 1965–1966 Crozier parent. This was also probably eaten in the rookery.

Cephalopods. Two beaks of immature *Nototodarus sloani* (Gray), a squid that attains a rather large size and is fairly common in waters adjacent to Antarctica, were found in the stomach contents. One was found in a 1964–1965 Crozier Adélie, the other in a Beaufort Island Adélie.

Copepods. A single Beaufort Island Adélie had 24 small (2–3 mm) unidentified copepods in its stomach. These were found either in or near partially digested fish stomachs, indicating they had been consumed by the fishes rather than the Adélies. Significantly a ros-

TABLE 10. Amphipoda Present in Adélie Penguin Stomach Contents

Family	
Lysianassidae	*Orchomenella plebs* Hurley
	Orchomenella rossi (Walker)
	Orchomenella pinguides Walker
	Uristes murrayi (Walker)
	Cheirimedon fougneri Walker
	Cyphocaris richardi Chevreux
Hyperiidae	*Hyperia macrocephala* (Dana)
	Hyperiella dilatata Stebbing
	Hyperia macronyx Walker
	Hyperoche medusarum (Kroyer)
Eusiridae	*Eusirus microps* Walker
	Eusirus perdentatus Chevreux
Paramphithoidae	*Epimeriella macronyx* Walker
Pontogeneiidae	*Paramoera walkeri* (Stebbing)

TABLE 11. Numbers of Amphipods Taken by Adélies (Based on Complete Counts)

	Cape Crozier 1964–1965		Cape Crozier 1965–1966		Beaufort Island 1965–1966		Franklin Island 1965–1966		Total	
	number	% of total	number	% of total	number	% of total	number	% of total	number	% of total
Orchomenella										
plebs	90	27	870	48	815	61	6	40	1781	51
rossi	70	21	136	8	53	4	2	13	261	7
pinguides	–	–	2	0*	–	–	–	–	2	0*
spp.	34	10	324	18	109	8	4	27	471	14
Sub-total	194	58	1332	74	977	73	12	80	2515	72
Uristes										
murrayi	1	0*	14	1	–	–	–	–	15	0*
spp.	2	1	9	0*	–	–	–	–	11	0*
Sub-total	3	1	23	1	–	–	–	–	26	1
Cyphocaris richardi	1	0*	–	–	–	–	–	–	1	0*
Cheirimedon fougneri	1	0*	–	–	–	–	–	–	1	0*
Hyperia macrocephala	40	12	44	2	168	13	1	7	253	7
Hyperiella dilatata	34	10	86	5	36	3	–	–	156	5
Hyperia macronyx	36	11	35	2	7	1	–	–	78	2
Hyperoche medusarum	4	1	–	–	–	–	–	–	4	0*
Eusirus perdentatus	1	0*	5	0*	3	0*	–	–	9	0*
Eusirus microps	15	5	76	4	72	5	–	–	163	5
Epimeriella macronyx	–	–	184	10	81	6	2	13	267	8
Paramoera walkeri	–	–	12	0*	–	–	–	–	12	0*
Unidentified	3	1	4	0*	–	–	–	–	7	0*
Totals for each sample group	332		1801		1344		15		3492	

* Less than 0.5%

trum of *Chionodraco kathleenae* was present in the same contents, for *Waite* [1916] recorded the stomach contents of the *C. kathleenae* he collected as containing copepods exclusively.

Hydrozoa. A single unidentified fragment of a sessile stage was found in the stomach of a Beaufort Island Adélie.

Discussion

During the chick-rearing period Adélie parents make periodic trips from land rookeries to the sea, where they consume 300–1100 grams of food (Figure 5) which, upon return to the rookery, they regurgitate to their chicks. In the Ross Sea region, stomach contents in this weight range usually contain 3500–7000 individual food items (Table 2), although as many as

12,500 were found in a single stomach. After the eggs hatch, the adults return to the rookery with their stomachs filled, consistent with the distance traveled and/or the availability of the prey items. They continue to do this as long as young are present (Figure 5). Thus, it appears that the mere presence of nestlings, regardless of age or size, provides the stimulus necessary for the adult to gather the maximum amount of food possible.

Only three groups of organisms, euphausiids, fishes, and amphipods, were found to be of importance at the three rookeries (Table 3). Numerically the Euphausiacea were the most abundant organisms in all groups of samples taken. They made up over 90% of the organisms taken at Cape Crozier in both 1964–1965 and 1965–1966; the five stomach contents collected at

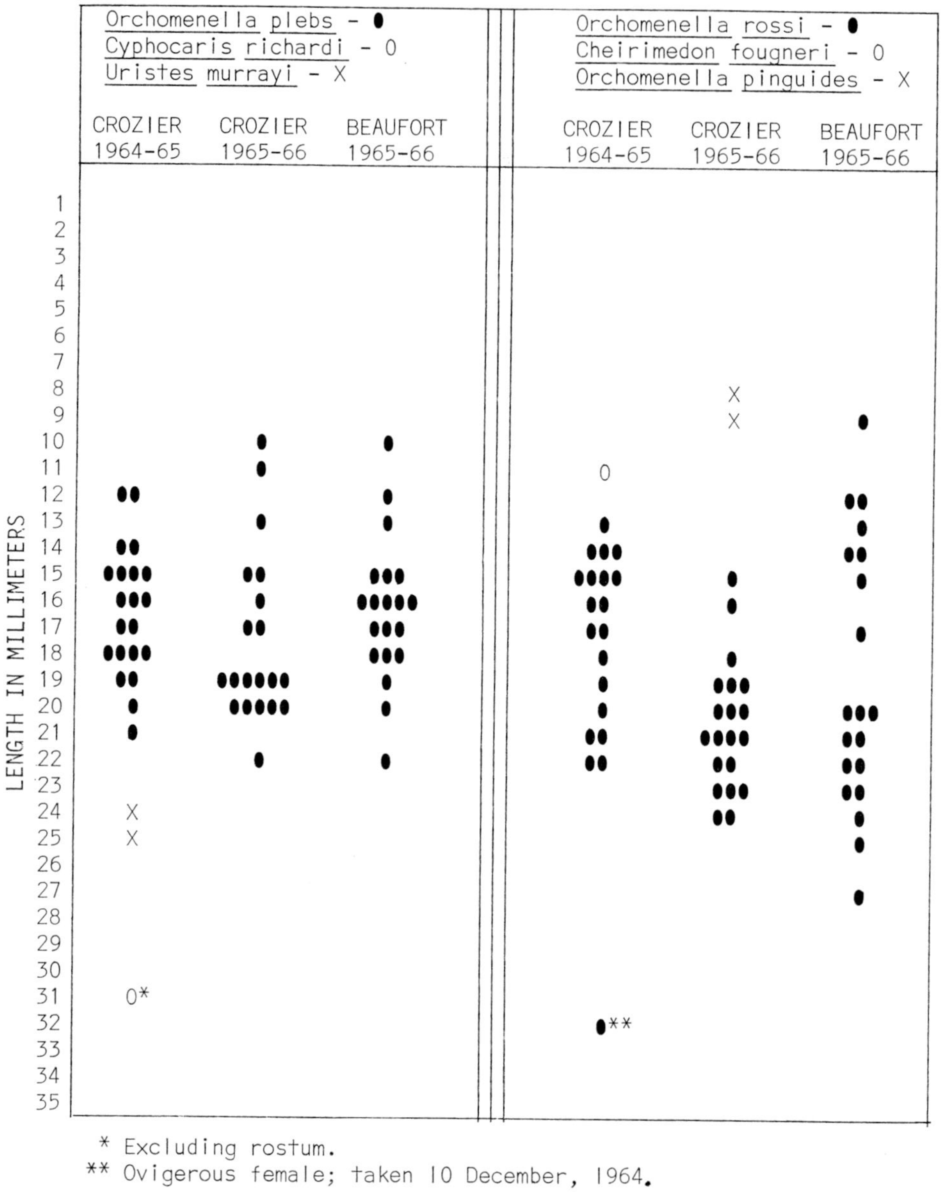

Orchomenella plebs - ●
Cyphocaris richardi - O
Uristes murrayi - X

Orchomenella rossi - ●
Cheirimedon fougneri - O
Orchomenella pinguides - X

	CROZIER 1964-65	CROZIER 1965-66	BEAUFORT 1965-66		CROZIER 1964-65	CROZIER 1965-66	BEAUFORT 1965-66

* Excluding rostum.
** Ovigerous female; taken 10 December, 1964.

Fig. 10. Lengths of amphipods found in Adélie penguin stomach contents. (Family Lysianassidae.)

Beaufort Island averaged less than 65% Euphausiacea (Figure 6). At Beaufort two stomachs contained the lowest percentages (less than 40%) of euphausiids found in any of the 37 complete stomach samples collected (none of the other 35 contained less than 60%). The other three Beaufort samples contained between 69% and 92% Euphausiacea. Because of the large variation in percentages and the small number of samples collected at Beaufort Island, comparisons of the percentages obtained at Cape Crozier with those from Beaufort were not statistically significant (Table 4). A more thorough examination of Beaufort Island feed-

ing habits may show that the parents are using two different foraging areas. On the other hand, the concentration of euphausiids may be low around Beaufort Island, and in their place the Adélies are consuming larger quantities of small fishes.

The importance of euphausiids in the diet of penguins was first reported in 1820 by *Bellingshausen* [1945]. Although most references to the euphausiids consumed by Adélies are general (euphausians, krill, opossum-shrimps, schizopods), a few authors [*Wilson*, 1907; *Eklund*, 1945; *Sladen*, 1955, 1964; *Marr*, 1962] refer specifically to *Euphausia superba*. This crustacean occurs in immense swarms in the antarctic waters; its circumpolar range extends from the antarctic convergence in the north to the shallow waters along the coast and the Ross Sea in the south, where it is replaced by *Euphausia crystallorophias* [*Marr*, 1962].

Marr suggests that the deep warm oceanic currents that carry *E. superba* eggs and larvae do not penetrate far into the relatively shallow Ross Sea, and the surface currents that enter this region from the east carry no larval, adolescent, or adult *E. superba*. He states (p. 124) "the combined observations of four well-equipped expeditions operating at intervals over a period of 35 years point fairly conclusively it would seem to the absence of *E. superba* from the greater

part of the shelf water of the Ross Sea region." But he later amends this statement (p. 125) to include only adolescent and adult *E. superba* and shows that 15 larval stage individuals have been collected in the Ross Sea shelf water. He suggests that these represent isolated stragglers that occasionally are carried into the shallow waters of the shelf from the larger oceanic populations. All other euphausiids collected in the Ross Sea between 175°W and the Victoria Land coast and southward of a line in 74°45'S, about 200 miles north of Cape Crozier have been identified as *E. crystallorophias*. Thus, Marr (p. 135) suggests that in the McMurdo Sound region of the Ross Sea, *E. superba* must be replaced by *E. crystallorophias* in the diets of

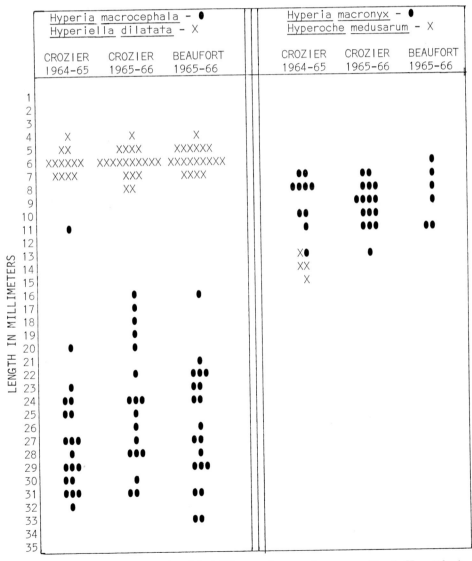

Fig. 11.   Lengths of amphipods found in Adélie penguin stomach contents. (Family Hyperiidae.)

the animals. Table 5 shows that this is partly the case with the Adélies, as more than 99% of all euphausiids taken were *E. crystallorophias*. However *E. superba* over 20 mm in length were also found in the stomach contents. Table 5 shows their numbers were small, but Table 6 shows that 30% of all the complete stomach samples collected contained at least one individual. *E. superba* were found in the Crozier stomach samples during both the 1964–1965 and 1965–1966 seasons, and were found in each of the 8 weeks of the 1965–1966 chick-rearing period; they were also consumed by some of the Beaufort and Franklin Island Adélies

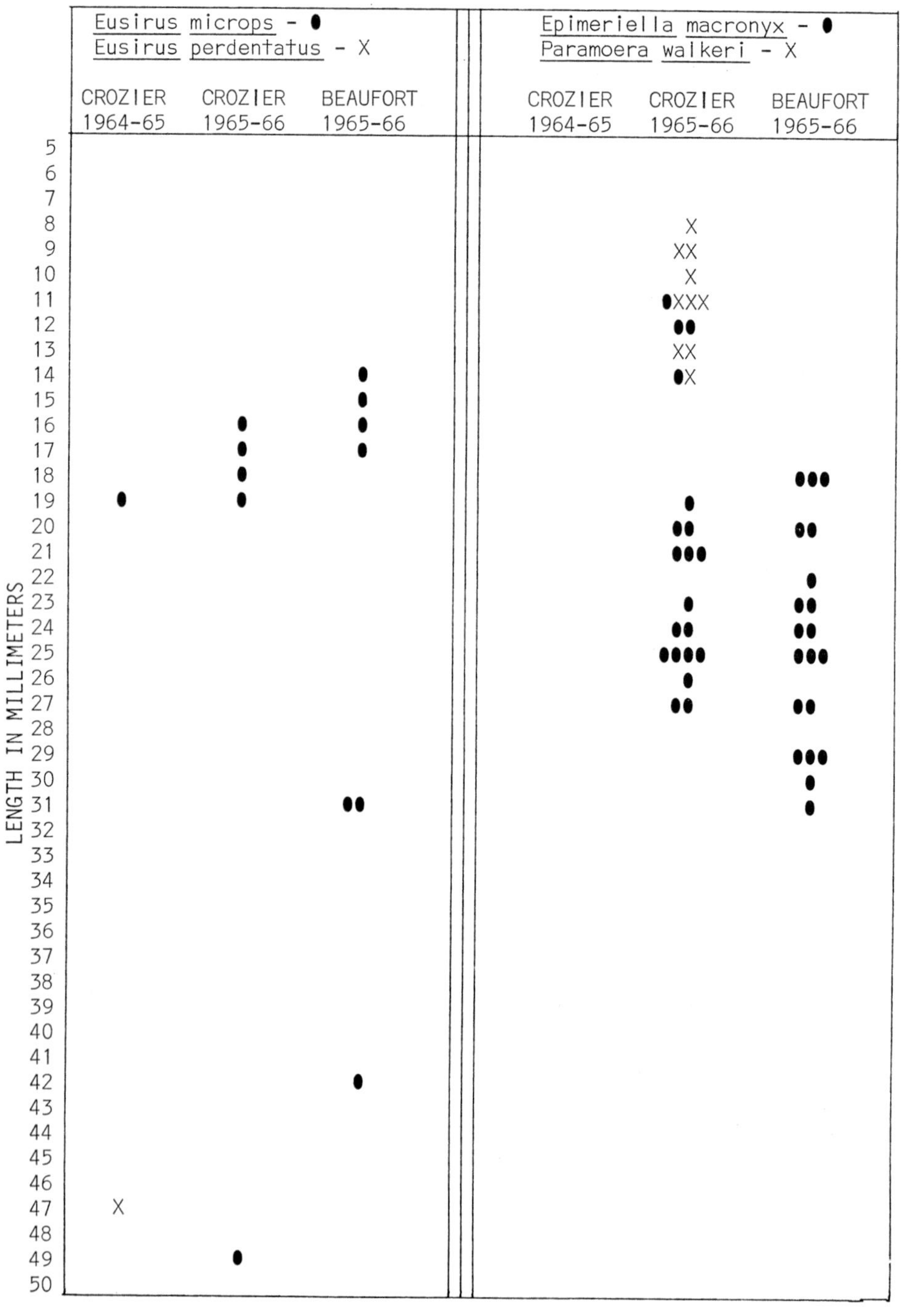

	Eusirus microps – ● / *Eusirus perdentatus* – X			*Epimeriella macronyx* – ● / *Paramoera walkeri* – X		
LENGTH (mm)	CROZIER 1964-65	CROZIER 1965-66	BEAUFORT 1965-66	CROZIER 1964-65	CROZIER 1965-66	BEAUFORT 1965-66
8					X	
9					XX	
10					X	
11					●XXX	
12					●●	
13					XX	
14			●		●X	
15			●			
16		●	●			
17		●	●			
18		●				●●●
19	●	●			●	
20					●●	●●
21					●●●	
22						●
23					●	●●
24					●●	●●
25					●●●●	●●●
26					●	
27					●●	●●
29						●●●
30						●
31		●●				●
42		●				
47	X					
49		●				

Fig. 12. Lengths of amphipods found in Adélie penguin stomach contents. (Families Eusiridae, Paramphithoidae, and Pontogeneiidae.)

TABLE 12. Frequency of Occurrence of Amphipod Species in 37 Complete Stomach Samples of Adélie Penguins

	Cape Crozier 1964–1965, 15 Samples, %	Cape Crozier 1965–1966, 16 Samples, %	Beaufort Island 1965–1966, 5 Samples, %	Franklin Island 1965–1966, 1 Sample, %	Totals, %
Orchomenella plebs	80	94	100	100	89
Orchomenella rossi	100	81	100	100	92
*Orchomenella pinguides**	–	–	–	–	–
Orchomenella sp.	60	63	100	100	70
Orchomenella totals	100	100	100	100	100
Uristes murrayi	7	25	–	–	14
Uristes sp.	7	25	–	–	14
Uristes	7	44	–	–	22
Cyphocaris richardi	7	–	–	–	13
Cheiremedon fougneri	7	–	–	–	3
Hyperia macrocephala	67	63	100	100	70
Hyperiella dilatata	60	31	60	–	46
Hyperia macronyx	47	25	40	–	35
Hyperoche medusarum	13	–	–	–	5
Eusirus microps	27	31	100	–	38
Eusirus perdentatus	7	19	40	–	16
Epimeriella macronyx	–	56	60	100	35
*Paramoera walkeri**	–	–	–	–	–
Unidentified	20	–	–	–	8

* Found only in the 170 partial samples collected at Cape Crozier during the 1965–1966 season.

(Table 5). The stomach samples were from chick-feeding adults; it is doubtful that they were making round trips of 400 miles (to the nearest area Marr records adolescent *E. superba* taken) every 1½ to 2 days during the chick-rearing period. Further, the undigested state of some of the specimens indicates that they were ingested shortly before the birds landed at the rookery.

The small number of *E. superba* found in any single sample (6 was the maximum) suggests that the Adélies were not encountering them in pure swarms, but rather as either isolated individuals or mixed in with the swarms of *E. crystallorophias*. These could well be the survivors of the larval stages, swept errantly into the Ross Sea from the larger oceanic populations, or perhaps a few adolescents may be carried in by the westward-flowing surface currents. In any event, they are being taken in the vicinity of the Cape Crozier rookery, although in very small numbers.

E. crystallorophias, the most abundant food species in the stomach contents of the Crozier Adélies (Table 5), is a swarming, circumpolar species found along the coasts of the antarctic continent and in the relatively shallow Ross Sea. Because of its neritic habitat, antarctic expeditions have seldom encountered it and comparatively few studies of it have been undertaken.

To date, the most informative publication on the life history of *E. crystallorophias* is a short discussion by *Tattersall* [1908]. He collected calyptopis larvae

"from the beginning of January to nearly the end of February" (p. 11), which suggests a relatively short spawning period. He also collected during January specimens 13 mm in length, which were "presumedly larvae of the preceding season" (p. 12). The adults he collected measured up to 32 mm in length, which corresponds well to the maximum lengths (28 mm less telson) taken by the Adélies in this study.

From Tattersall's work it appears that *E. crystallorophias* has a longevity at least equal to if not longer than some of the more thoroughly studied southern euphausiid species. Both *E. superba* [*Bargmann*, 1945; *Marr*, 1962] and *E. triacantha* [*Baker*, 1959] spawn at about 2 years of age. These two species have comparable growth rates for their first year of life; shortly after this *E. triacantha* stops growing, whereas *E. superba* continues until time of spawning. If *E. crystallorophias* also spawns at 2 years of age, then its pattern of growth would be similar to that of *E. superba*, although the rate would be much slower. The lengths of *E. crystallorophias* consumed by the Adélies (Figure 9), however, do little to clarify the question of its longevity. Most individuals measured (Figure 9A) fell within a length range of 14–25 mm (telson not included), which seems a rather wide range in view of the restricted spawning period suggested by Tattersall's collection of calyptopis larvae only during January and part of February. Possibly this wide range can be attributed partly to the degree of digestion of the individuals measured. The smaller size group, measuring 9–12 mm (less telson), being consumed by the Beaufort Island Adélies (Figure 9D) would correspond to the 13 mm specimens (probably 1 year of age) reported by Tattersall.

From the sizes shown in Figure 9, it appears that the Adélies are selecting the larger *E. crystallorophias*. The inclusion of the much smaller individuals in the diet of the Beaufort Adélies adds further support to the possibility that euphausiids may not be plentiful near the rookery and the Adélies are consuming all sizes encountered.

Probably the *E. crystallorophias* in the larger size group are 2 years of age, but the wide length range tends to make one wonder if another year group may not also be present. Further investigation into the life history of this important neritic species is needed.

An average of 5,300 euphausiids per complete stomach was being brought in by parents during the 1965–1966 chick-rearing period. Because of the relative small size of *E. crystallorophias*, this is a much larger average number of euphausiids than would be found in stomach contents in other parts of the Antarctic where the Adélies are feeding exclusively upon the larger *E. superba*. *Bagshawe* [1938:200] records a moderately full stomach of a gentoo penguin, *Pygoscelis papua* (slightly larger than the Adélie), as containing only 960 *E. superba*. Thus the success of the Adélie rookeries in the Ross Sea indicates that euphausiids are an important constituent in the Adélie diet mainly because of their swarming habits rather than individual size, although, when given a choice, the Adélies apparently select the larger euphausiids in the population.

Numerically the fishes were second in importance to the euphausiids. At Cape Crozier in both 1964–1965 and 1965–1966 and in the single Franklin Island sample fishes made up less than 10% of the individual food organisms consumed, while at Beaufort Island they made up over 30% of the individual food organisms consumed (Figure 6). Because of their rapid digestion, it was not possible to obtain an accurate estimation of biomass. A rough volumetric measurement of the 1964–1965 Crozier samples (Figure 8) shows that they were an important food group, making up nearly 40% by volume of the food consumed.

Pleuragramma antarcticum, a pelagic, circumpolar fish occurring in large shoals in the young stages, was the most abundant fish species found in the Adélie stomach contents. A few of these were calculated to be as much as 200 mm in length; 98% of all fishes

taken were less than 75 mm in length (Table 9). Of 171 small skulls examined, nearly 90% were *P. antarcticum* (Table 7), and probably most of these were 1- and 2-year-old individuals, which are less than 75 mm in length [*DeWitt and Tyler*, 1960]. The Adélies were apparently encountering the small *P. antarcticum* in shoals, as they occurred within the stomach contents in discrete masses and surrounded by remains composed almost entirely of euphausiids.

P. antarcticum seems to be an important constituent in the diets of a number of the birds and mammals inhabiting the Ross Sea. In addition to being important to Adélies, it has been recovered from the stomachs of a crabeater seal, *Lobodon carcinophagus* [*Schultz*, 1945], Weddell seals, *Leptonychotes weddelli* [*DeWitt and Tyler*, 1960; *Dearborn*, 1965], south polar skuas, *Catharacta maccormicki* [*Young*, 1962]. *Schultz* [1945] thought that 18 specimens in bad condition found in a snow petrel, *Pagodroma nivea*, were this species. The unidentified "silvery fish" that figures frequently in reported lists of emperor penguin food *Aptenodytes forsteri* (*Falla*, 1937:41) could well be this species also.

The fishes other than *P. antarcticum* all appeared to belong to the family Chaenichthyidae, the "white-blooded" or ice fishes, which possess the unique character among vertebrates of containing neither erythrocytes nor hemoglobin. They made up 10–15% of the total fishes consumed, both large and small (Tables 7 and 8). *Chaenodraco wilsoni* was the only chaenichthyid species identified from the contents of the Crozier Adélies.

Both *Chaenodraco wilsoni* and *Chionodraco kathleenae* were found in the Beaufort Island samples (Table 8), and although the numbers are small, the difference in chaenichthyid species composition between the Crozier and Beaufort rookeries suggests that different foraging areas are being utilized. The small (less than 75 mm in length) chaenichthyids could not be identified to species, and their distribution within the stomach contents gave no indication of their being encountered in large shoals.

Andriashev [1965:506] states, "the majority of chaenichthyid species live at a depth of not less than 100–200 m (and up to 600–700 m)." He points out, however, that *Nybelin* [1947:55] has found that some species of chaenichthyids are sometimes observed in the upper layers of water (often far from the coast) and in the stomachs of seals and whales. Thus the presence of the chaenichthyids in the stomachs of Adélies, along with large numbers of pelagic *Pleura-*

gramma and euphausiids, suggests that these are some of the chaenichthyid species Nybelin speaks of as occasionally inhabiting the upper layers of water.

It is noteworthy that *Trematomus borchgrevinki*, a pelagic, plankton-feeding fish taken in abundance from under the fast ice in McMurdo Sound [*Wohlschlag*, 1964], was not found in any of the Adélie stomachs examined. This suggests that small *T. borchgrevinki* (up to 200 mm), may be restricted to the fast ice area where they are not exposed to Adélie predation.

It is significant that the most important fish group in the stomach contents was the small shoaling *P. antarcticum*, whose habits are very similar to those of the swarming *Euphausia crystallorophias*. In addition to the large percentage of small fishes in some of the Beaufort Island Adélies in this study, *Eklund* [1945] reported that the food of Adélies at the East Base, Antarctic Peninsula, consisted chiefly of a small transparent fish (Nototheniidae) and *E. superba*. Apparently Adélies can sustain themselves and rear chicks successfully in waters containing few euphausiids if they are rich in small shoaling fishes such as *P. antarcticum*.

The amphipods, the only other group of any importance in the Ross Sea Adélie stomachs (Table 3), made up by number less than 5% of the individual food organisms found in all groups of samples taken (Figure 6). Their importance by volume appeared to be about the same as that by number (Figure 8).

The species composition, numbers, and sizes of the amphipods consumed bring out several important points in the feeding habits of the Ross Sea Adélies.

1. The preponderance of *Orchomenella plebs* and *O. rossi* in relation to the other amphipods ingested (Table 11) confirms the collections made by various expeditions into the Ross Sea in which *O. rossi* (which *Hurley* [1965] now considers two species, *O. plebs* and *O. rossi*) was the most abundant amphipod. The low percentage of amphipods in the stomachs suggests they are ingested incidentally while the Adélies feed on the swarming euphausiids and fishes.

2. Both *rossi* and *plebs* were found in about 90% of the complete stomach samples (Table 12), but *plebs* was usually far more numerous in any single sample. Less than 15% of the orchomenelids identified were *rossi*, the rest being *plebs*. *Hurley* [1965] examined a collection of amphipods taken from under the Ross Ice Shelf at White Island and found that *rossi* formed at most 10% of the collection and *plebs* made up the

bulk of the remainder. Thus the relative abundances of the two species taken by the Adélies and by sampling methods were fairly close, again indicating no selection of one species over the other by the Adélies.

3. None of the amphipods were bottom-dwelling species, although the lysianassids, *Orchomenella plebs*, *O. rossi*, *Cheirimedon fougneri* and *Uristes murrayi*, have mouthparts adapted to a carnivorous habit and occur in large numbers around suitable foods at all depths (Thurston, personal communication). Only two littoral species, *Paramoera walkeri* and *Orchomenella pinguides*, were found in the stomach contents. As these were only found in small numbers in three and two samples respectively and were on top of the contents, the Adélies had probably ingested them after they had returned from their normal deep-water feeding grounds just before they had landed at the rookery. Thus, it appears that most of the amphipods were being taken in the upper layers of deep water, along with the euphausiids and fishes.

4. The sizes of amphipods consumed are commensurate with those of euphausiids consumed. Figure 9A shows the majority of euphausiids were 14–25 mm in length (less telson), while Figures 10, 11, and 12 show that, except for *Hyperia macronyx* and *Hyperiella dilatata* (some of which had been initially consumed by the larger fishes), all the common amphipods fell within the same general size range.

Adélie penguins have not been recorded as feeding extensively upon amphipods in any part of the Antarctic, even though a number of species occur in swarms at certain seasons. *Ealey* [1954] found that the diets of most of the plankton-feeding birds on Heard Island, including the rockhopper penguin, *Eudyptes crestatus*, consisted primarily of swarming amphipods, but the gentoo penguins (closely related to the Adélie) on the island were eating mainly fishes of the genus *Notothenia* and taking only a few amphipods. Thus, the pygoscelid penguins apparently prefer euphausiids and fishes when available to the plentiful swarming amphipods.

The Ross Sea Adélies rarely took food organisms other than euphausiids, fishes, and amphipods (Table 3). The fragment of marine alga and of the sessile stage of a hydrozoan were the only bottom dwelling organisms recovered. These were probably both littoral and consumed by the Adélies near their rookeries after returning from feeding trips. The virtual absence of cephalopods from the Adélie stomachs is of interest because the stomachs of emperor penguin chicks found dead at the nearby Crozier emperor rookery contained the remains of numerous cephalopod beaks (Boyd, personal communication). The small number of copepods found, some of which have similar habits to the euphausiids, is probably a result of their much smaller size.

In brief the Adélie penguins in the Ross Sea region feed essentially upon small marine animals that occur in large swarms in the upper layers of water and select organisms between 15 and 75 mm long, with an occasional larger fish.

A number of competitors for this food source are present in the antarctic waters, and probably the most important are the baleen whales. The food preferences of the Adélie penguins and the baleen whales are strikingly similar. *Marr* [1962:142] concludes that the whales select the larger euphausiids in the population and notes that, for some unknown reason, they also do not prey on the large, swarming, pelagic amphipods. The small fishes in the Adélies' diet are the main feeding difference between the two species.

As Marr was concerned with whales feeding in waters inhabited by *E. superba*, most of his data on sizes and numbers of individuals consumed are not comparable to those of the smaller *E. crystallorophias* the Ross Sea Adélies were eating. However from his data on the smaller size groups of *E. superba* consumed (p. 145), he estimates that an average-sized whale, gorged to repletion on yearling *E. superba* ranging from 17 to 32 mm with a 30% mode at 24 mm (which is very close to the sizes of *E. crystallorophias* the Adélies ate), would consume in one meal more than 10 million individuals. Marr also presents evidence, although emphasizing that further studies are needed to confirm it, that the baleen whales fill their stomachs once a day. Thus the removal of a single average-sized whale from the nearby waters during the Adélie chick-feeding period (which at Crozier in 1965–1966 spanned 52 days) would make available an additional 520 million euphausiids, enough to rear more than 5,000 additional Adélie chicks.

In 1946 the International Whaling Commission set a limit for the annual catch at 16,000 blue whale units (1 blue = 2 fin = 2½ humpback = 6 sei). In 1962–1963 this was reduced to 15,000 units and by 1965–1966 it had been reduced to 4,500 units. Therefore, at least 15,000 blue whale units were being removed from the antarctic waters every year from 1946 until 1963. This removal of 15,000 whale units during the Adélie chick-rearing period would have resulted in the presence of enough additional food to rear 5,000 × 15,000

or 75 million Adélie chicks above the normal annual production. If one then considers that the baleen whales were present and feeding in the antarctic waters for 6 months rather than just the 2 months of chick rearing, this would further expand the figure to over 200 million additional Adélie chicks.

While this is speculative, and certainly the Adélie penguin shares these increased food resources with many other nonharvested krill feeders, it shows that whaling in antarctic waters could have a pronounced impact on the populations of a number of antarctic birds and mammals, assuming that their food is a major limiting factor. Probably certain species will benefit more than others; *Sladen* [1964] suggests it has caused an apparent increase in the pygoscelid penguin populations and ranges during the last few years.

Assessing the effects of large scale whale harvesting on antarctic biopopulations will be difficult. The close similarity this study reveals between the diets of the Adélie and the baleen whales shows the Adélie penguin would be an ideal indicator species for such researches. In addition the concentration of the Adélies in rookeries during the breeding season provides a convenient method of assessing their population changes from year to year. Adélie rookeries in regions where whales have been harvested can be censused either by ground counts of nesting birds, or preferably, by counting from aerial photographs of the rookeries. Sladen (personal communication) has found that the best time to make counts of breeding Adélies is during the week following egg laying (mid-November at Crozier); at this time the females have left the rookery, the males are incubating the eggs, and few nonbreeding birds are present. Counts of the birds present by either method give good estimates of nest numbers and breeding density. Aerial photographs of selected rookeries taken at this time each year over an extended period of years should provide a reliable measure of what impact the recent intensive whaling is having on the population of a major nonharvested krill-consuming species.

Acknowledgments. The support of this study by the following organizations is gratefully acknowledged: the National Science Foundation for field facilities and financial support, U.S. Navy Task Force 43 for logistic support, and Air Development Squadron VX-6 for helicopter support to Cape Crozier.

I am indebted to the following specialists who identified, confirmed identifications, and gave valuable comments on the organisms found in the stomach contents of the Adélies:

Euphausiacea and Isopoda, M. H. Thurston of the British Museum.

Fishes, H. H. DeWitt of the University of Southern California.[1]

Amphipoda, T. E. Bowman of the Smithsonian Institution identified the Hyperiidae and *Cyphocaris richardi*. D. E. Hurley of the New Zealand Oceanographic Institute, D.S.I.R., identified the remainder of the Lysianassidae and the Eusiridae, Paramphithoidae, and Pontogeneiidae. M. H. Thurston gave valuable comments on the distribution and ecology of the various amphipods taken.

Nematodes and cestodes, H. L. Holloway, Jr., of Roanoke College.

Hydrozoa and marine algae, J. S. Zaneveld of Old Dominion College.

Cephalopoda, G. L. Voss of the University of Miami.

I wish to express gratitude for the advice, encouragement, and patience of W. J. L. Sladen during the course of this study. The assistance received from the following individuals is also greatly appreciated: Admiral G. J. Dufek, U.S.N., and P. J. Burstall of the New Zealand Wildlife Research Division for their support and encouragement when I first entered the U.S. Antarctic Research Program; the 1964–1965 and 1965–1966 Johns Hopkins University field personnel at Cape Crozier for their assistance and unfailing good humor; R. C. Wood and H. R. Lowery for collecting the Beaufort Island Adélies, and C. Ray for collecting the Franklin Island Adélies; W. L. N. Tickell, F. C. Kinsky, G. S. Watson, J. H. Dearborn, J. C. Boyd, and D. H. Thompson for their comments and assistance on various aspects of this study.

This study has been supported by the National Science Foundation, Office of Antarctic Programs, from Grants GA-151 and GA-645 for Antarctic Avian Population Studies (USARP Bird-Banding Program) under the direction of W. J. L. Sladen.

---

[1] Dr. DeWitt has not examined the skeletal parts of *Chionodraco kathleenae* that I found and identified after his examinations of *Pleuragramma antarcticum* and *Chaenodraco wilsoni*.

REFERENCES

Andriashev, A. P., A general review of the antarctic fish fauna, in *Biogeography and Ecology in Antarctica*, edited by J. Van Mieghem and P. Van Oye, pp. 491–550, Dr. W. Junk publishers, The Hague, 1965.

Bagshawe, T. W., Notes on the habits of the gentoo and ringed or antarctic penguins, *Trans. Zool. Soc. Lond.*, 24, 185–307, 1938.

Baker, A. de C., The distribution and life history of *Euphausia triacantha* Holt and Tattersall, *Discovery Reports*, 29, 309–340, 1959.

Bargmann, H. E., The development and life-history of adolescent and adult krill, *Euphausia superba*, *Discovery Reports*, 23, 103–176, 1945.

Barnard, K. H., Amphipoda, *Brit. Antarctic ('Terra Nova') Expedition 1910, Zool.*, 8, 307–454, 1930.

Bellingshausen, F. G., *The Voyage of Captain Bellingshausen to the Antarctic Seas 1819–1821* (London), 1945.

Dearborn, J. H., Food of Weddell seals at McMurdo Sound, Antarctica, *J. Mammal.*, 46, 37–43, 1965.

DeWitt, H. H., and J. C. Tyler, Fishes of the Stanford antarctic biological research program, 1958, 1959, *Stanford Ichthyol. Bull.*, 7, 162–199, 1960.

Ealey, E. H. M., Analysis of stomach contents of some Heard Island birds, *Emu*, 54, 204–210, 1954.

Eklund, C. R., Condensed ornithology report, East base, Palmer Land, *Proc. Am. Phil. Soc.*, 89, 299–304, 1945.

Falla, R. A., Birds of the British, Australian, and New Zealand antarctic research expedition (1929–1931), and of the Australasian antarctic expedition (1911–1914), *B.A.N.Z. Antarctic Res. Expedition, 1929–1931, Ser. B. (Zool. & Bot.)*, 2, 1–304, 1937.

Foxton, P., The distribution of the standing crop of zooplankton in the southern ocean, *Discovery Reports*, 28, 191–236, 1956.

Hurley, D. E., A common but hitherto undescribed species of *Orchomenella* (Crustacea Amphipoda: Family Lysianassidae) from the Ross Sea, *Trans. Roy. Soc. N.Z. (Zool.)*, 6, 107–113, 1965.

John, D. D., The southern species of the genus *Euphausia*, *Discovery Reports*, 14, 193–324, 1936.

Marr, J. W. S., The natural history and geography of the antarctic krill (*Euphausia superba* Dana), *Discovery Reports*, 32, 33–464, 1962.

Nybelin, O., Antarctic fishes, *Sci. Res. Norweg. Antarctic Expedition*, 26, 1–76, 1947.

Schultz, L. P., Fishes of the United States antarctic service expedition 1939–1941, *Proc. Am. Phil. Soc.*, 89, 298, 1945.

Sladen, W. J. L., Some aspects of the behaviour of Adélie and chinstrap penguins, *Acta XI Congressus Internationalis Ornithologici*, Basel, 441–447, 1955.

Sladen, W. J. L., The pygoscelid penguins, 1, Methods of study; 2, The Adélie penguin, *Falkland Islands Depend. Surv. Sci. Rept.*, no. 17, 1–97, 1958.

Sladen, W. J. L., The distribution of the Adélie and chinstrap penguins, in *Biologie Antarctique*, edited by R. Carrick, M. Holdgate, and J. Prevost, pp. 359–365, Hermann, Paris, 1964.

Tattersall, W. M., Crustacea, 7, Schizopoda, *Nat. Antarct. Exped. 1901–1904. Nat. Hist.*, 4, 1–42, 1908.

Waite, E. R., Fishes, *Australasian Antarctic Expedition 1911–1914. Sci. Rept. Ser. C, Zool. & Bot.*, 3, 1–92, 1916.

Wilson, E. A., Aves, *Brit. Natl. Antarctic Expedition Rept., 1901–1904, Nat. Hist. 2*, 1–121, 1907.

Wohlschlag, D. E., Respiratory metabolism and ecological characteristics of some fishes in McMurdo Sound, Antarctica, in *Biology of the Antarctic Seas, Antarctic Res. Ser., 1*, edited by M. O. Lee, pp. 33–62, American Geophysical Union, Washington, D. C., 1964.

Young, E. C., Feeding habits of the south polar skua *Catharacta maccormicki*, *Ibis*, 105, 301–318, 1962.

Zaneveld, J. S., Vertical zonation of antarctic and subantarctic benthic marine algae, *Antarctic J. U.S.*, 1, 211–213, 1966.

THE USARP[1] BIRD BANDING PROGRAM, 1958–1965

W. J. L. Sladen, R. C. Wood, and E. P. Monaghan

Department of Pathobiology, Johns Hopkins University, Baltimore, Maryland

INTRODUCTION

Antarctica supports vast colonies of specially adapted, long-lived, polar sea birds that, because of their comparative tameness and easy accessibility, offer opportunities for research into population dynamics and social behavior difficult to match elsewhere in the world.

The USARP Bird Banding Program, started in 1958, has maintained three specific foci of activity: (1) international cooperation through providing bands to other countries in support of their own specialized ornithological studies; (2) support of USARP research workers following their own research problems; (3) support of our own Johns Hopkins University programs in three main locations: Bird Island, South Georgia; Cape Crozier, Ross Island; and the Falkland Islands. A list of the nations, U.S. universities, and individual scientists participating in these studies is given in Table 2.

This paper reports on birds banded and recovered away from the place of banding for the period 1958 to June 30, 1965. It gives no information on birds recaptured at or within 16 kilometers (10 miles) of the place of banding, which are to be dealt with elsewhere. The authors have been intimately involved in the continuity of this program: Sladen and Wood at Cape Crozier and Monaghan at Johns Hopkins. R. B. Napier has been responsible for the continuity maintained at Westpoint Island, Falkland Islands, and W. L. N. Tickell was at Bird Island, South Georgia.

REVIEW OF ANTARCTIC BIRD BANDING ACTIVITIES 1909–1965

"Gain has fastened rings of variously coloured celluloid, such as are used for fowls, round the legs of numerous penguins, both young and old, and of some cormorants. Thus it will perhaps be possible one day to get some certain information about the movements of these birds. Some writers claim, though I do not know upon what observations they found their statements, that the parents do not return to the old rookery a second year, and that

it is only inhabited by the young who were hatched there."

Thus did Jean Charcot, leader of the Second French Antarctic Expedition of 1908–1910, write about the first banding operations in Antarctica, which were, in fact, among the first anywhere [*Rydzewski*, 1951]. *Gain* [1914] marked his birds (Adélie and gentoo penguins and blue-eyed cormorants) on Petermann Island, 65°10′S, 64°10′W, in January and February 1909. The following spring he was able to show that, contrary to current belief, adult penguins return to the same breeding rookeries, whereas the young of the last season do not return, at least during their first year.

Twenty-six years passed before banding was again used as an aid to antarctic ornithological research. In 1935 and 1936 *Roberts* [1940] banded most of the birds in a 23-nest colony of Wilson's petrels in the Argentine Islands, 65°16′S, 64°16′W, during the British Graham Land Expedition. After another visit in February 1937, he verified an earlier conclusion that "each bird returns annually to the same burrow and mate." This discovery was especially interesting at that time for these, the smallest of antarctic birds, had all returned safely after their extensive migrations to the North Atlantic during the austral winter.

The first to use bands with addresses and reference numerals appear to be the U.S. biologists *Eklund* [1945] and Bryant of the U.S. Antarctic Service Expedition, 1939–1941. Their bands, provided by the U.S. Fish and Wildlife Service, were addressed *Notify Biological Survey, Washington, D. C.* A south polar skua, banded as an adult in March 1941, was recovered more than 8½ years later indicating this bird had lived for at least 10 years [*Sladen and Tickell*, 1958].

In 1947 the first continuing antarctic banding program was organized by Sladen under the name of the FIDS Bird Ringing Scheme (FIDS: Falkland Islands Dependencies Survey; now called British Antarctic Survey, BAS) [*Roberts and Sladen*, 1952; *Sladen*, 1952; *Sladen and Tickell*, 1958]. This program pro-

---

[1] United States Antarctic Research Program.

vided the first long-distance recovery of an antarctic bird when a giant petrel was caught alive in Fremantle harbor, Western Australia, in July 1948, about 9 weeks after fledging at the South Orkney Islands [*Serventy*, 1948; *Roberts and Sladen*, 1952]. After learning to fly, it had traveled an estimated 16,000 kilometers (10,000 miles) in this short time. In keeping with international cooperation, the address first used on the British bands, *Inform F.I.D.S. Colonial Office London*, was soon changed to *Inform F.I.D.S. British Museum Nat. Hist. London*. In the meantime other nations had started marking birds: the Australian National Antarctic Research Expeditions (ANARE) on Heard and Macquarie Islands used bands addressed *ANARE Melbourne Australia* [*Chittleborough and Ealey*, 1950; *Downes et al.*, 1954; *Gwynn*, 1955]; Expeditions Polaires Francaises (EPF) in Adélie Coast [*Sapin-Jaloustre*, 1952; *Sapin-Jaloustre and Bourliere*, 1951]; and the Norwegian *Brategg* expedition on Peter I Island [*Holgersen*, 1957].

Prior to and during the International Geophysical Year (IGY), U.S. Fish and Wildlife Service bands were again used, first by *Austin* [1957] and later by *Eklund* [1961]. Austin assessed the sizes of the Adélie colonies in the Ross Sea area in 1955–1956 and banded a significant sample of adult and young skuas that are still being recorded at Cape Royds and Cape Hallett. Eklund's researches during the IGY concentrated on the distribution and movements of the south polar skua. His international enquiry using tall plastic bands with different color combinations for each nation was the first of its kind in antarctic cooperation. The collaboration of 10 nations in this (Argentina, Australia, Belgium, Chile, Japan, New Zealand, Norway, United Kingdom, U.S.A., and U.S.S.R.) yielded new data on the hitherto unknown coastal and circumpolar movements of this species. This program yielded an interesting long-distance recovery; a south polar skua banded at Wilkes station (66°15'S, 110° 31'E) as a breeding adult was found dead 2 months later on a beach near Adelaide, Australia. The conspicuous green plastic band, indicating that it had been banded at Wilkes, caught the eye of the finder. This bird proved to be the first record of this species for Australia [*Eklund*, 1959].

The spirit of international cooperation has been encouraged by the Scientific Committee for Antarctic Research (SCAR) by the formation of a subcommittee of the SCAR Biology Working Group on Antarctic Bird Banding [International Council of Scientific Unions, *SCAR Manual*, 1966, p. 26]. Representatives from each nation engaged in antarctic ornithological research are exchanging data on banded and recovered birds, as well as cooperating in planning field research and techniques and the design of bands. Following the lead given by Eklund's international skua investigation, this SCAR subcommittee is coordinating a program whereby chicks and breeding adults (but not adults of unknown status) are given special color bands according to the region where they are caught. This will enable biologists to recognize instantly an immigrant visiting from another area in Antarctica.

Banding has now proved an indispensable tool for ecological and behavioral studies as well as for shorter-term and more specialized research in physiology and adaptation in the Antarctic and elsewhere. Thus banding has been used as a basis for detailed publications on the breeding biology of penguins [*Sladen*, 1953, 1958; *Sapin-Jaloustre*, 1960; *Stonehouse*, 1960; *Prevost*, 1961; *Taylor*, 1962; *Penney*, 1964, 1967], of albatrosses and petrels [*Roberts*, 1940; *Tickell*, 1962, 1967a, b; *Tickell and Pinder*, 1967; *Warham*, 1962; *Pinder*, 1966], and of skuas and sheathbills [*Stonehouse*, 1956; *Eklund*, 1961; *Young*, 1963a, b; *Jones*, 1963; *Reid*, 1966] and as a basis for our present knowledge of the long-distance movements of the great travelers of the southern oceans [*Sladen and Tickell*, 1958; *Ingham*, 1959; *Tickell and Scotland*, 1961; *Sladen*, 1965; *Tickell*, 1967a, 1968; *Tickell and Gibson*, 1967]. Of the specialized researches in physiology and adaptation in which the banding technique has been used, mention should be made of studies on navigation and orientation in the Adélie penguin [*Emlen and Penney*, 1964; *Penney and Emlen*, 1964], on the salt gland in the Adélie [*Douglas*, 1964, 1968], on feeding preferences in skuas, albatrosses, and the Adélie [*Young*, 1963b; *Tickell*, 1960, 1964; *Emison*, 1968], on temperature regulation and biotelemetry of Adélie and emperor penguins [*Goldsmith and Sladen*, 1961; *Sladen et al.*, 1966; *Boyd et al.*, 1967], and on circadian rhythms of the Adélie [*Müller-Schwarze*, 1968]. Many of these projects have been supported by the USARP Bird Banding Program.

THE USARP BIRD BANDING PROGRAM

Objectives and Organization

The USARP Bird Banding Program began operation in December 1958 on board the USS *Staten Island* [*Sladen and Goldsmith*, 1960] and at Bird Island, South Georgia [*Tickell and Cordall*, 1960]. This program, a collaborative effort between biologists of the

Fig. 1. Adélie penguins marked with individually numbered flipper bands.
(Photographed by R. L. Penney.)

Johns Hopkins University and the U.S. Fish and Wildlife Service, has as its main objectives: (1) to develop
specially designed bands for antarctic and subantarctic
sea birds that will last their lifetimes and aid studies
on local movements, long-distance migrations, longevity, breeding biology and ecology, social behavior, and
physiology; (2) to promote international cooperation
and interest in antarctic sea birds and to encourage
their conservation at their southern breeding grounds
and during their movements in the southern oceans
and along the shores of southern hemisphere countries.

The collaboration of the U.S. Fish and Wildlife
Service (USFWS) has been most helpful and productive. The USFWS has assumed responsibility for receiving and recording recovery and recapture data and
for incorporating them into their IBM system, for sup-
plying appropriate forms, and for providing limited
quantities of conventional USFWS bands.

For the period covered in this report, almost 100,000
birds of 35 different species have been banded (Table
4). A manual circulated to all stations and coopera-
tors outlines the objectives of the program, describes
the bands, gives instructions for the use and application of the bands, for filling in forms, and for reporting recovered birds. Special traps, tools for capturing
and marking birds, and literature have also been distributed to antarctic stations.

Band Designs and Special Markings

These will be discussed in greater detail elsewhere. All
bands have the usual USFWS 8-digit serial number
and bear one of two addresses: "Avise Fish and Wildlife Service, Write Washington D.C. USA" or "Avise
F. & W. Serv. Wash. USA."

Some of the bands distributed have been from the
regular issue of the Fish and Wildlife Service. The
majority, however, have been our own designs made
to suit the birds we are studying and purchased with
funds from the Office of Antarctic Programs of the
National Science Foundation. So far some fifteen designs have been made, starting from the basic designs
used by the FIDS Bird Ringing Scheme (see *Sladen
and Tickell*, 1958, Figure 1, p. 7). The greatest effort
has been put into penguin flipper bands [*Sladen and
Penney*, 1960; *Penney and Sladen*, 1966], double
inscription bands for albatrosses and the giant petrel,
and a band for skuas 1 inch (25 mm) tall (see Figures
1 and 2).

Our two aims with the band designs have been:
(1) to develop bands that will last the birds' lifetimes;
(2) to make the band numbers legible enough to be
read without catching and handling the bird.

While we have had considerable success with penguin and skua bands, we are still not satisfied that

our objectives have been fulfilled, for we have little idea as yet of how long these birds live. Evidence at present points to a life expectancy of from 20 to 30 and perhaps more years for most antarctic sea birds. The latest design in penguin flipper bands carries numerals ¼-inch (6 mm) high that can be read through field glasses from a distance of 25 meters. The skua design is a band 1-inch (25 mm) tall, with *duplicated* numerals reading vertically (not around the band as on the conventional USFWS bands), so that the band number can be read through a spotting scope 50 meters away no matter in what direction the standing bird faces (Figure 2). This distance can be increased somewhat by filling the digits with black paint and wrapping the band with transparent pressure-sensitive tape.

The position of the band is also important. Members of the SCAR subcommittee on Antarctic Bird Banding have agreed that metal bands (with reference numbers) be put on the left tarsus of all chicks and on the right tarsus of all birds banded as adults. Thus for long-term population studies that involve following birds of known age, observers need look only for bands on the left tarsus.

The USFWS issued the USARP Bird Banding Program species numbers for all antarctic species not listed in the *Check-list of North American Birds* of the American Ornithologists' Union (AOU), so that they could be recorded on IBM cards. These species numbers and the band sizes used are listed in Table 1.

BANDING RECOVERIES

Species Banded and Recovered
Away from Banding Stations

Table 3 lists the banding stations and their coordinates. Birds have been banded at 44 stations in the Antarctic and the Falkland Islands, and at one location in the Northern Hemisphere, Kent Island, New Brunswick, Canada. At the Canadian location, Charles E. Huntington banded 196 Leach's petrels to test a special monel overlap band designed for Wilson's petrels, and these bands are still in excellent condition after 7 years of wear on this burrow-nesting species. During the summer of 1965, 120 Wilson's petrels (not included in Table 4 because they were banded after June 30, 1965) were caught at sea in the Bay of Fundy, Canada [D. W. Gill, personal communication, 1967] by a recently developed technique: throwing a mist-net suspended on a hoop 152 cm (5 feet) in diameter over birds as they feed on bait thrown from a boat. This technique could be used very successfully

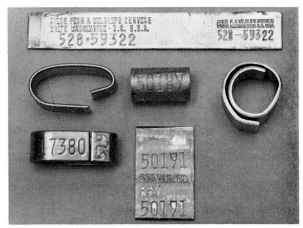

Fig. 2. Special band designs. *Top:* Double inscription band for albatrosses and giant petrels. *Center (l. to r.):* penguin flipper band, skua band, double inscription band. *Bottom (l. to r.):* penguin flipper band, skua band.

for capturing large numbers of Wilson's petrels and cape pigeons in the sheltered harbors of Antarctica where these birds often congregate.

The large numbers of birds banded during the period covered reflects the intensive population studies being carried on in the Antarctic.

System of Reporting Recoveries

The system outlined in Table 5 is modified for antarctic birds from *Rydzewski* [1954, 1955, 1960]. Rydzewski's first recommendations for an international system of reporting recoveries were discussed at a meeting of the International Coordinating Committee for Bird Ringing held at the Eleventh International Ornithological Congress in Basel, Switzerland, in 1954. This international system has been adopted with slight modifications in many countries in programs such as the Australian Bird Banding Scheme [*Carrick and Turnbull*, 1956; *Carrick et al.*, 1957], the New Zealand Bird Banding Scheme [*Kinsky*, 1963; *Robertson*, 1964], and the national program in Switzerland [*Schifferli*, 1965a, b]. So far as we know, this system has not been adopted as yet by any banding program in North America.

Results

Different types of results are to be expected from banding the various species of antarctic birds. We can describe two arbitrary groups. The first, group 1, are those species whose migrations to and from the pack ice, southern oceans, or even northern oceans (e.g. Wilson's petrel) make recoveries away from the Antarctic unlikely. Work on these species (penguins, most of the petrels except the giant petrel, cormorants, and skuas) is best confined to population studies in specially chosen breeding areas. Group 2 includes most of the albatrosses and the giant petrel that migrate long distances and feed off southern hemisphere continents and shores, thus offering greater chances of recovery.

The full value of the USARP Bird Banding Program will only be appreciated when the results of the various studies still in progress have been published. Population investigations of long-lived vertebrates must be long-term projects with annual marking and counting. The process of collecting data in these and other quantitative studies is not spectacular, and only as the study develops over a number of years is any indication gleaned of what is occurring. This applies particularly to group 1 birds, and at present we are concentrating on the Adélie penguin and the south polar skua at Cape Crozier. The longer this work continues, the more significant such marked populations will become. Recoveries from group 2 birds away from the breeding places tend to be rapid and spectacular, as is demonstrated below.

Adélie penguin. Of the 18 Adélie recoveries reported in Table 6, all but 2 have been from comparatively short distances within the Ross Island area. Thus 10 Adélies banded at Cape Crozier have been recovered at the Cape Bird rookery 74 kilometers away, and 2 have been recovered at the Cape Royds rookery more than 100 kilometers distant as the penguin must travel. So far we have no evidence that a bird banded as a chick has bred in a location other than its place of birth. However, good evidence of distant wanderings by nonbreeders is shown by the two longer distance recoveries in Table 6C. One of these, 509-82169, banded as a breeding bird at the Cape Hallett rookery in 1961 and observed in March 1963 molting at the Cape Crozier rookery some 565 kilometers away, indicates how far from their home rookeries Adélies may stray during the nonbreeding part of their annual cycle. Adélie 509-79109, banded as a chick at Wilkes Station and recovered by Spitchkin at Mirnyy 800 kilometers from its rookery of birth, also shows the extent of their travels. This 5-year-old should have been old enough to breed, but was apparently not seen again. These two are the first long-distance recoveries of Adélies that we know of.

The experimental Adélies *Penney* [1968] banded at Wilkes Station are not recorded in Table 6. In December 1959 5 Adélies were flown to McMurdo and released as part of a series of experiments on penguin orientation. None was seen again during that season, but 3 were recovered back at their original nest-sites at Wilkes at the beginning of the next season (1960–61). These spectacular recoveries of 3 flightless birds purposely displaced over 3200 kilometers from their home territory demonstrate that these birds are capable of interchanges between rookeries considerable distances apart.

In contrast to recoveries of other birds, recoveries of Adélies have without exception been sight records, the band numbers being read through field glasses or the birds captured briefly for checking and released with the band still intact. We can hope for many more long-distance recoveries, especially of Adélies of known age banded at Cape Crozier, but we are curious to know when we will find our first Adélie breeding away from its place of birth.

The albatrosses and giant petrel. These birds are dealt with together. They are being reported in greater detail by Tickell and others from recoveries of USARP, Australian, and British bands [*Tickell*, 1967a, 1968; *Tickell and Gibson*, 1967]. We are presenting data from all USARP Bird Banding Program recoveries in Tables 7–11 because we believe it important at this stage of our knowledge to have the original data available for future correlation with weather, mortality, distribution, and band survival.

Figure 3 shows two major areas of recovery of wandering albatrosses banded at South Georgia; 5 were recovered in the Mar del Plata area of South America, and 9 some 11,250 kilometers away off Bellambi, New South Wales, in Australia. The Bellambi recoveries reflect the work of the New South Wales Albatross Study Group [*Gibson and Sefton*, 1960; *Gibson*, 1963, 1967], a team of ornithologists who are capturing, banding, and releasing wandering albatrosses in one of this bird's regular winter feeding areas. Many more recoveries have been made since June 1965 (*Tickell*, 1967a] of birds banded at South Georgia and recovered in New South Wales and vice versa of birds banded in Australia and recovered back in their breeding places in South Georgia. Only 5 out of a total of 26 recoveries received have been of birds banded as chicks, and none has been reported from the South Pacific between New Zealand and the west coast of South America. Of the 176 wanderers Tufft banded on Îles Crozet, 2 have been recovered, one southwest of Madagascar, the other south of Bunbury, Western Australia.

SLADEN, WOOD, AND MONAGHAN

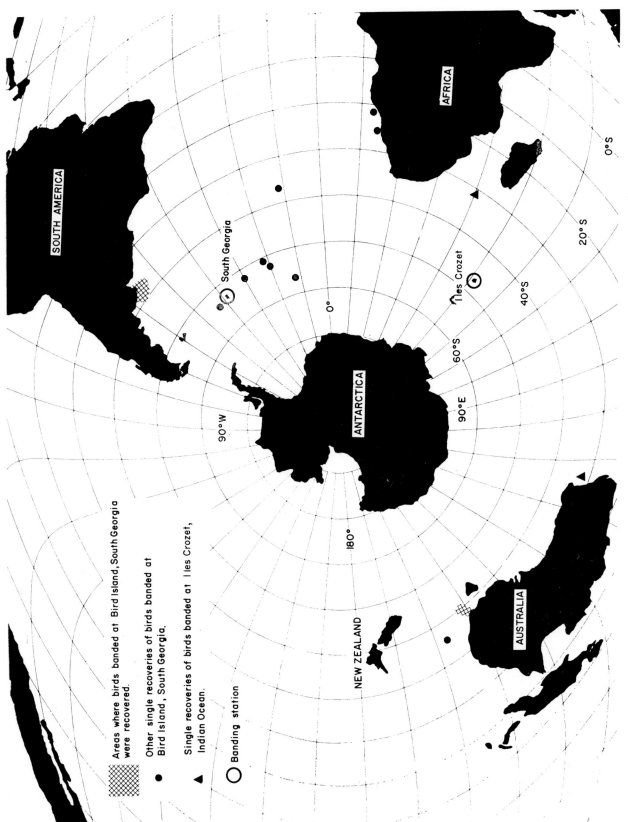

Fig. 3. Wandering albatross recoveries.

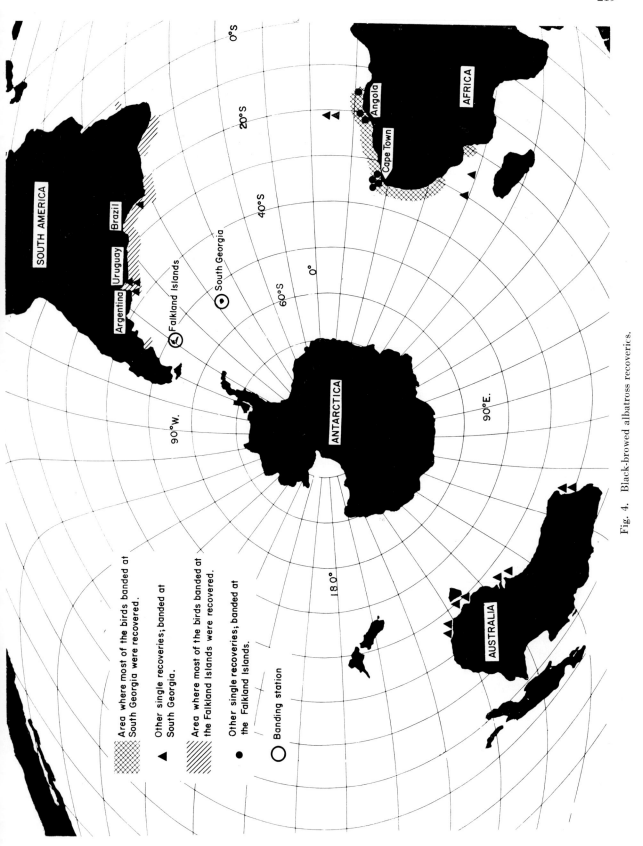

Fig. 4. Black-browed albatross recoveries.

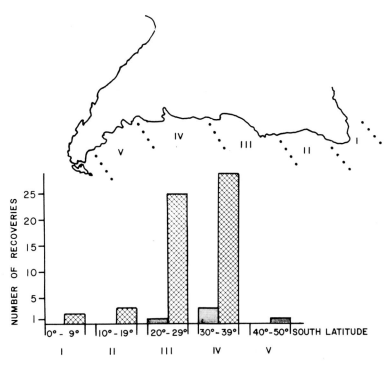

Fig. 5. Black-browed albatross recoveries off eastern South America. Cross-hatched areas represent birds banded at Westpoint Island, Falkland Islands. Shaded areas represent birds banded at Bird Island, South Georgia.

Our most extensive recovery data come from the banding of black-browed albatrosses in two locations: Bird Island, South Georgia, and Westpoint Island, Falkland Islands. Of the 226 recoveries (a 1.1% recovery rate), by far the most have been from the east coast of South America and from South Africa and Angola. These recoveries are too numerous to represent individually in Figure 4; therefore we have shaded the areas in which most of the birds have been recovered. This shows a clear pattern of distribution; the great majority of birds banded in the Falkland Islands were recovered along the east coast of South America, and those banded in South Georgia were reported off South Africa. Only 7 Falkland Islands birds have been recovered elsewhere, and these have been from South Africa and Angola. The South Georgia birds have shown a greater scatter, 4 along the east coast of South America, 4 off Africa, and 8 off the south coast of Australia. The quantitative representation of the distribution of the two groups of birds given in Figures 5 and 6 shows the center of distribution of the South Georgia birds to be off Cape Town and the Cape Point area, South Africa, while the Falkland Islands birds are concentrated off the coasts

of Brazil, Uruguay, and especially Mar del Plata, Argentina.

Like the wandering albatross, no black-browed albatross has been recovered in the South Pacific between New Zealand and the west coast of South America.

In contrast to the relatively plentiful recoveries of the black-browed albatross, the gray-headed albatross has yielded only 9 recoveries from a total of 14,276 banded (0.06% recovery). Figure 7 shows that the distribution follows much the same pattern as that of the South Georgia black-browed albatrosses, with 3 recoveries in South Africa, 1 in Angola, 4 in Australia, and 1 in New Zealand, farther east than any black-browed recovery. This albatross also has not been recovered in the South Pacific or South Atlantic between New Zealand and South Africa.

Table 10 presents recoveries of black-browed and gray-headed albatrosses banded during the first season on Bird Island when it was not possible to identify the species of the chicks positively. They are presented here because others of these birds are likely to be recaptured in South Georgia as adults. Most recoveries are likely to be of black-browed albatrosses.

Though fewer giant petrels have been banded than

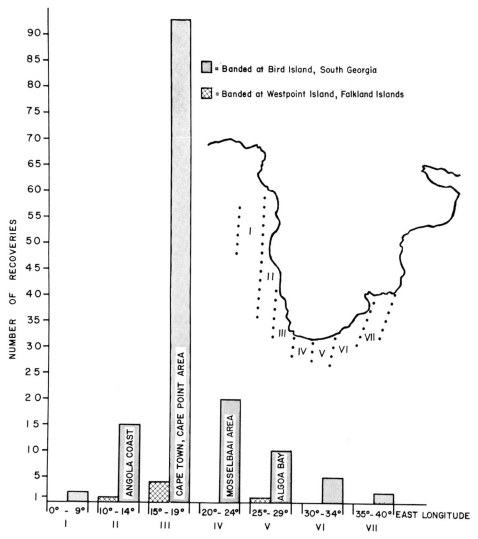

Fig. 6. Black-browed albatross recoveries off Africa.

any of the three species of albatross, they have produced, as expected from previous experience, a higher recovery rate. The total sample of 5,194 birds has yielded a recovery rate of 2.4%, but this has varied considerably from season to season and from area to area. For example, we had a 3.4% recovery rate from the 888 giant petrels banded as chicks on Bird Island in 1958–1959 and an 8% recovery rate from 76 chicks banded on the Frazier Islands near Wilkes station the same season. During the 1960–1961 season Bird Island chicks gave a recovery rate of 2.4%, while at Penguin Island, South Shetland Islands, the recovery rate was 3.4%.

The distribution of recoveries is truly circumpolar and generally in keeping with results from other banding programs [see *Sladen and Tickell, 1958; Stone-*

house, 1958; Ingham, 1959; Tickell and Scotland, 1961], except that no recoveries from the east coast of South America have been previously reported; 7 Bird Island birds were recovered there and one from the Frazier Islands.

The recoveries from 76 chicks banded at the Frazier Islands in February 1959 are of particular interest and warrant a more detailed analysis (see Figure 8). They were the first giant petrel chicks banded on the coast of Antarctica. Previously giant petrel chicks were banded only on such antarctic islands as South Georgia, South Orkney, Macquarie, Heard, and islands off the Antarctic Peninsula. The first recovery, 528-15098 (Table 11E), was reported on May 9, 1959, off Kaiteriteri in Tasman Bay, New Zealand. As this bird probably did not fledge fully and leave its nest

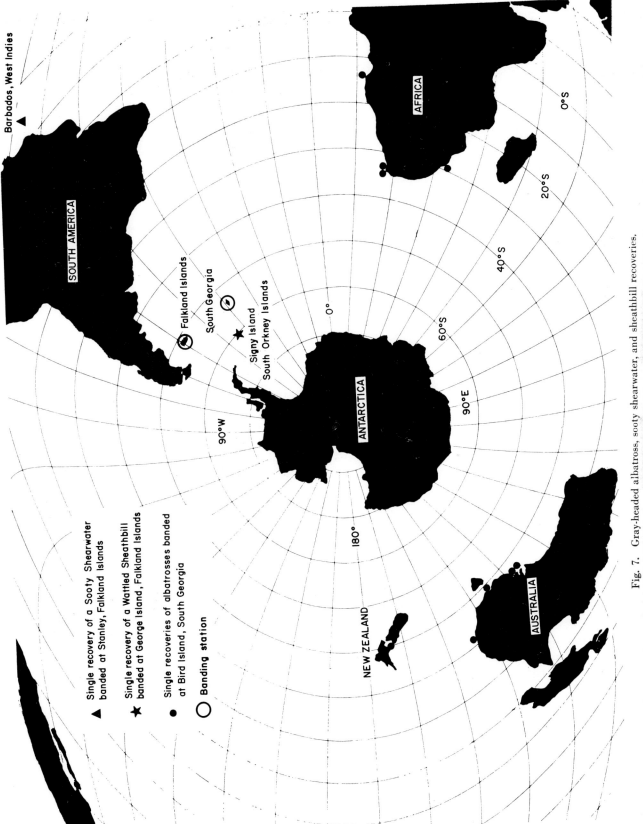

Fig. 7. Gray-headed albatross, sooty shearwater, and sheathbill recoveries.

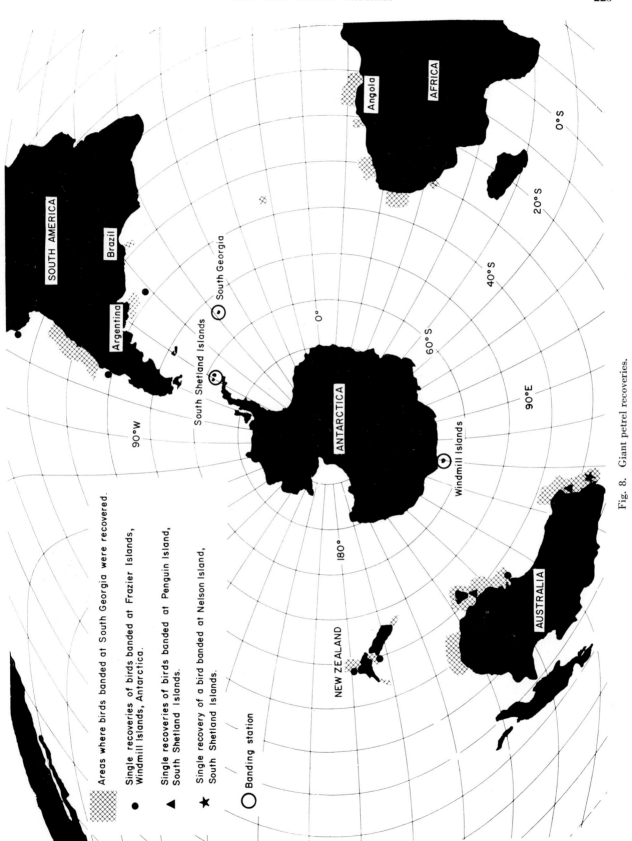

Fig. 8. Giant petrel recoveries.

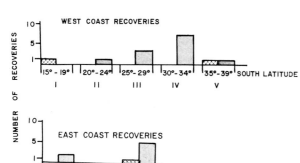

Fig. 9. Giant petrel recoveries off South America. Shaded areas represent birds banded at Bird Island, South Georgia, and cross-hatched areas represent birds banded at Frazier Island, Windmill Islands, Antarctica.

until the end of April, it presumably not only learned to fly, but covered a distance of more than 4,000 kilometers in approximately 2 weeks. The next recovery, 528-15035, was on June 5 near Te Araroa, North Island, New Zealand. Ten days later on June 15, 518-94904 (Table 11A) was found off north Valdivia in Chile, and on September 6 another, 528-15046, was found dead on San Juan beach in Peru. The Peruvian recovery is the most northerly of any of our giant petrel recoveries so far. The fifth recovery, 528-15025, was one of the rare ones on the east coast of South America, south of Puerto Quequén, Argentina, on October 19, 1959. Finally, on June 26, 1960, 16 months after banding, 518-94924 (only the leg with the band remained of the bird) was found in Louth Bay, north of Port Lincoln in South Australia. The dates of recovery succeeding each other circumpolarly provide strong evidence that these birds in their first and second years move in the west wind zone. The thoroughness of this circumpolar dispersion of young giant petrels is nicely demonstrated, for the young from this small colony of less than 100 nests on the fringe of the

Fig. 10. Giant petrel recoveries off Africa.

antarctic continent scattered all the way around the world and moved more than 50° latitude northward.

Figures 9 through 12 demonstrate, as do the Frazier Islands recoveries, a remarkably even distribution of recoveries around the world.

Skuas. In Table 4 we have combined the subspecies *Catharacta skua lonnbergi* from the Antarctic Peninsula area and *C. s. antarctica* from the Falkland Islands.

From a total of 3,149 banded *C. skua*, we have had but one distant recovery (Table 13), that of 617-13722 (almost certainly *C. s. lonnbergi*), banded on King George Island and recovered 20 days later 150 kilometers away at Deception Island. In contrast are 31 recoveries from a total of 4,681 banded south polar skuas, *C. maccormicki*. Many of these (20) are local movements within the Ross Island area (Table 12A, 12B, and 12C). However 5 birds (Table 12D) banded as nestlings at Cape Hallett have been found at Cape Crozier, 565 kilometers away, and 3 of these birds were sighted twice at Crozier, each in consecutive seasons.

Eklund [1961] presents considerable evidence from color-banded birds that *maccormicki* moves extensively along the antarctic coast, but this has not yet been shown for skuas banded in the southern Ross Sea area. Although 70% (3,332) of the total south polar skuas banded have been marked at Crozier, only 6 of these have been recorded away from the Ross Island

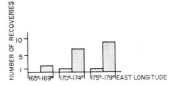

Fig. 11. Giant petrel recoveries off New Zealand. Shaded areas represent birds banded at Bird Island, South Georgia, and cross-hatched areas represent birds banded at Frazier Island, Windmill Islands, Antarctica.

☐ = Banded at Bird Island, South Georgia

▨ = Banded at Frazier Islands, Windmill Islands, Antarctica

▦ = Banded at Penguin Island, South Shetland Islands

▨ = Banded at Nelson Island, South Shetland Islands

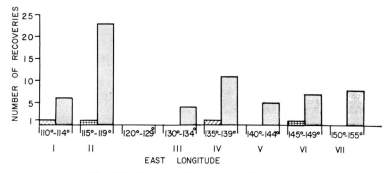

Fig. 12.   Giant petrel recoveries off Australia.

vicinity. These were recovered in New South Wales, Australia, and at Hallett and Dumont D'Urville stations in Antarctica, but too recently to be otherwise listed in this summary. It is apparent that the number of recoveries is in direct proportion to the frequency of visits by interested biologists.

We can, however, report on one very distant recovery during the period of this report; south polar skua[2] 647-27146, banded as an adult by R. and T. Leech at González Videla station (Table 13) March 5, 1961, was caught about 18,150 kilometers away in Udipi, Mysore, India [*Bombay Natural History Society*, 1966] on August 7, 1964, more than three years later. This recovery, like *Eklund*'s [1959] Australian recovery and more recent long-distance recoveries, demonstrates that *maccormicki* makes much more extensive movements away from Antarctica than either *lonnbergi* or *antarctica*.

As was the case with the Adélies, many of the skua recoveries were sight observations, indicating that there is a satisfactory band design for reference number reading (p. 229) and the desire to observe rather than

collect these components of long-term population studies. Within the next year or two, as our known-age population grows, we hope for more long-distance recoveries of breeding skuas in places other than of their birth.

Long-distance recoveries of two Falkland Islands birds, a sooty shearwater and a wattled sheathbill, are given in Table 14. Each of these deserves special comment.

The sooty shearwater was banded by R. Woods on May 4, 1962, in the harbor at Stanley, Falkland Islands. Less than a month later, on June 1, a fisherman found it dead, entangled in his fishing gear near the West Indian island of Barbados. This bird had traveled about 9,500 kilometers in 28 days, and its recovery constitutes one of the very few records of sooty shearwaters in the tropical Atlantic Ocean [*Watson*, 1966].

It is remarkable that the recovery of a sheathbill banded by R. Reid in the Falkland Islands should provide the first positive evidence of migration between the Falklands and the South Orkney islands. This reflects the small amount of banding done on this species and the inadequate design of the band used [see *Sladen and Tickell*, 1958]. The number of sight records of sheathbills far from land on icebergs or

---

[2] After confirmation of species by the U.S. National Museum, the specimen was returned to the Bombay Natural History Society, Bombay.

near whale catchers [*Murphy*, 1936; *Bierman and Voous*, 1950] suggested previously that this bird travels long distances.

Discussion

Our banding results are beginning to demonstrate some interesting patterns of distribution and some important isolating mechanisms. As expected, the Adélie penguins have given us very few recoveries other than the exchanges between neighboring rookeries, but we believe many more exchanges occur than we have detected. Every team member searching diligently for banded penguins in a new part of the Ross Sea area has found some. So far other priorities have prevented it, but in the next few years a team of trained observers should look simultaneously for marked birds in rookeries both near and far from the sites of intensive banding. This search will have to be planned carefully and centered around our knowledge of the species' breeding biology. *Sladen* [*unpublished data*] has compared measurements and weights of Adélie penguins collected at Hope Bay, Antarctic Peninsula (63°24'S, 56°59'W), at Signy Island, South Orkneys (60°43'S, 45°36'W), and on the opposite side of Antarctica at Cape Crozier, Ross Island (77°29'S, 169° 20'E). Their remarkable constancy suggests the occurrence of considerable interchange between the breeding populations of this circumpolar species. We should seek more evidence of this from banded birds.

Some interesting patterns of distribution have been revealed in the albatrosses and giant petrels. The wandering albatross appears to have an almost circumpolar distribution in its wanderings, with a gap in the South Pacific. Breeding wanderers as well as young birds travel to supposedly common feeding grounds annually for part of their winter.

The black-browed albatross recoveries have demonstrated a marked segmentary distribution, the Falkland Islands birds moving to the east coast of South America, the South Georgia birds moving to South Africa and a few to Australia. Several insular subspecies have been described [*Murphy*, 1936]. Our banding program in the Falklands and South Georgia should be repeated with other insular populations.

The gray-headed albatross, which nests side by side with the black-browed, has a completely different pattern of pelagic feeding and movements, confining itself to the open sea and rarely approaching land. This indicates an isolating mechanism that separates these closely related species [*Tickell*, 1968]. The differ-

ences in feeding preferences between these two species have already been reported by *Tickell* [1964].

We have been able to gather more evidence of the truly circumpolar travels of the young giant petrels and have added to existing knowledge on the routes and speed with which these birds travel around the world.

It is of considerable biological interest to compare the ages of the different species recovered. Tables 15–18 analyze the age at recovery of these albatrosses and giant petrels banded as nestlings. There are discrepancies between the totals in these tables and the recovery totals for these species in Table 4 because the dates of some of the recoveries (14 black-brows and 3 giant petrels) are too uncertain to put into age groups. We are following the U.S. Fish and Wildlife Service Bird Banding Laboratory's (1961–1967) age classification system, with modifications for the Southern Hemisphere breeding season. Thus, for the wandering albatross with its 9-month chick period, the first year extends from the mean date of hatching to June 30 of the following year, a period of 15½ months; the second and subsequent years extend from July 1 to June 30. With the other albatrosses and the giant petrel the first year extends from the mean date of hatching to June 30, and, as for the wanderer, the second and subsequent years extend from July 1 to June 30. This classification somewhat changes our previous concept of age groups, for we have talked of the "first year" giant petrels as the great travelers. We feel this classification, which reckons age from hatching, more accurately represents the age of the bird.

The recoveries of known-age wanderers (Table 15) are too few for drawing any conclusions, but they indicate that all age groups are being recovered, as evidenced particularly by the many adults recovered in Australian waters. The few gray-headed albatross recoveries (Table 17) fall into about the same pattern as those of the wanderers. In contrast, and confirming previous studies, the giant petrel recoveries are largely confined to their 1st and 2nd years, there being only 3 recoveries of 3rd year birds out of 3,732 banded and one 4th year recovery out of 2,932 banded.

The black-browed albatrosses (Table 16) show an intermediate pattern, the largest number of recoveries being in their 2nd year (0.5%); there are, however, significant recoveries in their 3rd (0.2%), 4th (0.04%), and 5th years (0.06%).

We urge continued large-scale banding of giant petrel chicks at selected breeding places, for banding programs are still too young to produce really meaningful results. Despite an accumulating amount of data on the worldwide movements of these young birds, increasing evidence [*Warham*, 1962] indicates that the young tend eventually to return to their place of birth to breed. One nestling Sladen banded in 1951 at Signy Island, South Orkneys, was found breeding in the same area at the age of 8 years [*Tickell and Scotland*, 1961]. We know almost nothing about the movements of the mature breeding birds. This challenge we will have to pass to the electronics experts, for we believe tracking these large birds by satellite as suggested by *Warner* [1963] and others would be very profitable.

Acknowledgments. This report is concerned with field work done by many antarctic biologists from the U.S. and other nations, and with recoveries reported by a large number of people from all over the world. We are most grateful to all these and especially to W. L. N. Tickell for the part he played in the banding at South Georgia and his helpful criticism of the text. The work has been supported in part by National Science Foundation grants G-6327, G-9990, G-17924, G-24067, GA-63, and GA-151 from the Office of Antarctic Programs. We are grateful for help and good counsel from NSF, especially from George Llano, and for logistic support from the U.S. Navy. Richard Penney and the late Carl Eklund have at one time during this period been joint project directors of the USARP Bird Banding Program, and we thank them for help and advice. We deeply regret that Eklund did not survive to be a joint author of this publication, for many of his early ideas on antarctic ornithology have matured in this and other recent publications. Much of this work would have been impossible without help from Margaret Gambrill at Johns Hopkins, our close collaboration with the Banding Laboratory of the U.S. Fish and Wildlife Service, and especially Alan Duvall, Earl Baysinger, and Chandler Robbins. To all these we extend our most sincere thanks.

REFERENCES

Austin, O. L., Jr., Notes on banding birds in Antarctica, and on the Adélie penguin colonies of the Ross Sea sector, *Bird-Banding, 28,* 1–26, 1957.

Bierman, W. H., and K. H. Voous, *Birds Observed and Collected during the Whaling Expeditions of Willem Barendsz in the Antarctic, 1946–1947 and 1947–1948,* 123 pp., Brill, Leiden, 1950.

Bombay Natural History Society (Editors), Recovery of ringed birds, *J. Bombay Nat. Hist. Soc., 62,* 564–565, 1966.

Boyd, J. C., W. J. L. Sladen, and H. A. Baldwin, Biotelemetry of penguin body temperature, 1966–1967, *Antarctic J. U. S., 2,* 97–99, 1967.

Carrick, R., N. Keith, and K. Keith, Third annual report of the Australian Bird-Banding Scheme, July 1956 to June 1957, *CSIRO Wildlife Research, 2,* 145–163, 1957.

Carrick, R., and N. Turnbull, First annual report of the Australian Bird-Banding Scheme, October 1953 to June 1955, *CSIRO Wildlife Research, 1,* 31–39, 1956.

Charcot, J., *The Voyage of the "Why Not" in the Antarctic,* p. 244, viii + 315 pp., Hodder and Stoughton, New York, 1911.

Chittleborough, R. G., and E. H. M. Ealey, Bird ringing at Heard Island during 1949, *Emu, 50,* 102–104, 1950.

Douglas, D. S., Extra-renal salt excretion in the Adélie penguin chick, in *Biologie Antarctique,* edited by R. Carrick, M. Holdgate, and J. Prevost, pp. 503–508, Hermann, Paris, 1964.

Douglas, D. S., Salt and water metabolism of the Adélie penguin, *this volume,* 1968.

Downs, M. C., A. M. Gwynn, and P. F. Howard, Banding of giant petrels at Heard and Macquarie islands, *Emu, 54,* 257–262, 1954.

Eklund, C. R., Condensed ornithology report, East Base, Palmer Land, *Proc. Am. Phil. Soc., 89,* 299–304, 1945.

Eklund, C. R., South polar skua recovery in Australia, *Emu, 59,* 158, 1959.

Eklund, C. R., Distribution and life history studies of the south polar skua, *Bird-Banding, 32,* 187–223, 1961.

Emison, W. E., Feeding preferences of the Adélie penguin at Cape Crozier, Ross Island, *this volume,* 1968.

Emlen, J. T., and R. L. Penny, Distance navigation in the Adélie penguin, *Ibis, 106,* 147–431, 1964.

Gain, L., Oiseaux antarctiques, *Deuxieme Expedition Antarctique Francaise, 1908–10,* 200 pp., Paris, 1914.

Gibson, J. D., Third report of the New South Wales Albatross Study Group (1962), *Emu, 63,* 215–223, 1963.

Gibson, J. D., The wandering albatross (*Diomedea exulans*): Results of banding and observations in New South Wales coastal waters and the Tasman Sea, *Notornis, 114,* 47–57, 1967.

Gibson, J. D., and A. R. Sefton, Second report of the New South Wales Albatross Study Group, *Emu, 60,* 125–130, 1960.

Goldsmith, R., and W. J. L. Sladen, Temperature regulation of some antarctic penguins, *J. Physiol., 157,* 251–262, 1961.

Gwynn, A. M., Penguin marking at Heard Island, 1953, *Australian National Antarctic Expeditions, Interim Reports No. 8,* pp. 8–12, Melbourne, 1955.

Holgerson, H., *Ornithology of the "Brategg" Expedition,* Norwegian Research Council for Science and the Humanities, 80 pp., A. S. Griegs, Bergen, 1957.

Ingham, S. E., Banding of giant petrels by the Australian National Antarctic Research Expeditions 1955–1958, *Emu, 59,* 189–200, 1959.

International Council of Scientific Unions, SCAR, *Scientific Committee on Antarctic Research, Manual,* 84 pp., Scott Polar Research Institute, Cambridge, England, 1966.

Jones, N. V., The sheathbill, *Chionis alba* (Gmelin), at Signy Island, South Orkney Islands, *Brit. Antarct. Surv. Bull.,* no. 2, 53–71, 1963.

Kinsky, F. C., *13th Annual Report of the New Zealand Bird Banding Scheme for the Year Ending 31st March 1963,* pp. 1–32, Dominion Museum, Wellington, N. Z., 1963.

Müller-Schwarze, D., Circadian rhythms in the Adélie penguin, *this volume,* 1968.

Murphy, R. C., *Oceanic Birds of South America,* vols. 1 and 2, 1,245 pp., Am. Mus. Nat. Hist., New York, 1936.

Penney, R. L., The Adélie penguin's faithfulness to territory and mate, in *Biologie Antarctique,* edited by R. Carrick,

M. Holdgate, and J. Prevost, pp. 401–406, Hermann, Paris, 1964.

Penney, R. L., Territorial and social behavior in the Adélie penguin, *this volume*, 1968.

Penny, R. L., and J. T. Emlen, Further experiments on distance navigation in the Adélie penguin *Pygoscelis adeliae*, *Ibis, 109*, 99–109, 1967.

Penney, R. L., and W. J. L. Sladen, The use of Teflon for banding penguins, *J. Wildlife Management, 30*, 847–850, 1966.

Pinder, R., The cape pigeon, *Daption capensis* Linnaeus, at Signy Island, South Orkney Islands, *Brit. Antarct. Surv. Bull.*, no. 8, 19–47, 1966.

Prevost, J., Ecologie du manchot empereur, *Expeditions Polaires Francaises*, no. 222, 204 pp., 1961.

Pryor, M. S., The avifauna of Haswell Island, Antarctica, *this volume*, 1968.

Reid, B., The growth and development of the south polar skua (*Catharacta maccormicki*), *Notornis, 13*, 71–89, 1966.

Roberts, B. B., The life cycle of Wilson's petrel (*Oceanites oceanicus*), *Brit. Graham Land Exped., 1934–37, Sc. Reports, 1*, 141–194, 1940.

Roberts, B. B., and W. J. L. Sladen, Preliminary note on bird ringing by the Falkland Islands Dependencies Survey, 1945–51, *Ibis, 94*, 538–540, 1952.

Robertson, C. J. R., *14th Annual Report of the New Zealand Bird Banding Scheme for the Year Ending 31st March 1964*, pp. 1–32, Dominion Museum, Wellington, N. Z., 1964.

Rydzewski, W., A historical review of bird marking, *Dansk Ornithologisk Forenings Tidsskrift, 45*, 61–95, 1951.

Rydzewski, W., *The Ring*, no. 1, 2–4, 1954.

Rydzewski, W., *The Ring*, no. 2, 17–26, 1955.

Rydzewski, W., *The Ring*, no. 22, 203–207, 1960.

Sapin-Jaloustre, J., Medecine-Biologie, in *Expedition en Terre Adélie* 1949–51, *Rop. prel.*, serie scient., no. 14, pp. 76–86, 1952.

Sapin-Jaloustre, J., Ecologie du manchot Adélie, *Expeditions Polaires Francaises*, no. 208, 211 pp., 1960.

Sapin-Jaloustre, J., and F. Bourliere, Incubation et developpement du poussin chez le manchot Adélie, *Pygoscelis adeliae*, *Aluada, 19*, 65–83, 1951.

Schifferli, A., Bericht der Schweizerischen Vogelwarte Sempach fur die Jahre 1963 und 1964, *Orn. Beob., 62*, 125–173, 1965a.

Schifferli, A., Schweizerische Ringfundmeldung fur 1963 und 1964, *Orn. Beob., 62*, 141–169, 1965b.

Serventy, D. L., Recovery of a ringed giant petrel from South Orkney Islands, *Emu, 48*, 158–159, 1948.

Sladen, W. J. L., Notes on methods of marking penguins, *Ibis, 94*, 541–543, 1952.

Sladen, W. J. L., The Adélie penguin, *Nature, 171*, 952–955, 1953.

Sladen, W. J. L., The Pygoscelid penguins, 1, Methods of study, 2, The Adélie penguin, *Falkland Islands Dependencies Surv. Sci. Rept.*, no. 17, 97 pp., 1958.

Sladen, W. J. L., Ornithological research in Antarctica, *Bio Science, 15*, 264–268, 1965.

Sladen, W. J. L., J. C. Boyd, and J. M. Pedersen, Biotelemetry studies on penguin body temperatures, *Antarctic J. U. S., 1*, 142–143, 1966.

Sladen, W. J. L., and H. Friedmann, Antarctic ornithology, in *Science in Antarctica*, vol. 1, pp. 62–76, Nat. Acad. Sci., Nat. Res. Council, Publ. 839, 1961.

Sladen, W. J. L., and R. Goldsmith, Biological and medical research based on USS *Staten Island*, Antarctic, 1958–59, *Polar Record, 10*, 146–148, 1960.

Sladen, W. J. L., and R. L. Penney, Penguin flipper bands used by the USARP Bird Banding Program, 1958–60, *Bird-Banding, 31*, 79–82, 1960.

Sladen, W. J. L., and W. L. N. Tickell, Antarctic bird banding by the Falkland Islands Dependencies Survey, 1945–57, *Bird-Banding, 29*, 1–26, 1958.

Spencer, R., Report on bird-ringing for 1961, *British Birds, 55*, 493–556, 1962.

Spencer, R., Report on bird-ringing for 1962, *British Birds, 56*, 477–540, 1963.

Spencer, R., Report on bird-ringing for 1963, *British Birds, 57*, 525–596, 1964.

Stonehouse, B., The brown skua of South Georgia, *Falkland Islands Dependencies Surv. Sci. Rept.*, no. 14, 25 pp., 1956.

Stonehouse, B., Notes on the ringing and breeding distribution of the giant petrel *Macronectes giganteus*, *Ibis, 100*, 204–208, 1958.

Stonehouse, B., The king penguin of South Georgia, *Falkland Islands Dependencies Surv. Sci. Rept.*, no. 23, 81 pp., 1960.

Taylor, R. H., The Adélie penguin at Cape Royds, *Ibis, 104*, 176–204, 1962.

Tickell, W. L. N., Chick feeding in the wandering albatross, *Nature, 185*, 116–117, 1960.

Tickell, W. L. N., The dove prion, *Pachyptila desolata* Gmelin, *Falkland Islands Dependencies Surv. Sci. Rept.*, no. 33, 55 pp., 1962.

Tickell, W. L. N., Feeding preferences of the albatrosses *Diomedea melanophris* and *D. chrysostoma* at South Georgia, in *Biologie Antarctique*, edited by R. Carrick, M. Holdgate, and J. Prevost, pp. 383–387, Hermann, Paris, 1964.

Tickell, W. L. N., Movements of black-browed and gray-headed albatrosses in the South Atlantic, *Emu, 66*, 357–367, 1967a.

Tickell, W. L. N., Biology of the great albatrosses, *Diomedea exulans* and *D. melanophris*, *this volume*, 1968.

Tickell, W. L. N., and P. A. Cordall, South Georgia Biological Expedition, 1958–1959, *Polar Record, 10*, 145–146, 1960.

Tickell, W. L. N., and J. D. Gibson, Movements of wandering albatrosses, *Diomedea exulans*, *Emu* (in press), 1967.

Tickell, W. L. N., and R. Pinder, Two-egg clutches in albatrosses, *Ibis, 108*, 126–129, 1966.

Tickell, W. L. N., and R. Pinder, Breeding frequencies in the albatrosses *Diomedea melanophris* and *D. chrysostoma*, *Nature, 213*, 315–316, 1967.

Tickell, W. L. N., R. Pinder, and H. B. Clagg, Biological studies on Bird Island, South Georgia, 1962–64, *Polar Record, 12*, 601–602, 1965.

Tickell, W. L. N., and C. Scotland, Recoveries of ringed giant petrels, *Ibis, 103a*, 260–266, 1961.

U.S. Fish and Wildlife Service, *Bird Banding Manual*, Loose leaf, n.p., 1961–1967.

Warham, J., The biology of the giant petrel, *Auk, 79*, 130–160, 1962.

Warner, D. W., Space tracks, *Natural History, 72*, 8–15, 1963.

Watson, G. E., *Seabirds of the Tropical Atlantic Ocean*, 120 pp., Smithsonian Press, Washington, D.C., 1966.

Young, E. C., The breeding behaviour of the south polar skua, *Ibis, 105*, 203–233, 1963a.

Young, E. C., Feeding habits of the south polar skua, *Ibis, 105*, 301–318, 1963b.

TABLE 1. Common and Scientific Names of Species Banded with U. S. Fish and Wildlife Service Numbers and Band Sizes

Species Number*	Species	Band Size
8502	Emperor Penguin, *Aptenodytes forsteri*	9†
8503	Gentoo Penguin, *Pygoscelis papua*	9†
8504	Adélie Penguin, *P. adeliae*	9†
8505	Chinstrap Penguin, *P. antarctica*	9†
8508	Rockhopper Penguin, *Eudyptes crestatus*	9†
8510	Macaroni (or Royal) Penguin, *E. chrysolophus*	9†
8518	Wandering Albatross, *Diomedea exulans*	8§
0822	Black-browed Albatross, *D. melanophris*	8
8522	Gray-headed Albatross, *D. chrysostoma*	8
8551	Black-browed or Gray-headed Albatross‡	8
8524	Light-mantled Sooty Albatross, *Phoebetria palpebrata*	8
8525	Giant Petrel, *Macronectes giganteus*	8
1020	Pintado (or Cape) Petrel, *Daption capensis*	4
8526	Silver-gray Fulmar, *Fulmarus glacialoides*	5
8533	White-chinned Petrel (or Shoemaker), *Procellaria aequinoctialis*	5
0890	Greater Shearwater, *Puffinus gravis*	5
0950	Sooty Shearwater, *P. griseus*	5
8537	Snow Petrel, *Pagodroma nivea*	4

TABLE 1 (*continued*)

Species Number*	Species	Size Band
1060	Leach's Petrel, *Oceanodroma leucorhoa*	1A
1090	Wilson's Petrel, *Oceanites oceanicus*	1B
8538	Gray-backed Petrel, *Garrodia nereis*	1B
8541	Blue-eyed Cormorant, *Phalacrocorax atriceps*	8
8543	Wattled Sheathbill, *Chionis alba*	7A
0350	Great Skua, *Catharacta skua*	7A or 7B
8545	South Polar Skua, *C. maccormicki*	7A
8546	Southern Black-backed Gull, *Larus dominicanus*	7A

* Species numbers that begin with the numeral 8 were assigned by the U. S. Fish and Wildlife Service specifically for antarctic species that are not included in the American Ornithologists' Union *Check-list of North American Birds*.

† Special designs of flipper bands were used on these species.

§ Bands used for *D. exulans* were double inscription and 4½ inches (115 mm) long. All other size-8 bands were 4-inch (101 mm) long, double-inscription bands.

‡ During 1958–59 it was not possible to distinguish positively between the chicks of these species; therefore a separate species number was used.

TABLE 2. USARP Banders and Affiliating Organizations

Banders

BA	B. Araya (IAC)	SH	S. Haven (SU)	CR	C. J. R. Robertson (NZARP)
CB	C. Bailey (NZARP)	TK	T. Kitamura (JARE)	RR	R. Reid (JHU*)
JB	J. Bromley (JHU*)	GL	G. Lowry (JHU)	BS	B. Stonehouse (NZARP)
WB	W. Bryden (ANARE)	GEL	G. E. Lippert (BBM)	IS	I. Strange (JHU*)
GC	G. Caughley (NZARP)	PL	P. LeMorvan (EPF)	IFS	I. F. Spellerberg (NZARP)
JC	J. Cranfield (NZARP)	RL	R. E. Leech (BBM)	KS	K. Salmon (NZARP)
PC	P. A. Cordall (JHU)	TL	T. Leech (BBM)	PS	P. M. Smith (USARP)
DD	D. Davidson (JHU)	CM	C. Martell (JHU*)	WS	W. J. L. Sladen (JHU)
DSD	D. S. Douglas (DU)	RM	R. C. Murphy (AMNH)	CT	C. B. Taylor (NZARP)
HD	H. Dollman (JHU)	WM	W. J. Maher (UC)	RT	R. Tufft (TE)
JD	J. Drabkin	LN	L. Napier (JHU*)	RBT	R. B. Thomson (NZARP)
JHD	J. H. Dearborn (SU)	RN	R. B. Napier (JHU*)	RHT	R. H. Taylor (NZARP)
WE	W. B. Emison (JHU)	NO	M. N. Orton (ANARE)	WT	W. L. N. Tickell (JHU)
CF	C. F. le Feuvre (JHU)	AP	A. G. H. Parker (JHU)	EV	E. M. van Zinderen Bakker, Jr. (SANAE)
VF	V. F. Flyger (JHU)	JP	J. Poltock (JHU*)	AW	A. Woods (JHU*)
PG	P. Gladstone (JHU*)	MP	M. E. Pryor (OSU)	RW	R. C. Wood (JHU)
DG	D. M. Galloway (JHU*)	RP	R. Pinder (JHU)	RWW	R. W. Woods (JHU*)
CH	C. E. Huntington (BC)	RLP	R. L. Penney (UW)	EY	E. C. Young (NZARP)
RH	R. Haga (JARE)	BR	B. E. Reid (NZARP)		

Affiliating Organizations

ANARE	Australian National Antarctic Research Expeditions
EPF	Expéditions Polaires Françaises
IAA	Instituto Antartico Argentino
IAC	Instituto Antartico Chileno
JARE	Japanese Antarctic Research Expedition
NZARP	New Zealand Antarctic Research Programme
SANAE	South African National Antarctic Expedition
TE	H. W. Tillman Expedition to Îles Crozet and Kerguélen
USARP	United States Antarctic Research Program
AMNH	American Museum of Natural History[a]
BBM	Bernice P. Bishop Museum[a]
BC	Bowdoin College[a]
DU	Duke University[a]
JHU	Johns Hopkins University[a]
JHU*	Johns Hopkins collaborators in Falkland Islands[a]
OSU	Ohio State University[a]
SU	Stanford University[a]
UC	University of California[a]
UW	University of Wisconsin[a]

[a] Included in USARP.

TABLE 3. Principal Banding Stations and Their Coordinates

Ardley Island, King George Island, South Shetland Islands	62.13S, 58.54W
Beaufort Island, Ross Sea	76.53S, 167.00E
Bellingshausen Sea	72S, 85W
Bird Island, South Georgia	54.00S, 38.02W
Bleaker Island, Falkland Islands	52.10S, 58.50W
Bluff Cove, Falkland Islands	51.47S, 58.10W
Cape Adare, Victoria Land, Antarctica	71.17S, 170.15E
Cape Barne, Ross Island, Antarctica	77.35S, 166.13E
Cape Bird, Ross Island, Antarctica	77.16S, 166.20E
Cape Crozier, Ross Island, Antarctica	77.27S, 169.13E
Cape Evans, Ross Island, Antarctica	77.38S, 166.24E
Cape Royds, Ross Island, Antarctica	77.33S, 166.07E
Carcass Island, Falkland Islands	51.15S, 60.33W
Chartres, Falkland Islands	51.43S, 60.01W
Coulman Island, Victoria Land, Antarctica	73.21S, 170.40E
Darwin, Falkland Islands	51.49S, 58.58W
Deception Island, South Shetland Islands	62.57S, 60.38W
Dunbar Island, Falkland Islands	51.21S, 60.22W
Ellsworth station, Filchner Ice Shelf, Antarctica	77.43S, 41.08W
Frazier Islands, Windmill Islands, Antarctica	66.13S, 110.10E
Géologie Archipelago, Adélie Coast, Antarctica	66.39S, 139.55E
George Island, Falkland Islands	52.20S, 59.44W
González Videla station, Danco Coast, Antarctic Peninsula	64.49S, 62.51W
Hallett station, Victoria Land, Antarctica	72.18S, 170.13E
Îles Crozet, Indian Ocean	46.25S, 51.40E
Kent Island, New Brunswick, Canada	44.35N, 66.45W
Kidney Island, Falkland Islands	51.38S, 57.45W
Marble Point, Victoria Land, Antarctica	77.26S, 163.48E
Marion Island, Indian Ocean	46.55S, 37.45E
McMurdo Sound, Ross Sea	77.38S, 166.05E
McMurdo station, Ross Island, Antarctica	77.51S, 166.37E
Mirnyy station, Queen Mary Coast, Antarctica	66.33S, 93.01E
Nelson Island, South Shetland Islands	62.17S, 59.02W
Ongulkalven Island, Prince Olav Coast, Antarctica	69.02S, 39.25E
Palmer station, Anvers Island, Antarctic Peninsula	64.45S, 64.05W
Penguin Island, South Shetland Islands	62.08S, 57.55W
Prince Harald Coast, Antarctica	68.25S, 38.47E
Duthiers Point (Punta Canelo), Danco Coast, Antarctic Peninsula	64.48S, 63.49W
Sabrina Island, Balleny Islands	66.55S, 163.20E
Sparrow Cove, Falkland Islands	51.39S, 57.48W
Stanley, Falkland Islands	51.45S, 57.56W
Ongul Island, Prince Olav Coast, Antarctica	69.03S, 39.30E
Westpoint Island, Falkland Islands	51.21S, 60.41W
Wilkes station, Budd Coast, Antarctica	66.15S, 110.32E
Windmill Islands, Budd Coast, Antarctica	66.20S, 110.28E

TABLE 4*A*.   Totals of Each Species Banded and Recovered, 1958 through June 1965

Species	Banding Station	Year	Bander	Totals Banded			Totals Recovered		
				P*	A†	Total	P*	A†	Total
Emperor Penguin[a]	McMurdo Sound	1958–1959	WS		8	8			
(*Aptenodytes*	Coulman Island	1958–1959	WS,RLP	30	73	103			
forsteri)	Hallett station	1958–1959	BR		2	2			
		1959–1960	RBT		4	4			
		1959–1960	BR		84	84			
		1960–1961	BR		18	18			
	Wilkes station	1959–1960	RLP		4	4			
		1960–1961	RLP		3	3			
		1961–1962	NO		1	1			
			Totals	30	197	227			
Gentoo Penguin	Penguin Island	1959–1960	RM,PS		2	2			
(*Pygoscelis papua*)	Sparrow Cove	1959–1960	RM,PS		33	33			
			Totals		35	35			
Adélie Penguin	Cape Bird	1958–1959	GC	53		53			
(*P. adeliae*)	Cape Royds	1958–1959	WS		33	33			
		1959–1960	RHT	312	188	500		2	2
		1961–1962	RW	109		109			
		1963–1964	IFS	952		952			
		1964–1965	BS		357	357		2	2
	Cape Crozier	1958–1959	WS		198	198			
		1961–1962	WS,RW	2,310	1,347	3,657	3	2	5
		1962–1963	RW,VF	4,110	304	4,414	7		7
		1963–1964	WS,AP	5,000	136‡	5,136			
		1964–1965	AP,GL	5,000	124	5,124			
	Hallett station	1958–1959	WS	207	150	357			
		1958–1959	BR	181		181			
		1959–1960	BR	75	938	1,013			
		1960–1961	DSD		246	246			
		1960–1961	CB	100		100			
		1961–1962	RW		3,580	3,580		1	1
		1962–1963	BR		59	59			
		1962–1963	JC		120	120			
		1964–1965	JC		486	486			
	Cape Adare	1958–1959	WS	110	15	125			
	Beaufort Island	1958–1959	RLP	30		30			
	Sabrina Island	1958–1959	WS		22	22			
	Wilkes station	1958–1959	WS	5	16	21			
		1958–1959	RLP	15	481	496			
		1959–1960	RLP	200	1,020	1,220	1		1
		1960–1961	RLP		97	97			
	Mirnyy station	1964–1965	RLP		39	39			
	Géologie Archipelago	1964–1965	PL	153	172	325			
	Ongulkalven Island	1959–1960	TK	1	11	12			
	Ongul Island	1959–1960	TK	5	12	17			
	Prince Harald Coast	1959–1960	RH		30	30			
	Bellingshausen Sea	1959–1960	RM,PS		3	3			
	Penguin Island	1959–1960	RM,PS		1	1			
			Totals	18,928	10,185‡	29,113	11	7	18

TABLE 4A (*continued*)

Species	Banding Station	Year	Bander	Totals Banded P*	A†	Total	Totals Recovered P*	A†	Total
Chinstrap Penguin	Sabrina Island	1958–1959	WS		3	3			
(*P. antarctica*)		1963–1964	RW		1	1			
		1964–1965	CR	6	17	23			
	Deception Island	1959–1960	RM,PS		9	9			
	Penguin Island	1959–1960	RM,PS		83	83			
	González Videla station	1960–1961	RL,TL	21	229	250			
	Géologie Archipelago	1964–1965	PL		5	5			
	Cape Crozier	1962–1963	RW		1	1			
		1963–1964	WE		1	1			
		1964–1965	RW		3	3			
			Totals	27	352	379			
Rockhopper Penguin	Kidney Island	1961–1962	RWW		54	54			
(*Eudyptes crestatus*)	Westpoint Island	1961–1962	RN		21	21			
			Totals		75	75			
Macaroni Penguin	Penguin Island	1959–1960	RM,PS		1	1			
(*E. chrysolophus*)	Kidney Island	1961–1962	RWW		1	1			
	Géologie Archipelago	1964–1965	PL		1	1			
			Totals		3	3			
Wandering Albatross	Bird Island	1958–1959	WT,PC		89	89		2	2
(*Diomedea exulans*)		1960–1961	WT,HD		1,415	1,415		10	10
		1961–1962	HD,CF		1,500	1,500		7	7
		1962–1963	WT,RP	400	758	1,158	4		4
		1963–1964	WT,RP	1,000	331	1,331	1		1
	Îles Crozet	1959–1960	RT		176	176		2	2
	Marion Island	1964–1965	EV	225		225			
			Totals	1,625	4,269	5,894	5	21	26
Black-browed Albatross	Westpoint Island	1959–1960	RN	199		199	3		3
(*D. melanophris*)		1960–1961	RN,LN	600	100	700	5		5
		1961–1962	RN,RWW, AW,IS	3,000		3,000	30		30
		1962–1963	RN,RWW, AW	3,000		3,000	17		17
		1963–1964	RN,LN,IS	1,994	6	2,000	11		11
		1964–1965	RN,LN, PG,CM	2,000		2,000	3		3
	Bird Island	1960–1961	WT,HD	3,251	260	3,511	54	1	55
		1961–1962	HD,CF	5,300	480	5,780	40		40
		1962–1963	WT,RP	7,622	40	7,662	54		54
		1963–1964	WT,RP	2,781	40	2,821	8		8
			Totals	29,747	926	30,673	225	1	226
Gray-headed Albatross	Bird Island	1960–1961	WT,HD	3,529	40	3,569	4		4
(*D. chrysostoma*)		1961–1962	HD,CF	2,750	670	3,420	2		2
		1962–1963	WT,RP	5,600	95	5,695	3		3
		1963–1964	WT,RP	1,400	192	1,592			
			Totals	13,279	997	14,276	9		9

TABLE 4*A* (*continued*)

Species	Banding Station	Year	Bander	Totals Banded			Totals Recovered		
				P*	A†	Total	P*	A†	Total
Black-browed or Gray-headed Albatross (*D. melanophris* and *chrysostoma* chicks)	Bird Island	1958–1959	WT,PC	3,550		3,550	39		39
Light-mantled Sooty Albatross (*Phoebetria palpebrata*)	Bird Island	1963–1964	WT,RP		20	20			
Giant Petrel	Frazier Islands	1958–1959	WS	76	7	83	6		6
		1959–1960	RLP	46		46			
(*Macronectes giganteus*)	Bird Island	1958–1959	WT,PC	888	19	907	30		30
		1960–1961	WT,HD	1,207	93	1,300	39		39
		1961–1962	HD,CF	633	117	750	18		18
		1962–1963	WT,RP	800	1	801	15		15
		1963–1964	WT,RP	547	1	548	16		16
	Îles Crozet	1959–1960	RT	24		24			
	Ardley Peninsula	1959–1960	IAA		10	10			
	Penguin Island	1960–1961	RL,TL	58		58	2		2
	Nelson Island	1964–1965	BA	320	121	441	1		1
	Marion Island	1964–1965	EV	2		2			
	Palmer station	1964–1965	GEL	224		224			
			Totals	4,825	369	5,194	127		127
Pintado Petrel (*Daption capensis*)	Windmill Islands	1958–1959	WS	2	3	5			
		1958–1959	WB		10	10			
		1959–1960	RLP	10	24	34			
	Bird Island	1960–1961	WT,HD	17		17			
	Mirnyy station	1962–1963	MP		25	25			
	Nelson Island	1964–1965	BA	4		4			
			Totals	33	62	95			
Silver-gray Fulmar (*Fulmarus glacialoides*)	Windmill Islands	1958–1959	WS		5	5			
		1959–1960	RLP	30	29	59			
		1960–1961	RLP		39	39			
			Totals	30	73	103			
White-chinned Petrel (*Procellaria aequinoctialis*)	Kidney Island	1960–1961	RWW		1	1			
		1961–1962	RWW		45	45			
		1962–1963	RWW,DD		20	20			
			Totals		66	66			
Greater Shearwater (*Puffinus gravis*)	Kidney Island	1961–1962	RWW		2	2			
Sooty Shearwater (*P. griseus*)	Stanley	1961–1962	RWW		1	1		1	1
	Kidney Island	1961–1962	RWW	2	48	50			
		1962–1963	RWW,DD, IS		39	39			
		1963–1964	DD		95	95			
			Totals	2	183	185		1	1

TABLE 4*A* (*continued*)

Species	Banding Station	Year	Bander	Totals Banded			Totals Recovered		
				P*	A†	Total	P*	A†	Total
Snow Petrel	Hallett station	1959–1960	BR		31	31			
(*Pagodroma nivea*)	Mirnyy station	1962–1963	MP		25	25			
			Totals		56	56			
Leach's Petrel (*Oceanodroma leucorhoa*)	Kent island	1960–1961	CH		196	196			
Wilson's Petrel (*Oceanites oceanicus*)	Windmill Islands	1960–1961	NO		1	1			
Gray-backed Petrel	Kidney Island	1961–1962	RWW		5	5			
(*Garrodia nereis*)		1962–1963	RWW		3	3			
			Totals		8	8			
Blue-eyed Cormorant (*Phalacrocorax atriceps*)	Duthiers Point (Punta Canelo)	1960–1961	RL,TL	37	63	100			
Wattled Sheathbill	George Island	1961–1962	RR		5	5	1		1
	Stanley	1962–1963	DD		1	1			
(*Chionis alba*)	Bird Island	1962–1963	WT,RP		34	34			
		1963–1964	WT,RP		74	74			
			Totals		114	114	1		1
Great Skua	Ardley Island	1959–1960	(IAA)		26	26	1		1
	Bird Island	1960–1961	WT,HD	95	191	286			
		1961–1962	HD,CF	297	480	777			
(*Catharacta skua*)		1962–1963	WT,RP	255	827	1,082			
		1963–1964	WT,RP	229	306	535			
	Dunbar Island	1960–1961	RN,LN	81		81			
		1961–1962	RN	34	41	75			
		1962–1963	RN	36	21	57			
	Kidney Island	1961–1962	RWW		2	2			
	Stanley	1961–1962	RWW		3	3			
	Carcass Island	1961–1962	DG	50		50			
	Nelson Island	1964–1965	BA	45	1	46			
	Marion Island	1964–1965	EV		10	10			
	Palmer station	1964–1965	GEL	119		119			
			Totals	1,241	1,908	3,149		1	1
South Polar Skua	Hallett station	1958–1959	WS	21		21			
(*C. maccormicki*)		1958–1959	KS		18	18			
		1958–1959	BR	35	3	38	1		1
		1960–1961	WM	85	32	117	2		2
		1961–1962	RW	8	243	251			
		1961–1962	CT	56		56	5		5
	McMurdo station	1958–1959	JHD		48	48		5	5
		1958–1959	JD		2	2			
		1959–1960	SH		6	6			
		1963–1964	RW		1	1			
		1963–1964	IFS		4	4			
	Windmill Islands	1958–1959	WS	1	1	2			
		1959–1960	RLP	9	3	12	1		1
		1960–1961	RLP		16	16			
		1960–1961	NO	12		12			
	Marble Point	1961–1962	RW		4	4			
		1964–1965	RW	13	9	22			

Species	Banding Station	Year	Bander	Totals Banded			Totals Recovered		
				P*	A†	Total	P*	A†	Total
	Cape Crozier	1961–1962	RW	224	208	432	1		1
		1962–1963	RW	501	1,361	1,862	3	4	7
		1963–1964	RW	237	197	434			
		1964–1965	RW	151	453	604			
	Cape Royds	1959–1960	SH		13	13			
		1959–1960	EY	28	4	32	2		2
		1961–1962	RW	3	3	6	1		1
		1963–1964	IFS	84	17	101			
	Cape Barne	1959–1960	EY	9		9			
		1963–1964	IFS	19		19			
	Cape Evans	1961–1962	RW	13	17	30			
		1963–1964	IFS	54		54			
	Cape Bird	1964–1965	RW	183	148	331		2	2
	Ellsworth station	1959–1960			1	1			
	González Videla station	1960–1961	RL,TL		97	97		2	2
	Mirnyy station	1962–1963	MP		26	26			
			Totals	1,746	2,935	4,681	16	13	29
Southern Black-backed Gull (*Larus dominicanus*)	Stanley	1960–1961	RWW		12	12			
		1961–1962	RWW,AW		96	96			
		1962–1963	RWW,AW		424	424			
		1962–1963	DD		4	4			
		1963–1964	DD		137	137		1	1
		1964–1965	DD		269	269			
	Bluff Cove	1960–1961	RWW	48		48			
		1961–1962	RWW	84		84	1		1
		1962–1963	RWW,DD	12		12			
	Bleaker Island	1961–1962	RR		1	1			
	George Island	1961–1962	RR		1	1			
	Carcass Island	1961–1962	DG	50		50			
		1962–1963	JB	55		55			
	Darwin	1961–1962	JP		72	72			
		1962–1963	JP		48	48			
	Westpoint Island	1962–1963	RN		25	25			
	Chartres	1962–1963	DG		1	1			
	Nelson Island	1964–1965	BA	8		8			
			Totals	257	1,090	1,347			
Totals, all species				75,357	24,185	99,542¶	432	45	477

ᵃ 90 Teflon flipper bands with reference numbers and "USARP" put on Emperors at Cape Crozier in Oct. and Dec. 1964 [*Penney and Sladen*, 1966] are not listed here.

* Adult plumage (age uncertain).

† Nestling not yet capable of extended flight (or swimming).

‡ 23 of these were juveniles, first year.

§ A total of 854 of these were banded as juveniles.

¶ The total of all species (99,542) in the adult column includes 877 birds banded as juveniles.

TABLE 4B. Additional Species Banded in the
Falkland Islands

These have not been given species numbers, and therefore have not been recorded on the FWS IBM cards.

Species	Number of Birds
Magellan Cormorant (*Phalacrocorax magellanicus*)	74
Yellow-billed Teal (*Anas flavirostris*)	1
Black Oystercatcher (*Haematopus ater*)	2
Fuegian Oystercatcher (*H. leucopodus*)	7
Falkland Plover (*Charadrius falklandicus*)	1
Caracara (*Phalcobaenus australis*)	1
Magellan Gull (*Leucophaeus scoresbii*)	231
Short-eared Owl (*Asio flammeus*)	1
South Georgia Teal (*Anas georgica*) *	1

* Banded on Bird Island, South Georgia.

TABLE 5. System of Reporting Recoveries, with Signs and Abbreviations Used in Tables 6–14
[Modified from *Rydzewski*, 1954, 1955, 1960]

1. Band number, in italics if band has been returned. See sections 10 and 19 for replaced bands. Two band numbers indicate double banding.

2. Resident status:

   o banded in nest (or on territory either as nestling or as adult; or in creche for penguin nestlings).

   (o) experimental birds banded in nest (or on territory either as nestling or adult).

   　　(Experimental purpose explained after symbol; for example (o) (flipper-notched).)

   * adults or juveniles banded as nonterritorial birds.

   (*) experimental birds, adults, or juveniles banded as nonterritorial birds.

   　　(Experimental purpose explained after symbol; for example (*) (blood taken).)

3. Age:

   P nestling, incapable of extended flight (or swimming, as in penguins).

   Juv young or immature plumage; in case of birds having successive annual plumages, I—1st year, II—2nd year, etc.

   A Adult plumage (age uncertain).

4. Sex (omitted if unknown):

   M male　　F female

   (M + F confirmed by dissection)

5. Banding date:

   day, month, year (spell out month as a 3-letter abbreviation, as 1Jan60).

6. Banding site.

7. Coordinates of banding site.

8. Administrative unit (example, Ross Island).

9. Bander's name omitted. (Information, especially on international cooperation, is listed in Table 2.)

10. Manner of recovery and disposition:

    + shot or killed by man.

    x found dead, ill, or exhausted, and eventually died. (Cause of death in bracket after symbol; for example x (fish hook).)

    (x) found long dead. (State of remains after symbol; example (x) (decomposed) or (skeleton).)

    v caught or trapped alive and released with band intact or replaced. (If band has been replaced, put new band number in parentheses after symbol; example v (528–01019).)

    (v) caught or trapped alive and not released (eventually killed after trapping) or released with band removed.

    s sight record (band number read with glasses, etc.).

    /?/ manner of recovery unknown.

    f fate of bird and/or band unknown.

10a. Breeding status (omitted if not applicable):

    br. known to be breeding at time of recovery.

11. Recovery date:

    day, month, year (spell out month as a 3-letter abbreviation); example, 1Jan60.

    16Jun60—exact date of recovery.

    c.16Jun60—approximate date, within a few days.

    e.Jun60, m.Jun60, l.Jun60—early, middle, late, June 1960.

    _.Jun60—sometime during June 1960.

    (16)Jun60—recovery date not stated by finder; correspondence or postmark date taken instead.

12. Recovery site; parentheses indicate the origin of the letter of recovery, when the place of recovery is unknown. Loco is used when the bird was recovered on or in close proximity to the banding site, i.e. within the same coordinates; this does not apply to recoveries reported in this paper.

13. Coordinates of recovery site are given for the first recovery; they are not repeated in the same table for subsequent recoveries.

14. Administrative unit.

15. Country, in boldface type.

16. Finder or reporter's name omitted (see 9.).

17. Distance from place of banding omitted.

18. Direction from place of banding omitted.

19. Length of time banded, given in years, months, days; for example, 1-8-14. For birds whose bands have been replaced, the length of time is given for the 1st band, 2nd band, 3rd band, etc. If the datum in section 11 is uncertain, the length of time banded is estimated to the nearest month (for example, 1-4-?) or left indefinite (?-?-?), according to the information from the original letter.

EXAMPLE

1	2	3	4	5	6	7	8	10	11
528–66237	o	P	M	1Jan60	Bird Island	54.00S, 38.02W	South Georgia	+	16Jun60

12	13	14	15	19
East London	33.00S, 27.54E	Cape of Good Hope	**South Africa**	0–5–15

TABLE 6. Adélie Penguin Recoveries

Band Number	Resident Status Manner of Recovery	Age Breeding Status	Banding Data Recovery Data	Banding Site Recovery Site	Length of Time Banded
6A. Banded at Cape Crozier, 77.27S, 169.13E, Ross Island, Antarctica					
509–67723	o	P	4Feb63		
	s		5Jan65	Cape Bird, 77.13S, 166.20E, Ross Island, Antarctica	1–11–1
509–70434	o	P	9Feb63		
	s		5Jan65	Cape Bird	1–10–27
509–70499	o	P	10Feb63		
	s		5Jan65	Cape Bird	1–10–26
519–01553	o	P	6Feb62		
	s		5Jan65	Cape Bird	2–10–30
519–03232	o	P	13Feb62		
	s		5Jan65	Cape Bird	2–10–23
519–03801	*	A	17Feb62		
	s		5Jan65	Cape Bird	2–10–19
509–68146	o	P	4Feb63		
	s		6Jan65	Cape Bird	1–11–2
519–03204	o	P	13Feb62		
	s		6Jan65	Cape Bird	2–10–24
519–03570	*	A	14Feb62		
	s	br.	6Jan65	Cape Bird	2–10–23
509–67296	o	P	30Jan63		
	s		21Jan65	Cape Bird	1–11–22
509–68812	o	P	6Feb63		
	s		25Jan65	Cape Royds, 77.33S, 166.07E, Ross Island, Antarctica	1–11–19
509–69717	o	P	8Feb63		
	s		8Feb65	Cape Royds	2–0–0
6B. Banded at Cape Royds, 77.33S, 166.07E, Ross Island, Antarctica					
509–75508	o	A	28Dec59		
	s		3Dec62	On sea ice about 10 mi. off Cape Royds	2–11–5
509–75529	o	A	31Dec59		
	s		5Jan65	Cape Bird	5–0–5
519–16816	*	A	7Nov64		
	s		5Jan65	Cape Bird	0–1–29
519–16852	*	A	8Nov64		
	s		5Jan65	Cape Bird	0–1–28
6C. Banded Elsewhere					
509–82169	o	A	18Dec61	Hallett station, 72.18S, 170.13E, Victoria Land, Antarctica	
	s		1Mar63	Cape Crozier, 77.27S, 169.13E, Ross Island, Antarctica	1–2–11
509–79109	o	P	21Feb60	Wilkes station, 66.15S, 110.32E, Antarctica	
	s		26Oct64	Mirnyy station, 66.33S, 93.01E, Antarctica	4–8–5

TABLE 7. Wandering Albatross Recoveries

All birds reported in this table, unless otherwise stated, were banded on Bird Island, 54.00S, 38.02W, South Georgia. This banding station is therefore not repeated below.

Band Number	Resident Status Manner of Recovery	Age Breeding Status	Banding Data Recovery Data	Banding Site Recovery Site	Length of Time Banded
7A. Recovered in South America or the South Atlantic					
528-10018	o (v)	A M	11Jan59 12Jun60	At sea, 36.55S, 54.10W	1–5–1
528–71478	(o) (dyed) +	A	6Mar61 2Apr61	At sea, 55S, 12W	0–0–27
528–71014	o (v)	A	29Jan61 2Jan62	At sea, 52.40S, 21.37W	0–11–4
528–73145	o v	A	8Feb62 25Sep62	At sea, 37.42S, 55.08W	0–7–17
528–10039	o f/?/	A F	13Jan59 e.Nov62	Off Tristan da Cunha, 37.15S, 12.30W	3–10–?
528–72303	o +	A	16Mar61 29Dec62	At sea, 54.42S, 42.12W	1–9–13
568–25081	o (v)	P	4Dec62 16Sep63	At sea, 130 mi. NE of Mar del Plata, 38.00S, 57.32W, **Argentina**	0–9–12
528–72111	o v	A	12Mar61 30Dec63	At sea, approximately 52S, 30W	2–9–18
528–71433	(o) (dyed) +	A	1Mar61 11Mar64	At sea, 53.30S, 20.00W	3–0–10
528–74632	o (v)	P	6Aug63 11Apr64	At sea, 39.00S, 55.01W, SE of Mar del Plata, **Argentina**	0–8–5
568–25343	o v	P	5Dec62 12Nov64	At sea, 38.10S, 55.00W, near Mar del Plata, **Argentina**	1–11–7
7B. Recovered in Africa					
528–73899	o +	A	26Feb62 (6)Sep62	70 miles W. of Lüderitz, 26.38S, 15.10E, **South West Africa**	0–?–?
568–25384	o x	P	5Dec62 22Jan63	20 miles N. of Swakopmund, 22.40S, 14.34E, **South West Africa**	0–1–17
528–43483	o (v)	A	6Jan60 8Sep63	Îles Crozet, 46.25S, 51.40E, Indian Ocean Approximately 31S, 40E, SW of **Madagascar**	3–8–2
7C. Recovered in Australia					
528–71119	o v	A	1Feb61 22Jul61	Off Bellambi, 34.22S, 150.56E, New South Wales	0–5–21
528–72064	o v	A	12Mar61 6Aug61	Off Bellambi	0–4–25
528–72453	o v	A	20Mar61 6Aug61	Off Bellambi	0–4–17
528–71217	(o) (dyed) v	A	28Feb61 9Sep61	Off Bellambi	0–6–12

TABLE 7 (*continued*)

Band Number	Resident Status Manner of Recovery	Age Breeding Status	Banding Data Recovery Data	Banding Site Recovery Site	Length of Time Banded
528–73061	o	A	8Feb62		
	v		22Jul62	Off Bellambi	0–5–14
528–72651	o	A	4Feb62		
	v		28Jul62	Off Bellambi	0–5–24
528–72598	o	A	26Feb62		
	v		25Aug62	Off Bellambi	
	v		17Aug63	Off Bellambi	1–5–22
528–72725	o	A	27Feb62		
	v		3Aug63	Off Bellambi	1–5–7
528–43279	o	A	9Jan60	Îles Crozet, 46.25S, 51.40E, Indian Ocean	
	x		15Sep63	S of Bunbury, 33.20S, 115.34E, Western Australia	3–8?6
568–25044	o f/?/	P	4Dec62 10Oct63	At sea, 35.40S, 156.30E	0–10–6
528–71575	o	A	8Mar61		
	v		8Aug64	Off Bellambi	3–5–0

TABLE 8. Black-browed Albatross Recoveries

8A. Recovered in South America

All birds in section 8A, unless otherwise stated, were banded on Westpoint Island, 51.21S, 60.41W, Falkland Islands. This banding station is therefore not repeated below.

Band Number	Resident Status Manner of Recovery	Age Breeding Status	Banding Data Recovery Data	Banding Site Recovery Site	Length of Time Banded
528–58774	o (v)	P	8Apr60 _.Aug60	110 miles NE of Mar del Plata, 36.54S, 56.02W, **Argentina**	0–4–?
528–58900	o x	P	9Apr60 11Jan61	30 miles S of Mar del Plata, 38.00S, 57.32W, **Argentina**	0–9–2
528–59081	o (v)	P	4Mar61 (1)Jul61	Ponta de Juatinga, 23.17S, 44.30W, **Brazil**	0–?–?
528–59644	o x	P	6Feb61 30Jun61	Praia Brava, 22.45S, 41.52W, **Brazil**	0–4–22
528–59091	o (v)	P	4Mar61 19Jul61	At sea, 29.30S, 49.45W, **Brazil**	0–4–15
528–59755	o x	P	4Mar61 11Aug61	Pôrto Belo, 27.09S, 48.35W, **Brazil**	0–5–7
568–12558	o f	P	1Mar62 11May62	Punta Mogotes, 38.05S, 57.33W, **Argentina**	0–2–10
568–11404	o v	P	25Feb62 14May62	NNE of Mar del Plata, **Argentina**	0–2–19
568–12394	o (v)	P	1Mar62 18May62	At sea, 37.10S, 55.00W, **Argentina**	0–2–17
568–10572	o x	P	22Feb62 5Jun62	On the coast, Estado do Rio Grande do Sul (Porto Alegre, 30.03S, 51.10W) **Brazil**	0–3–14

TABLE 8 (*continued*)

Band Number	Resident Status Manner of Recovery	Age Breeding Status	Banding Data Recovery Data	Banding Site Recovery Site	Length of Time Banded
568–10049	o	P	20Feb62		
	x		19Jun62	Montevideo, 34.55S, 56.10W, **Uruguay**	0–3–30
568–11897	o	P	26Feb62		
	v		28Jun62	SE of Puerto Quequén, 38.30S, 58.44W, **Argentina**	0–4–2
568–12422	o	P	1Mar62		
	(v)		28Jun62	Near Cabode Santa Marta Grande, 28.36S, 48.49W, **Brazil**	0–3–27
568–11916	o	P	26Feb62		
	x		9Jul62	Near Playa Portezuelo, approximately 34.53S, 56.10W, **Uruguay**	0–4–13
528–18450	o	P	21Feb62		
568–10274	v		15Jul62	São Sebastião, 23.48S, 45.26W, **Brazil**	0–4–25
	x		31Jul62	Cabo Frio, 22.51S, 42.03W, **Brazil**	0–5–10
528–18405	o	P	21Feb62		
568–10278	x		17Oct62	Praia de Juréia, near Iguape, 24.44S, 47.31W, **Brazil**	0–7–26
568–03660	o	P	13Feb62	Bird Island, 54.00S, 38.02W, South Georgia	
	(v)		16Aug62	off Ilha Grande, 23.07S, 44.16W, **Brazil**	0–6–3
568–11714	o	P	26Feb62		
	f/?/		(17)Aug62	At sea, 36.05S, 54.10W, **Argentina**	0–?–?
568–12239	o	P	1Mar62		
	x		13Sep62	Near Garopaba, 28.03S, 48.40W, **Brazil**	0–6–12
568–11863	o	P	26Feb62		
	f		c.15Sep62	Mar del Plata, **Argentina**	0–7–?
568–10987	o	P	24Feb62		
	x		16Sep62	São Paulo, 23.33S, 46.39W, **Brazil**	0–6–23
568–12375	o	P	1Mar62		
	x		e.Oct62	Itajai, 26.50S, 48.39W, **Brazil**	0–7–?
568–10777	o	P	23Feb62		
	f/?/		(17)Oct62	Armção do Itapocoróia, 22.54S, 43.08W, **Brazil**	0–?–?
568–12869	o	P	4Mar62		
	x		19Nov62	São Francisco do Sul, 26.17S, 48.39W, **Brazil**	0–8–15
568–12970	o	P	5Mar62		
	(v)		5Dec62	Barra de Ribeira, 30.20S, 51.15W, **Brazil**	0–9–0
568–49649	o	P	16Feb63		
	(v)		28May63	Mar del Plata, **Argentina**	0–3–12
568–47666	o	P	11Feb63		
	v		5Jun63	At sea, 37.24S, 54.55W, **Argentina**	0–3–25
568–47098	o	P	10Feb63		
	f		29Jun63	Off Santos, 23.56S, 46.22W, **Brazil**	0–4–19
568–47347	o	P	11Feb63		
	x		_.63	San Luis beach, Dept. of Canelones, 34.32S, 56.17W, **Uruguay**	0–?–?
568–48165	o	P	13Feb63		
	x		8Jul63	District of Tramandaí, 29.58S, 50.06W, **Brazil**	0–4–25

TABLE 8 (*continued*)

Band Number	Resident Status Manner of Recovery	Age Breeding Status	Banding Data Recovery Data	Banding Site Recovery Site	Length of Time Banded
568–49744	o f/?/	P	16Feb63 15Jul63	Araranguá, 28.55S, 49.25W, **Brazil**	0–4–29
568–48093	o f/?/	P	13Feb63 14Aug63	60 miles SE of Ponta do Bai Light, 23.59S, 45.15W, **Brazil**	0–6–1
568–47980	o x (exhaustion)	P	11Feb63 (20)Aug63	100 miles N of Pôrto Alegre, 30.03S, 51.10W, **Brazil**	0–?–?
568–47772	o x	P	11Feb63 8Oct63	Maçambaba, near village of Arraial do Cabo, 22.58S, 42.02W, **Brazil**	0–7–27
528–58834	o v	P	9Apr60 27Apr64	SE of Puerto Quequén, Buenos Aires, **Argentina**	4–0–18
568–08050	o v	P	17Feb62 _.Mar64	Bird Island, South Georgia Mar del Plata, **Argentina**	2–?–?
568–11277	o v	P	25Feb62 (4)May64	Bird Island, South Georgia Mar del Plata, **Argentina**	2–?–?
568–48450	o /?/	P	15Feb63 24May64	At sea, 34.27S, 50.26W, **Argentina**	1–3–9
568–12150	o x	P	26Feb62 29May64	At sea, 37.22S, 54.48W, **Brazil**	2–3–3
568–10971	o v	P	24Feb62 16Jun64	Between 35 & 41S, **Argentina**	2–3–23
568–51471	o (v)	P	1Mar64 20Jun64	Off Canavieiras, 15.44S, 38.58W, **Brazil**	0–3–19
568–48781	o (v)	P	15Feb63 21Jun64	SE of Ilha das Couves (Rio de Janeiro, 22.53S, 43.17W) **Brazil**	1–4–6
568–10508	o v	P	22Feb62 29June64	Between 35S & 41S, **Argentina**	2–4–7
568–26445	o v	P	13Mar63 29Jun64	Bird Island, South Georgia Between 35S & 41S, **Argentina**	1–3–16
568–11074	o v	P	25Feb62 1Jul64	Between 35S & 41S, **Argentina**	2–4–5
568–49496	o v	P	16Feb63 9Jul64	Bird Island, South Georgia	1–4–23
568–50221	o x	P	18Feb64 9Jul64	Mouth of Inhambupe River (Inhambupe, 11.52S, 38.22W) **Brazil**	0–4–21
568–51094	o v	P	26Feb64 _.Jun64	Cabo Frio, 22.59S, 42.00W, **Brazil**	0–4–?
568–50490	o x	P	20Feb64 10Jul64	S of Maceió, 9.40S, 35.44W, **Brazil**	0–4–20
568–51453	o x	P	1Mar64 12Jul64	N of Salvador, 12.58S, 38.29W, **Brazil**	0–4–11
568–48373	o v	P	15Feb63 13Jul64	Between 35S & 41S	1–4–28
568–12714	o v	P	4Mar62 16Jul64	Between 35S & 41S	2–4–12

TABLE 8 (*continued*)

Band Number	Resident Status Manner of Recovery	Age Breeding Status	Banding Data Recovery Data	Banding Site Recovery Site	Length of Time Banded
568–51883	o f/?/	P	15Mar64 18Jul64	 SE of Ilha de Santana, 2.20S, 43.40W, **Brazil**	0–4–3
568–50100	o x (drowned)	P	18Feb64 22Jul64	 Off Victoria Island, on N coast of São Paulo State (São Paulo, 23.33S, 46.39W) **Brazil**	0–5–4
568–50643	o v x	P	20Feb64 7Aug64 (16) Aug64	 Montevideo, 34.55S, 56.10W, **Uruguay** Off Montevideo, 34.55S, 56.10W, **Uruguay**	0–5–18 0–6–?
568–50302	o +	P	20Feb64 (20) Aug64	 Vitoria, 20.19S, 40.21W, **Brazil**	0–?–?
568–47120	o (v)	P	10Feb63 7Sep64	 Off Rio de Janeiro, 22.53S, 43.17W, **Brazil**	1–6–28
568–47679	o x	P	11Feb63 (28) Dec64	 Near Caleta Cordova, approximately 46S, 67W, **Argentina**	?–?–?
568–07835	o f	P	17Feb62 20May65	Bird Island, South Georgia Río de la Plata (La Plata, 36.15S, 57.55W) **Argentina**	3–3–3
568–53316	o (v)	P	20Feb65 24May65	 Off Rio Grande do Sul, between Solidad and Mostardas, 31.02S, 50.51W, **Brazil**	0–3–4
568–51044	o f	P	26Feb64 25May65	 São Francisco do Sul, 26.17S, 48.39W, **Brazil**	1–2–29
568–53951	o x	P	21Mar65 16Jun65	 Caraguatatuba, 23.37S, 45.25W, **Brazil**	0–2–26

8B. *Recovered in South Africa*

All birds in sections 8B, 8C, 8D, unless otherwise stated, were banded on Bird Island, 54.00S, 38.02W, South Georgia. This banding station is therefore not repeated below.

Band Number	Resident Status Manner of Recovery	Age Breeding Status	Banding Data Recovery Data	Banding Site Recovery Site	Length of Time Banded
528–66237	o +	P	25Feb61 (7) Jun61	 East London, 33.00S, 27.54E	?–?–?
528–14728	o (v)	A	30Jan61 7Jul61	 Cape Point, 34.12S, 18.29E	0–5–7
528–63721	o (v)	P	18Feb61 (17) Jul61	 Cape Point	?–?–?
528–67108	o (v)	P	2Mar61 20Jul61	 Simonstown, 34.12S, 18.26, Cape Point	0–4–18
528–63623	o (v)	P	18Feb61 23Jul61	 Cape Point	0–5–5
528–64734	o f	P	21Feb61 15Jul61	 Port Shepstone, 30.44S, 30.28E	0–4–24
528–64602	o /?/	P	21Feb61 (27) Jul61	 Off Natal (Durban, 29.53S, 31.00E)	0–?–?

TABLE 8 (*continued*)

Band Number	Resident Status Manner of Recovery	Age Breeding Status	Banding Data Recovery Data	Banding Site Recovery Site	Length of Time Banded
528–64377	o x	P	21Feb61 (4)Aug61	False Bay coast, Kenilworth, Cape Province (Cape Point)	0–?–?
528–64261	o (v)	P	21Feb61 5Aug61	WSW of Hout Bay, Cape peninsula, near Cape Town, 33.56S, 18.28E	0–5–15
528–64238	o (v)	P	19Feb61 _.Aug61	Munster, 100 miles S of Durban, Natal	0–6–?
528–63429	o (v)	P	18Feb61 c.7Sep61	Off Kalkbaai near Simonstown, Cape peninsula	0–7–?
528–64598	o (v)	P	21Feb61 c.7Sep61	Off Kalkbaai near Simonstown, Cape peninsula	0–7–?
528–59117	o (v)	P	2Mar61 1Oct61	Westpoint Island, 51.21S, 60.41W, Falkland Islands At sea, 32.25S, 17.40E	0–6–29
528–66209	o f	P	25Feb61 22Oct61	Off Durban	0–7–27
528–64217	o f/?/	P	19Feb61 (22)Nov61	SW of Cape Agulhas, 34.50S, 20.00E	0–9–?
528–63601	o f	P	18Feb61 (24)Nov61	50 miles SSW of Cape Point	0–?–?
528–64269	o x	P	21Feb61 24Nov61	Cape Infanta, 34.27S, 20.51E	0–9–3
528–64398	o (v)	P	21Feb61 (27)Nov61	(Port Alfred, 33.36S, 26.54E)	0–?–?
528–67267	o f	P	2Mar61 (6)Mar62	Danger Point, 34.37S, 19.17E	0–?–?
528–62771	o (v)	P	17Feb61 12Mar62	Mosselbaai, 34.12S, 22.08E	1–0–23
528–66234	o f/?/	P	25Feb61 14Mar62	W of Cape Town	1–0–17
528–62688	o (v)	P	16Feb61 1Apr62	Cape Point	1–1–16
528–63582	o (v)	P	18Feb61 29Apr62	SE of Cape of Good Hope, approximately 33.6S, 25.3E	1–2–11
528–63682	o (v)	P	18Feb61 9May62	NW of Cape Point	1–2–21
528–64164	o +	P	19Feb61 11May62	Off Witsandsbaai, Cape peninsula (Cape Point)	1–2–22
528–63484	o +	P	18Feb61 20May62	Bird Island, 33.49S, 26.17E, Algoa Bay	1–3–2
528–63074	o +	P	17Feb61 25May62	Bird Island, Algoa Bay	1–3–8
528–64368	o (v)	P	21Feb61 6Jun62	W of Cape St. Blaize, 34.12S, 22.10E	1–3–16

SLADEN, WOOD, AND MONAGHAN

TABLE 8 (*continued*)

Band Number	Resident Status Manner of Recovery	Age Breeding Status	Banding Data Recovery Data	Banding Site Recovery Site	Length of Time Banded
568–07758	o (v)	P	17Feb62 10Jun62	 SE of Mosselbaai	 0–3–24
568–12067	o v	P	26Feb62 14Jun62	Westpoint Island, Falkland Islands NW of Cape Columbine, at 32.10S, 17.00E	 0–3–19
528–66830	o (v)	P	26Feb61 (27)Jun62	 Near Cape Point	 ?–?–?
528–63860	o (v)	P	19Feb61 29Jun62	 In the Atlantic near Hermanus, 34.25S, 19.14E	 1–4–10
528–64533	o (v)	P	21Feb61 29Jun62	 In the Atlantic near Hermanus	 1–4–8
568–06698	o v x	P	15Feb62 17Jul62 (27)May63	 Port Edward, 31.03S, 30.14E Lambert's Bay, 32.04S, 18.20E, Cape of Good Hope	 0–5–2 ?–?–?
568–06214	o (v)	P	15Feb62 31Jul62	 Off Malgas Island near Saldanhabaai 33.00S, 17.56E	 0–5–16
568–11140	o (v)	P	25Feb62 1Aug62	Westpoint Island, Falkland Islands At sea, 70 miles SW of Cape Town	 0–5–7
528–63239	o (v)	P	21Feb61 12Aug62	 NW of Dasseniland, 33.25S, 18.05E	 1–5–22
568–09504	o f	P	18Feb62 14Aug62	 100 miles W of Lambert's Bay	 0–5–27
528–66980	o x (fishing gear)	P	26Feb61 29Aug62	 At sea, 34.53S, 18.37E	 1–6–3
528–63808	o v	P	19Feb61 _.Sep62	 WNW of Cape Town	 1–6–?
528–65439	o /?/	P	23Feb61 _.Sep62	 At sea, 50 miles W of Table Bay (near Cape Town)	 1–7–?
568–07881	o /?/	P	17Feb62 _.Sep62	 At sea, 50 miles W of Table Bay (near Cape Town)	 0–7–?
568–12318	o f/?/	P	1Mar62 12Sep62	Westpoint Island, Falkland Islands 60 miles SE of Port Elizabeth, 33.58S, 25.36E	 0–6–11
568–05759	o f	P	15Feb62 (20)Sep62	 W of Cape Infanta	 0–?–?
568–11636	o x	P	26Feb62 2Oct62	Westpoint Island, Falkland Islands NW of Mouille Point, Cape Town	 0–7–6
528–65202	o v	P	23Feb61 4Oct62	 W of Cape Point	 1–7–11
528–63844	o x	P	19Feb61 5Oct62	 NNW of Cape Town	 1–7–16
568–07803	o (v)	P	17Feb62 23Oct62	 Table Bay harbor, Cape Town	 0–8–6
528–66489	o f	P	25Feb61 25Oct62	 At sea, 60 miles WNW of Cape Town	 1–8–0

TABLE 8 (*continued*)

Band Number	Resident Status Manner of Recovery	Age Breeding Status	Banding Data Recovery Data	Banding Site Recovery Site	Length of Time Banded
568–04363	o f	P	13Feb62 25Oct62	At sea, 60 miles NW of Cape Town	0–8–12
568–05254	o +	P	14Feb62 13Nov62	Near Cape Point, Cape peninsula	0–8–30
568–09600	o +	P	18Feb62 28Nov62	SSE of Cape Infanta	0–9–10
568–03831	o +	P	13Feb62 1Dec62	SSW of Cape Town	0–9–18
528–63944	o x (drowned)	P	19Feb61 (26)Dec62	At sea, 70 miles W of Cape Town	?–?–?
528–67385	o (v)	P	2Mar61 1Jan63	N of Cape Columbine, 32.50S, 17.51E	1–9–30
528–67396	o (v)	P	2Mar61 (3)Jan63	At sea, 50 miles SSE of Cape Agulhas	?–?–?
528–63503	o (v)	P	18Feb61 14Jan63	At sea, 32.10S, 16.55E	1–10–27
568–07790	o (v)	P	17Feb62 24Apr63	W of Cape Town	1–2–7
528–63812	o f	P	19Feb61 3May63	St. Helena Bay, Vredenburg, 32.55S, 18.00E	2–2–14
528–64121	o +	P	19Feb61 6May63	Near Cape Town	2–2–17
528–70915	o (v)	P	9Mar63 17May63	N of Lambert's Bay	0–2–8
528–63249	o (v)	P	17Feb61 23May63	Off Hout Bay, 34.03S, 18.21E	2–3–6
568–08859	o +	P	18Feb62 27May63	St. Helenafontein, 32.31S, 18.20E	1–3–9
568–04932	o x (drowned)	P	14Feb62 __.Jun63	At sea, approximately 33.10S, 17.20E	1–4–?
568–27741	o (v)	P	14Mar63 16Jun63	SE of Mosselbaai	0–3–2
528–70655	o (v)	P	8Mar63 21Jun63	Mosselbaai	0–3–13
568–31589	o f	P	13Apr63 21Jun63	Off Hout Bay	0–2–8
568–03681	o (v)	P	13Feb62 __.Jul63	NW of Cape Columbine	1–5–?
568–30359	o +	P	27Mar63 8Jul63	At sea, 32.25S, 17.29E	0–3–11
568–30301	o +	P	27Mar63 8Jul63	At sea, 32.25S, 17.29E	0–3–11
528–63562	o +	P	18Feb61 14Jul63	Off Cape Point	2–4–26
568–26552	o +	P	13Mar63 16Jul63	Off Maitland River, near Port Elizabeth, 33.58S, 25.36E	0–4–3
528–69871	o x	P	7Mar63 __.Aug63	At sea, 25.30S, 38.00E	0–5–2

TABLE 8 (*continued*)

Band Number	Resident Status Manner of Recovery	Age Breeding Status	Banding Data Recovery Data	Banding Site Recovery Site	Length of Time Banded
568–08180	o (v)	P	17Feb62 7Aug63	Stilbaai, 34.23S, 21.24E, Riversdale	1–5–21
568–26261	o (v)	P	13Mar63 8Aug63	At sea, *ca.* 31S, 40E, SW of **Madagascar**	0–4–26
528–70594	o x (fishing nets)	P	8Mar63 19Aug63	Near Lambert's Bay	0–5–11
568–27085	o x (storm)	P	14Mar63 (31)Aug63	Bird Island	0–?–?
568–27942	o x (storm)	P	14Mar63 (31)Aug63	Bird Island	0–?–?
568–29289	o x (storm)	P	23Mar63 (31)Aug63	Bird Island	0–?–?
568–29675	o +	P	26Mar63 7Sep63	St. Helena Bay, 60 miles N of Cape Columbine	0–5–12
568–29552	o (v)	P	23Mar63 21Sep63	At sea, 34.51S, 18.30E	0–5–29
568–27637	o (v)	P	14Mar63 25Sep63	At sea, W of Cape of Good Hope, 34.20S, 18.15E	0–6–11
528–62805	o (v)	P	17Feb61 (24)Oct63	Jeffreys Bay, 34.05S, 24.55E, Cape of Good Hope	?–?–?
568–02283	o f	P	18Feb62 (31)Oct63	N of Cape Town	?–?–?
568–02482	o (v)	P	19Feb62 (31)Oct63	N of Cape Town	?–?–?
568–29883	o (v)	P	26Mar63 _.Nov63	70 miles NW of Saldanhabaai	0–8–?
568–04395	o x (drowned)	P	13Feb62 (12)Nov63	50 miles NW of Saldanha	?–?–?
568–08398	o f/?/	P	17Feb62 22Dec63	At sea, NW of Cape Town, 33.35S, 17.27E	1–10–5
528–66839	o x	P	26Feb61 _.63	Off Struisbaai, 34.48S, 20.03E, near Cape Agulhas	?–?–?
568–26291	o (x) (decomposed)	P	13Mar63 22Dec63	Port Elizabeth	0–9–9
568–28870	o x (fishing net)	P	23Mar63 (7)Jan64	NE of Cape Town	?–?–?
568–29665	o x (fishing net)	P	26Mar63 (7)Jan64	NE of Cape Town	?–?–?
568–08473	o v	P	17Feb62 11Jan64	At sea, 33.55S, 17.25E	1–10–25
568–07100	o (v)	P	16Feb62 24Jan64	S of Cape Point	1–11–8

TABLE 8 (*continued*)

Band Number	Resident Status Manner of Recovery	Age Breeding Status	Banding Data Recovery Data	Banding Site Recovery Site	Length of Time Banded
568–09051	o (v)	P	18Feb62 (18)Feb64	At sea, W of Saldanhabaai	?–?–?
528–70317	o /?/	P	8Mar63 _.Mar64	Cape Columbine	1–0–?
568–26262	o (v)	P	13Mar63 m.Apr64	S of Cape Infanta	1–1–?
568–30990	o (v)	P	29Mar63 10Apr64	Between Dasseneiland and Saldanhabaai	1–0–12
528–70522	o (v)	P	8Mar63 26Apr64	NNW of Robbeneiland, 33.48S, 18.22E	1–1–18
568–27470	o +	P	14Mar63 10May64	Off Hout Bay	1–1–26
568–30321	o f	P	27Mar63 30May64	Kalkbaai	1–2–3
568–26695	o (v)	P	13Mar63 (1)Jun64	At Sea, 34.10S, 17.52E	?–?–?
568–31251	o (v)	P	5Apr63 1Jun64	Danger Point	1–1–27
568–32912	o v	P	22Feb64 10Jun64	Off Cape Agulhas, at sea, 35.10S, 21.05E	0–3–19
568–30081	o x (injury)	P	26Mar63 12Jun64	Off Hout Bay fishing harbor	1–2–17
568–30400	o +	P	27Mar63 17Jun64	180 miles SSW of Cape Town	1–2–21
568–09271	o (v)	P	18Feb62 22Jun64	St. Helena Bay (Cape Columbine)	2–4–4
568–25716	o +	P	9Mar63 29Jun64	At sea, between Mosselbaai, and Agulhas Bank	1–3–20
528–70341	o +	P	8Mar63 _.Jul64	Off Cape Point	1–4–?
568–29193	o f	P	23Mar63 _.Jul64	Off Cape Point	1–3–?
568–31379	o /?/	P	5Apr63 (10)Jul64	At sea, 34.15S, 17.32E	?–?–?
528–62992	o f	P	17Feb61 11Jul64	60 miles from Cape Town	3–4–24
568–26417	o +	P	13Mar63 10May64	Off Hout Bay	1–1–27
568–27450	o f/?/	P	14Mar63 13Aug64	At sea, 31.10S, 16.00E	1–4–30
568–32260	o f/?/	P	5Mar64 13Aug64	At sea, 31.10S, 16.00E	0–5–8
528–64521	o x (drowned)	P	21Feb61 19Aug64	At sea, 34.56S, 23.15E	3–5–29
568–33874	o f/?/	P	2Mar64 _.Oct64	At sea, 50 miles S of Cape Agulhas	0–7–?
568–33921	o (v)	P	2Mar64 10Oct64	At sea, 34.40S, 24.54E	0–7–8

TABLE 8 (*continued*)

Band Number	Resident Status Manner of Recovery	Age Breeding Status	Banding Data Recovery Data	Banding Site Recovery Site	Length of Time Banded
568–34229	o (v)	P	13Mar64 10Oct64	At sea, 34.40S, 24.54E	0–6–27
568–26365	o x	P	13Mar63 2Nov64	At sea, approximately 110 miles NNW of Cape Town	1–7–20
568–33150	o (v)	P	22Feb64 (4)Dec64	Off SW coast near Cape Columbine and Dasseneiland	0–?–?
568–31798	o (v)	P	22Feb64 (4)Dec64	Off SW coast near Cape Columbine and Dasseneiland	0–?–?
568–26902	o x (rigging)	P	14Mar63 3Feb65	At sea, 55 miles WNW of Dasseneiland	1–10–20
568–29271	o (v)	P	23Mar63 4Feb65	Quoin Point lighthouse, near Hermanus	1–10–12
568–28820	o f/?/	P	20Mar63 8Mar65	At sea, 32.30S, 17.25E	1–11–16
568–08565	o f	P	17Feb62 10May65	St. Helena Bay	3–2–23
568–23426	o v	P	1Apr63 30May65	At sea, 31.14S, 6.06E	2–1–29
568–28871	o f	P	23Mar63 e.Jun65	At sea, 31.14S, 6.06E	2–2–?
568–30019	o (v)	P	26Mar63 (3)Jun65	At sea, 34S, 17.40E	?–?–?
568–31825	o (v)	P	16Feb64 (3)Jun65	At sea, 34S, 17.40E	?–?–?
528–70222	o x	P	8Mar63 5Jun65	Cape Point nature reserve, 45 miles from Cape Town	2–2–28

8C. *Recovered in Other African Nations*

Band Number	Resident Status Manner of Recovery	Age Breeding Status	Banding Data Recovery Data	Banding Site Recovery Site	Length of Time Banded
528–66979	o f/?/	P	26Feb61 8Sep61	15 miles S of Moçâmedes, 15.10S, 12.10E, Angola, **Portuguese West Africa**	0–6–13
528–64304	o f	P	21Feb61 18Sep61	Off Swakopmund, at approximately 22.40S, 14.22E, **South West Africa**	0–6–28
528–67363	o f/?/	P	2Mar61 10Jun62	At sea, 19.00S, 11.57E	1–3–8
568–05666	o f	P	15Feb62 20Oct62	Porto Alexandre, 15.50S, 11.51E, Angola, **Portuguese West Africa**	0–7–17

TABLE 8 (*continued*)

Band Number	Resident Status Manner of Recovery	Age Breeding Status	Banding Data Recovery Data	Banding Site Recovery Site	Length of Time Banded
568–09738	o (v)	P	19Feb62 27Oct62	Baía dos Tigres, 16.38S, 11.46E, Angola, **Portuguese West Africa**	0–8–8
568–02459	o x	P	19Feb62 (3)May63	20 miles N of Lüderitz, 26.38S, 15.01E, **South West Africa**	?–?–?
528–70755	o (v)	P	8Mar63 (6)Jun63	Conception Bay, 23.55S, 14.30E, **South West Africa**	0–?–?
568–47843	o f/?/	P	11Feb63 13Jul63	Westpoint Island, 51.21S, 60.41W, Falkland Islands 3 miles off Porto Alexandre, Angola, **Portuguese West Africa**	0–5–2
528–69780	o (v)	P	7Mar63 (17)Jul63	Walvisbaai, 22.59S, 14.31E, **South West Africa**	?–?–?
568–48153	o v	P	13Feb63 18Jul63	Westpoint Island, Falkland Islands Near Walvisbaai, **South West Africa**	0–5–5
568–31319	o (v)	P	5Apr63 (22)Jul63	Pelican Point, 22.54S, 14.25E, 3 miles from Walvisbaai, **South West Africa**	0–?–?
568–27768	o (v)	P	14Mar63 24Jul63	Near Pelican Point, at 22.50S, 14.25E, Walvisbaai, **South West Africa**	0–4–10
528–70129	o (v)	P	8Mar63 24Jul63	Near Pelican Point, at 22.50S, 14.25E, Walvisbaai, **South West Africa**	0–4–16
568–27972	o (v)	P	14Mar63 (26)Oct63	At sea, 17.24S, 10.51E, off **South West Africa**	?–?–?
568–29742	o f/?/	P	26Mar63 28Jun64	At sea, 20.00S, 12.30E, off **South West Africa**	1–3–2
568–10616	o (v)	P	22Feb62 _.Sep64	Westpoint Island, Falkland Islands Walvisbaai, **South West Africa**	2–7–?
568–02118	o x	P	19Feb62 (4)Sep64	At sea, 29.55S, 16.30E, off **South West Africa**	?–?–?
568–28933	o x	P	23Mar63 (4)Sep64	At sea, 29.55S, 16.30E, off **South West Africa**	?–?–?
568–30331	o x	P	27Mar63 (4)Sep64	At sea, 29.55S, 16.30E, off **South West Africa**	?–?–?
528–62737	o x	P	16Feb61 (4)Sep64	At sea, 29.55S, 16.30E, off **South West Africa**	?–?–?
528–64184	o x	P	19Feb61 (4)Sep64	At sea, 29.55S, 16.30E, off **South West Africa**	?–?–?
568–53385	o f/?/	P	20Feb65 7Jun65	Westpoint Island, Falkland Islands At sea, 23.34S, 13.55E, off **South West Africa**	0–3–18

8D. Recovered in Australia

Band Number	Resident Status Manner of Recovery	Age Breeding Status	Banding Data Recovery Data	Banding Site Recovery Site	Length of Time Banded
528–64172	o x (foxes)	P	19Feb61 11Jul61	Between Port Moorowie and Edithburgh, 35.05S, 137.44E, South Australia	0–4–22
568–07799	o x	P	17Feb62 e.Jun62	Quinn's R⸴ck, between Wanneroo and Yanchep Beaches (Midland Junction 31.54S, 116E), Western Australia	0–3–?
568–08979	o x	P	18Feb62 17Jun62	Badger Head beach, 41S, 146.5E, N. Tasmania	0–3–30
568–05629	o (x) (tarsus bone only)	P	15Feb62 6Aug63	Near Perth, 31.58S, 115.49E, Western Australia	1–5–22
568–08345	o (v)	P	17Feb62 23Sep62	E of Sydney, 33.55S, 151.10E, New South Wales	0–7–6
568–06283	o x	P	15Feb62 30Sep62	Elliston, 33.37S, 134.54E	0–7–15
568–07382	o x	P	16Feb62 (15)Oct62	Burgess Bay, S of Currie, 39.56S, 143.55E, King Island, Tasmania	0–?–?
568–06986	o x	P	16Feb62 e.Nov62	Ulladulla beach, 35.21S, 150.25E, New South Wales	0–9–?

TABLE 9. Gray-Headed Albatross Recoveries

All birds reported in this table, unless otherwise stated, were banded on Bird Island, 54.00S, 38.02W, South Georgia.

Band Number	Resident Status Manner of Recovery	Age Breeding Status	Banding Data Recovery Data	Banding Site Recovery Site	Length of Time Banded
			9A. Recovered in South Africa		
568–06466	o f	P	15Feb62 11Jul62	Durban Harbor, 29.53S, 31.00E	0–3–27
568–03182	o x	P	12Feb62 22Dec63	Near Cape Point, 34.21S, 18.29E, Cape peninsula	1–10–10
568–35214	o (v)	P	29Mar63 21Jul63	At sea, 25 miles W of Cape Point, Cape peninsula	0–3–22
			9B. Recovered in Other African Nations		
528–61454	o + (fish hook)	P	13Feb61 3Jul61	At sea, 15.11S, 12.09E, Angola, **Portuguese West Africa**	0–4–20
			9C. Recovered in Australia		
528–66713	o (x) (skeleton)	P	26Feb61 30Jul61	Cape Jaffa, 36.58S, 139.39E	0–5–4

TABLE 9 (*continued*)

Band Number	Resident Status Manner of Recovery	Age Breeding Status	Banding Data Recovery Data	Banding Site Recovery Site	Length of Time Banded
528–67762	o	P	4Mar61		
	x		17Jan63	Pennington Bay (Cape Willoughby, 35.54S, 138.04E), Kangaroo Island	1–10–13
568–35606	o	P	30Mar63		
	x		23Jun63	W coast of Wilson's Promontory, SE of Melbourne, 37.45S, 144.58E	0–2–24
568–13910	o	P	18Mar63		
	x (drowned)		25Jun65	Stockton, 32.54S, 151.48E, Newcastle	2–3–7
		9D. Recovered in New Zealand			
528–61468	o	P	13Feb61		
	x		26Dec63	Otaki Beach, 40.45S, 175.09E, North Island	2–10–13

TABLE 10. Recoveries of Black-Browed or Gray-Headed Albatrosses Banded as Nestlings in 1958–1959

(Banders were unable to distinguish positively between the young of the two species.) All birds reported in this table were banded on Bird Island, 54.00S, 38.02W, South Georgia.

Band Number	Resident Status Manner of Recovery	Age Breeding Status	Banding Data Recovery Data	Banding Site Recovery Site	Length of Time Banded
		10A. Recovered in South America			
528–14060	o	P	1Mar59		
	+		2Jan62	At sea, 52.40S, 21.37W	2–10–1
528–12421	o	P	21Feb59		
	+		26Nov63	At sea, 55.26S, 21.20W	4–9–5
		10B. Recovered in South Africa			
528–12883	o	P	22Feb59		
	+		__.Jun59	Great Fish River, 33.5S, 27.2E, near Port Elizabeth	0–4–?
528–13451	o	P	26Feb59		
	(v)		c.22Jul59	At sea, 32.46S, 16.33E	0–5–?
528–12935	o	P	22Feb59		
	(v)		16Aug59	At sea, between Saldanha, 33.00S, 17.56E, and Cape Columbine, 32.50S, 17.56E	0–5–25
528–12176	o	P	21Feb59		
	f/?/		27Aug59	Coast of Durban, 29.53S, 31.00E, Natal	0–6–6
528–11640	o	P	16Feb59		
	+		6Sep59	At sea, False Bay, off Cape Point, 34.21S, 18.29E	0–6–21
528–14598	o	P	3May59		
	v		19Sep59	At sea, 20 miles SW of Cape Agulhas, 34.50S, 20.00E	0–4–16
528–13390	o	P	26Feb59		
	+		13Nov59	Danger Point, 34.37S, 19.17E	0–8–18
528–12438	o	P	21Feb59		
	(x) (decomposed)		18Jan60	At sea, Algoa Bay near Port Elizabeth, 33.58S, 25.36E	0–10–28
528–14591	o	P	3Mar59		
	x (drowned)		30May60	At sea, 33.30S, 17.35E, off Dasseneiland	1–2–27

TABLE 10 (*continued*)

Band Number	Resident Status Manner of Recovery	Age Breeding Status	Banding Data Recovery Data	Banding Site Recovery Site	Length of Time Banded
528–14432	o v x	P	5Mar59 1Jun60 (4)Sep64	At sea, 20.50S, 16.20E At sea, 29.55S, 16.30E	1–2–27 ?–?–?
528–13564	o +	P	3Jan59 24Jun60	Hout Bay, 34.03S, 18.21E	1–5–21
528–14526	o +	P	5Mar59 2Aug60	At sea, ESE of Cape Point, 34.21S, 18.29E	1–4–28
528–11531	o (x) (decomposed)	P	16Feb59 11Aug60	At sea, Smitswinkelbaai, 8 miles off Cape Town, 33.56S, 18.28E	1–5–26
528–13215	o f	P	1Mar59 22Aug60	S of Danger Point	1–5–21
528–13064	o (v)	P	25Feb59 (27)Sep60	At sea, 34.15S, 22.10E, near Mosselbaai	?–?–?
528–13181	o (v)	P	26Feb59 8Feb61	At sea, 50 miles W of Cape peninsula, near Cape Town	1–11–13
528–13397	o (v)	P	26Feb59 14Feb61	At sea, 35 miles WSW of Cape Point	1–11–19
528–13297	o (v)	P	26Feb59 13Mar61	At sea, 34.15S, 17.53E, off SW coast	2–0–15
528–14573	o (v)	P	3May59 28Mar61	At sea, 34.10S, 17.45E, off Cape of Good Hope	1–10–25
528–13258	o v	P	3Jan59 17Apr61	At sea, WNW of Cape Town	2–3–14
528–13006	o x	P	25Feb59 (30)May61	120 miles NW of Cape Town	?–?–?
528–11996	o +	P	25Feb59 26Aug61	At sea, 34.21S, 17.43E, off Cape Town	2–6–1
528–13378	o (v)	P	26Feb59 27Aug61	At sea, 22.40S, 14.27E, off Swakopmund	2–6–1
528–13078	o (v)	P	25Feb59 24Mar62	At sea, 34.13S, 18.00E, off Slangkop light house	3–0–27
528–13927	o +	P	2Mar59 (27)Jun62	Near Cape Point	?–?–?
528–13820	o (v)	P	28Feb59 _.Jun63	At sea, off Hout Bay, 34.03S, 18.21E	4–3–?
528–13783	o +	P	28Feb59 (19)Jun63	Near Saint Helena Bay (St. Helenafontein, 32.31S, 18.20E)	?–?–?
528–13594	o x	P	28Feb59 e.Jul63	At sea, 31.50S, 18.10E	4–4–?
528–13977	o x	P	2Mar59 (4)Sep64	At sea, 29.55S, 16.30E	4–?–?
528–14588	o x	P	5Mar59 (4)Sep64	At sea, 29.55S, 16.30E	4–?–?
528–14015	o x	P	1Mar59 15Sep64	Mosselbaai, 34.12S, 22.08E	5–6–14

TABLE 10 (*continued*)

Band Number	Resident Status Manner of Recovery	Breeding Status	Banding Data Recovery Data	Banding Site Recovery Site	Length of Time Banded
			10C. Recovered in Other African Nations		
528–13291	o (v)	P	28Feb59 _.Sep59	N of Lobito, 12.20S, 13.34E, Angola, **Portuguese West Africa**	0–?–?
528–13100	o x	P	26Feb59 c.30Sep59	Sandfisch Bay, 30 miles S of Walvisbaai, 22.59S, 14.31E, **South West Africa**	0–7–?
528–13081	o (v)	P	26Feb59 20Oct59	At sea, 13.18S, 12.10E, off **Portuguese West Africa**	0–7–24
			10D. Recovered in Australia		
528–12241	o x	P	21Feb59 3Feb65	Albany, 34.57S, 117.54E	5–11–13
			10E. Recovered in New Zealand		
528–12065	o x	P	31Jan59 18Jul59	Waitarere beach, 40.33S, 175.12E, North Island	0–5–18

TABLE 11. Giant Petrel Recoveries

All birds reported in this table, unless otherwise stated, were banded on Bird Island, 54.00S, 38.02W, South Georgia.

Band Number	Resident Status Manner of Recovery	Age Breeding Status	Banding Data Recovery Data	Banding Site Recovery Site	Length of Time Banded
			11A. Recovered in South America		
518–94904 528–15047	o f/?/	P	3Feb59 15Jun59	Frazier Islands, 66.13S, 110.10E, Windmill Islands, Antarctica N of Valdivia, 39.26S, 73.15W, **Chile**	0–4–12
528–10921	o x	P	31Jan59 26Aug59	N of Bahía de Quintero (Quintero, 32.47S, 71.12W), **Chile**	0–6–26
528–15046	o x	P	3Feb59 6Sep59	Frazier Islands, Windmill Islands, Antarctica San Juan beach, 15.22S, 75.07W, **Peru**	0–7–3
528–15025	o v	P	3Feb59 19Oct59	Frazier Islands, Windmill Islands, Antarctica S of Puerto Quequén, 38.30S, 52.44W, **Argentina**	0–8–16
528–10493	o x	P	31Jan59 5Feb60	S of Valparíso, 33.05S, 71.40W, **Chile**	1–0–5
528–61050	o x	P	8Feb61 6Aug61	N of Chañaral, 26.15S, 70.40W, Atacama, **Chile**	0–5–29
528–60059	o f	P	1Feb61 10Aug61	Bay of Guayacán, 30.00S, 71.26W, Coquimbo, **Chile**	0–6–9
528–60053	o x	P	1Feb61 (15)Nov61	Maitencillo, 30.58S, 71.43W, **Chile**	0–?–?
528–60122	o f	P	3Feb61 (16)Nov61	San Antonio, 33.25S, 71.39W, **Chile**	0–?–?

SLADEN, WOOD, AND MONAGHAN

TABLE 11 (*continued*)

Band Number	Resident Status Manner of Recovery	Age Breeding Status	Banding Data Recovery Data	Banding Site Recovery Site	Length of Time Banded
568–01316	o x	P	3Feb62 10Aug62	Topcopilla, 22.05S, 70.10W, **Chile**	0–6–7
568–01456	o v	P	3Feb62 (20)Aug62	Garopaba, 28.03S, 48.40W, Santa Catarina, **Brazil**	0–?–?
	(v)		25Aug62	Garopaba, **Brazil**	0–6–22
568–01119	o (v)	P	2Feb62 12Oct62	Puerto Flamenco, 26.34S, 70.43W, Atacama, **Chile**	0–8–10
528–61164	o v v	P	13Feb61 13Nov62 17Nov62	Mar del Plata, 38.00S, 57.32W, **Argentina** Mar del Plata, **Argentina**	1–9–0 1–9–4
528–16852	o f	P	21Feb63 _.Sep63	San Antonio, **Chile**	0–7–?
528–68548	o +	P	21Feb63 (4)Sep63	Talcahuano, 36.40S, 73.10W, **Chile**	0–6–?
528–16934	o +	P	21Feb63 22Nov63	33.40S, 71.45W, off San Antonio, **Chile**	0–9–1
528–16781	o v	P	20Feb63 7Dec63	At sea, 39.00S, 56.00W, **Argentina**	0–9–17
528–61083	o v	P	13Feb61 9Jun64	Along coast, between 35S and 41S, **Argentina**	3–3–27
528–69062	o f/?/	P	15Feb64 3Nov64	N of Caldera, 27.05S, 70.41W, Copiapó, **Chile**	0–8–19
528–68963	o v	P	12Feb64 3Dec64	At sea, 39.20S, 55.50W, **Argentina**	0–9–21

11B. Recovered in South Africa

Band Number	Resident Status Manner of Recovery	Age Breeding Status	Banding Data Recovery Data	Banding Site Recovery Site	Length of Time Banded
528–10178	o x	P	31Jan59 7Jun60	Cintsa, NE of East London, 33.00S, 27.54E, Cape Province	1–4–7
528–61122	o x	P	13Feb61 20Jun61	SE of Mosselbaai, 34.12S, 22.08E, Cape of Good Hope	0–4–7
528–60231	o +	P	3Feb61 21Jul61	At sea (Port Elizabeth, 33.58S, 25.36E)	0–5–18
523–10955	o v	P	31Jan59 6Sep59	W of Dasseneiland, 33.25S, 18.05E	0–7–6
528–60717	o x (oiled)	P	7Feb61 26Oct61	Durban beach, 29.53S, 31.00E, Natal	0–8–19
528–60023	o f/?/	P	1Feb61 (1)Nov61	(Mosselbaai)	0–?–?
568–01484	o x	P	3Feb62 4Jul62	Lambert's Bay, 32.04S, 18.20E	0–5–1
568–01472	o v	P	3Feb62 26Aug62	Cape Town, 33.56S, 18.28E	0–6–23
568–01846	o (v)	P	4Feb62 (7)Oct63	N of Port Nolloth, 29.17S, 16.51E, Cape of Good Hope	?–?–?
528–68240	o x	P	22Feb63 c.26Oct63	Approximately 33S, 27E, between Port Elizabeth and East London	0–8–?

TABLE 11 (*continued*)

Band Number	Resident Status Manner of Recovery	Age Breeding Status	Banding Data Recovery Data	Banding Site Recovery Site	Length of Time Banded
528–68851	o x	P	13Feb64 _.Oct64	Near Walvisbaai, at 22.58S, 14.32E	0–8–?

11C. Recovered in Other African Nations and the South Atlantic

Band Number	Resident Status Manner of Recovery	Age Breeding Status	Banding Data Recovery Data	Banding Site Recovery Site	Length of Time Banded
528–10196	o f/?/	P	31Jan59 27Jun59	Moçâmedes, 15.10S, 12.10E, Angola, **Portuguese West Africa**	0–4–27
528–10961	o v	P	31Jan59 10Jul59	Baía dos Tigres, 16.38S, 11.46E, Angola, **Portuguese West Africa**	0–5–10
528–10123	o x	P	31Jan59 (16)Jul59	Benguela, 12.34S, 13.24E, Angola, **Portuguese West Africa**	0–?–?
528–61407	o x	P	8Feb61 16Aug61	Tristan da Cunha, 37.03S, 12.18W, **South Atlantic**	0–6–8

11D. Recovered in Australia

Band Number	Resident Status Manner of Recovery	Age Breeding Status	Banding Data Recovery Data	Banding Site Recovery Site	Length of Time Banded
528–10157	o x	P	31Jan59 26May59	Between Robe, 37.13S, 139.46E, and Beachport, 37.29S, 140.01E, S.A.	0–3–26
528–10428	o x	P	31Jan59 26May59	Barwon Heads, Aberfeldie, Vict.	0–3–26
528–10830	o x	P	31Jan59 19Jun59	Kwinana (Medina) beach, 32.15S, 115.46E, S of Fremantle, W.A.	0–4–19
528–10845	o (v)	P	31Jan59 21Jun59	Woodman Pt., 32.08S, 115.45E, S of Fremantle, W.A.	0–4–21
528–10835	o x	P	31Jan59 22Jun59	Ampol beach, 32.31S, 115.45E, N of Fremantle, W.A.	0–4–22
528–10468	o v	P	31Jan59 24Jun59	Fremantle, 32.07S, 115.44E, W.A.	0–4–24
528–10703	o x	P	31Jan59 25Jun59	At 34.01S, 115.59E, in vicinity of Margaret River, W.A.	0–4–25
528–10378	o x	P	31Jan59 (26)Jun59	Near Stanley, Tasmania, 40.46S, 145.20E	0–?–?
528–10194	o (v)	P	31Jan59 26Jun59	Safety Bay, Fremantle, W.A.	0–4–26
528–10529	o x	P	31Jan59 (29)Jun59	25 miles S of Fremantle, W.A.	0–?–?
528–10930	o v	P	31Jan59 29Jun59	S of Breaksea Island, off Albany, 34.57S, 117.54E, W.A.	0–4–29
528–10679	o x	P	31Jan59 1Jul59	Sturt Bay, *ca.* 35S, 137.30E, south Yorke Peninsula, S.A.	0–5–1
528–10725	o v	P	31Jan59 6Jul59	At sea, 20 miles S of Port Fairy, 38.23S, 142.17E, Vict.	0–5–6

TABLE 11 (*continued*)

Band Number	Resident Status Manner of Recovery	Age Breeding Status	Banding Data Recovery Data	Banding Site Recovery Site	Length of Time Banded
528–10985	o	P	31Jan59		
	v		9Jul59	Near Coffin Island, at 35.00S, 118.21E, E of Albany, W.A.	0–5–9
528–10682	o	P	31Jan59		
	x		22Jul59	Near Murwillumbah, 28.20S, 153.24E, N.S.W.	0–5–22
528–10667	o	P	31Jan59		
	x		24Jul59	Goolwa, 35.31S, 138.45E, S.A.	0–5–24
528–10936	o	P	31Jan59		
	x (foxes)		24Jul59	Near Mt. Drummond, at 34.8S, 135.15E, Coffin Bay area, S.A.	0–5–24
528–10527	o	P	31Jan59		
	x (storm)		16Aug59	Mouth of Powlett R., near Wonthaggi, 38.38S, 145.37E, Vict.	0–6–16
528–10688	o	P	31Jan59		
	x		14Sep59	Near Point Sir Isaac, 34.24S, 135.11E, Coffin Bay, S.A.	0–7–14
518–94924 *528–15084*	o	P	3Feb59	Frazier Islands, 66.13S, 110.10E, Windmill Islands, Antarctica	
	tarsus only		26Jun60	Louth Bay, 16 miles N of Port Lincoln, 34.43S, 135.49E, S.A.	1–4–16
528–61033	o	P	8Feb61		
	x		8Jun61	Cottesloe beach, Perth, 31.58S, 115.49E, W.A.	0–4–0
528–60829	o	P	8Feb61		
	x		15Jun61	Capri beach, 38S, 145E, Mornington peninsula, Vict.	0–4–7
528–60545	o	P	6Feb61		
	v		22Jun61	Off Waldegrave Island, at 33.35S, 134.48E, Anxious Bay, S.A.	0–4–17
528–60392	o	P	3Feb61		
	v		25Jun61	In Gulf St. Vincent, off Glenelg, 34.59S, 138.31E, Adelaide, S.A.	0–4–22
528–61029	o	P	8Feb61		
	(v)		30Jun61	Between Garden Island, 32.15S, 115.09E, and Mewstone, near Fremantle, W.A.	0–4–22
528–60274	o	P	3Feb61		
	(v)		_.Jun Jul61	Shark Bay, 25.31S, 113E, W.A.	0–?–?
528–60310	o	P	3Feb61		
	x		1Jul61	S of Geraldton, 28.49S, 114.36E, W.A.	0–4–28
528–60655	o	P	5Feb61		
	x		(1)Jul61	Wonthaggi, Vict.	0–?–?
528–60098	o	P	1Feb61		
	x		1Jul61	Waterloo Bay, at 33.39S, 134.52E, Elliston, S.A.	0–5–0
528–43540	o	P	18Mar61	Penguin Island, 62.08S, 57.55W, South Shetland Islands	
	(v)		(5)Jul61	Fremantle, W.A.	0–?–?
528–60683	o	P	5Feb61		
	x		22Jul61	Beachport, 37.29S, 140.01E, S.A.	0–5–17
528–60380	o	P	3Feb61		
	f/?/		28Jul61	8 miles off Port Victoria, 34.30S, 137.30E, S.A.	0–5–25
528–60303	o	P	3Feb61		
	x		6Aug61	SW of Port Fairy, 38.23S, 142.17E, Vict.	0–6–3

TABLE 11 (*continued*)

Band Number	Resident Status Manner of Recovery	Age Breeding Status	Banding Data Recovery Data	Banding Site Recovery Site	Length of Time Banded
528–60546	o (v)	P	5Feb61 6Aug61	Off Rottnest Island, 32.01S, 115.28E, W.A.	0–6–1
528–60286	o x	P	3Feb61 25Aug61	Cape Liptrap, 38.56S, 145.56E, Vict.	0–6–22
528–60499	o x	P	5Feb61 5Sep61	11 miles S of Fremantle, W.A.	0–7–0
528–60715	o x	P	7Feb61 (7)Sep61	On beach at Capel, 130 miles S of Perth, 31.58S, 115.49E, W.A.	0–?–?
528–43557	o v	P	18Mar61 1Oct61	Penguin Island, South Shetland Islands Cooee, near Burnie, 41.03S, 145.55E, Tasmania	0–6–13
528–60494	o (x) (decomposed)	P	5Feb61 7Oct61	Corny Point, 34.53S, 137.00E, S.A.	0–8–2
528–60050	o x	P	1Feb61 15Oct61	Streaky Bay, 32.50S, 134.15E, S.A.	0–8–14
528–60187	o x	P	3Feb61 4Nov61	Barrack Pt., 34.34S, 150.52E, N.S.W.	0–9–1
528–60474	o x	P	5Feb61 10Nov61	Near Batemans Bay, 35.45S, 150.09E, N.S.W.	0–9–5
528–61223	o x	P	13Feb61 3Jun62	Streaky Bay, S.A.	1–3–21
568–01022	o v	P	1Feb62 7Jul62	Between Sydney Harbour, 33.55S, 151.10E, and Botany Bay, 34.04S, 151.08E, N.S.W.	0–5–6
568–01400	o (v)	P	4Feb62 17Jul62	Off Fremantle, W.A.	0–5–13
568–01466	o x	P	3Feb62 22Jul62	Point Lookout, 27.28S, 153.29E, Stradbroke Island, Qnsld.	0–5–19
568–01509	o v	P	3Feb62 6Aug62	Abrolhos Islands, approximately 28S, 113E, Pelsart group, W.A.	0–6–3
568–01107	o x	P	2Feb62 11Aug62	Port Campbell, 38.37S, 143.04E, Vict.	0–6–9
568–01065	o x	P	2Feb62 (15)Aug62	Off Portland, 38.21S, 141.38E, Vict.	0–?–?
568–01588	o x	P	3Feb62 11Sep62	N of Perth, W.A.	0–7–8
568–01172	o x	P	2Feb62 20Oct62	Pipers Head, 16 miles E of mouth of River Tamar (Georgetown, 41.04S, 146.48E) Tasmania	0–8–18
568–01522	o x	P	3Feb62 1Nov62	E side of Pelsart Island, 28.55S, 113.55E, W.A.	0–8–29
528–16840	o x	P	20Feb63 —.Jun63	S of Smoky Cape, 30.55S, 153.05E, N.S.W.	0–4–?
528–16897	o (v)	P	21Feb63 —.Jun63	Neptune Islands, 35.20S, 136.07E, S.A.	0–4–?

SLADEN, WOOD, AND MONAGHAN

TABLE 11 (*continued*)

Band Number	Resident Status Manner of Recovery	Age Breeding Status	Banding Data Recovery Data	Banding Site Recovery Site	Length of Time Banded
528–16873	o	P	21Feb63		
	x		(2) Jul63	Bunbury, 33.20S, 115.34E, W.A.	0–?–?
528–16902	o	P	21Feb63		
	v		7Jul63	Rottnest Island, Vict.	0–4–16
528–16962	o	P	21Feb63		
	v		c.5Aug63	Koks island, Shark Bay area, approximately 25S, 113E, W.A.	0–5–?
528–16809	o	P	20Feb63		
	x		5Sep63	S of Ulladulla, 35.21S, 150.25E, N.S.W.	0–6–16
528–68275	o	P	22Feb63		
	x		(18) Sep63	S of Adelaide, 34.55S, 138.35E, S.A.	0–?–?
528–68932	o	P	12Feb64		
	(v)		14Jun64	Cockburn Sound, approximately 32.10S, 114E, W.A.	0–4–2
528–68922	o	P	12Feb64		
	x (fox)		16Jun64	Eaton, Bunbury, W.A.	0–4–4
528–68845	o	P	13Feb64		
	v		21Jun64	Harvey, 33.06S, 115.50E, W.A.	0–4–8
528–68879	o	P	13Feb64		
	x		(22) Jun64	NW of Port Lincoln, S.A.	0–?–?
528–68996	o	P	12Feb64		
	v		4Jul64	S of Albany, W.A.	0–4–22
	v		13Jul64	E of Breaksea Island, Albany, W.A.	0–5–1
528–68921	o	P	12Feb64		
	x		9Jul64	Off Cape Howe, 37.30S, 149.59E, N.S.W.	0–4–27
528–68800	o	P	13Feb64		
	x		18Jul64	Coogee beach near Sydney, N.S.W.	0–5–5
528–68747	o	P	12Feb64		
	x		19Sep64	Swanbourne, S of Perth, W.A.	0–7–7
528–16494	o	P	31Jan65	Nelson Island, 62.17S, 59.02W, South Shetland Islands	
	v		9Jun65	60 miles S of Geraldton, W.A.	0–4–9

11E. *Recovered in New Zealand*

Band Number	Resident Status Manner of Recovery	Age Breeding Status	Banding Data Recovery Data	Banding Site Recovery Site	Length of Time Banded
528–15098	o	P	4Feb59	Frazier Islands, 66.13S, 110.10E, Windmill Islands, Antarctica	
	(v)		9May59	Off Kaiteriteri, at 41.05S, 173.00E, in Tasman Bay	0–3–5
528–10943	o	P	31Jan59		
	x		24May59	Te Akau near Ngaruawahia, 37.41S, 175.10E, North Island	0–3–24
528–15035	o	P	3Feb59	Frazier Islands, Windmill Islands, Antarctica	
	v		5Jun59	Near Te Araroa, 37.37S, 178.23E, North Island	0–4–2
528–10599	o	P	31Jan59		
	x		30Jul59	Manganui Bluff, 35.43S, 173.32E, North Island	0–5–30
528–10964	o	P	31Jan59		
	x		7Aug59	Near Kaitaia, 35.12S, 173.10E, North Island	0–6–7
528–10494	o	P	31Jan59		
	(v)		3Sep59	ESE of Gisborne, 38.41S, 178.02E, in Poverty Bay, North Island	0–7–3

TABLE 11 (*continued*)

Band Number	Resident Status Manner of Recovery	Age Breeding Status	Banding Data Recovery Data	Banding Site Recovery Site	Length of Time Banded
528–60168	o x	P	3Feb61 (18) Jul61	Near Kaeo, 35.05S, 173.48E, North Island	0–?–?
528–60087	o (v)	P	1Feb61 3Sep61	Dargaville, 35.57S, 173.53E, North Island	0–7–2
528–61217	o x	P	13Feb61 11Sep61	Near Nuhaka, 39.02S, 177.50E, Bay of Plenty, North Island	0–6–29
528–60186	o v v	P	3Feb61 14Oct61 14Nov61	Off Flat Point, 41.15S, 175.58E, North Island 30 miles S of Castlepoint, 40.54S, 176.15E, North Island	0–8–11 0–9–11
568–01012	o x	P	1Feb62 (5) Aug62	Dargaville, North Island	0–?–?
528–68176	o x	P	22Feb63 10Aug63	Princess Cove, Wellington, 41.17S, 174.47E	0–5–19
568–01211	o x	P	2Feb62 c.15Sep63	South Cape, 47.17S, 167.13E, Stewart Island	1–7–?
528–16945	o x	P	21Feb63 16Sep63	South Cape, Stewart Island	0–6–26
528–68497	o v	P	5Mar63 27Oct63	E of Table Cape, 39.06S, 178.02E, Hawke Bay	0–7–22
528–68917	o x	P	12Feb64 29Aug64	Muriwai, 38.45S, 177.56E, N of Auckland, North Island	0–6–17
528–68796	o x	P	13Feb64 15Sep64	Te Puke, 37.47S, 176.22E, North Island	0–7–2
528–68798	o (v)	P	13Feb64 28Sep64	NE of North Cape, 34.23S, 173.04E, North Island	0–7–15
528–69056	o v	P	15Feb64 12Oct64	NNE of Mt. Maunganui, Tauranga, 37.42S, 176.11E, North Island	0–7–27

11F. Recovered in Tahiti

Band Number	Resident Status Manner of Recovery	Age Breeding Status	Banding Data Recovery Data	Banding Site Recovery Site	Length of Time Banded
528–60756	o f/?/	P	7Feb61 4Aug61	Off Teahupoo, 17.51S, 149.15W, Tairapu peninsula, **Tahiti**	0–5–28

TABLE 12. Recoveries of South Polar Skuas Banded in the Ross Sea Area

Band Number	Resident Status Manner of Recovery	Age Breeding Status	Banding Data Recovery Data	Banding Site Recovery Site	Length of Time Banded
12A.		*Banded at Cape Crozier, 77.27S, 169.13E, Ross Island, Antarctica*			
657–63538	*	A	2Jan63		
	x		14Jan64	Cape Royds, 77.33S, 166.07E, Ross Island, Antarctica	1–0–12
657–66069	*	A	11Jan63		
	s		9Jan65	Cape Bird, 77.16S, 166.20E, Ross Island, Antarctica	1–11–29
657–66953	o	P	19Feb63		
	v		13Jan65	Cape Royds	1–10–25
657–60257	o	A	9Nov62		
	v		20Jan65	Cape Royds	2–2–11
657–66692	o	P	5Feb63		
	s		22Jan65	Cape Bird	1–11–17
657–66729	o	P	5Feb63		
	s		22Jan65	Cape Bird	1–11–17
657–63887	o	P	4Feb62		
	x		29Jan65	McMurdo station, 77.51S, 166.37E, Ross Island, Antarctica	2–11–25
657–63548	*	A	2Jan63		
	v		6Feb65	Cape Royds	2–1–4
12B.		*Banded at McMurdo Station, 77.51S, 166.37E, Ross Island, Antarctica*			
617–12644	*	A	3Feb59		
	s		22Jan62	Cape Evans 77.38S, 166.24E, Ross Island, Antarctica	2–11–19
617–12621	*	A	23Jan59		
	v	br.	6Jan65	Cape Bird	5–11–14
617–12637	*	A	25Jan59		
	s	br.	20Jan65	Cape Bird	5–11–26
	v		2Feb65	Cape Royds	6–0–8
617–12627	*	A	23Jan59		
	s		21Jan65	Cape Bird	5–11–29
12C.		*Banded at Other Stations on Ross Island*			
617–12437	o	P	16Feb60	Cape Royds	
	s		17Feb62	Cape Crozier	2–0–1
617–12428	o	P	16Feb60	Cape Royds	
	s		17Jan64	Cape Crozier	3–11–1
657–63687	o	P	24Jan62	Cape Royds	
	s		28Jan64	Cape Crozier	2–0–4
657–64386	o	A	5Jan65	Cape Bird	
	v		28Jan65	Cape Royds	0–0–23
657–64463	o	A	6Jan65	Cape Bird	
	v		2Feb65	Cape Royds	0–0–27

TABLE 12 (*continued*)

Band Number	Resident Status Manner of Recovery	Age Breeding Status	Banding Data Recovery Data	Banding Site Recovery Site	Length of Time Banded
			12D. Banded at Hallett Station, 72.18S, 170.13E, Victoria Land, Antarctica		
647–26213	o	P	24Jan59		
	s		25Feb63	Cape Crozier	4–1–1
657–63258	o	P	18Jan62		
	s		15Jan64	Cape Crozier	1–11–28
657–63250	o	P	18Jan62		
	s		17Jan64	Cape Crozier	1–11–30
	s		17Jan65	Cape Crozier	2–11–30
657–63216	o	P	17Jan62		
	s		16Feb64	Cape Crozier	1–10–29
	s		16Dec64	Cape Crozier	2–10–29
657–61149	o	P	25Jan61		
	s		16Dec64	Cape Crozier	3–10–21
	s		20Jan65	Cape Bird	3–11–26

TABLE 13. Recoveries of Skuas[a] Banded in the Antarctic Peninsula and Budd Coast Areas

Band Number	Resident Status Manner of Recovery	Age Breeding Status	Banding Data Recovery Data	Banding Site Recovery Site	Length of Time Banded
617–13722	*	A	31Dec59	Ardley Island, 62.13S, 58.54W, King George Island, South Shetland Islands	
	v		20Jan60	Deception Island, 62.57S, 60.38W, South Shetland Islands	0–0–20
647–27111	*	A	4Mar61	González Videla station, 64.49S, 62.51W, Antarctic Peninsula	
	+		1Feb62	Hope Bay, 63.24S, 57.00W, Antarctic Peninsula	0–11–28
617–11413	o	P	25Jan60	Windmill Islands, 66.20S, 110.28E, Budd Coast, Antarctica	
	/?/	f	9Jan64	Hallett station, 72.18S, 170.13E, Victoria Land, Antarctica	3–11–15
647–27146	*	A	5Mar61	González Videla station	
	x		7Aug64	Udipi, 13.23N, 74.45E, Mysore, **India**	3–5–2

[a] 617–13722 was almost certainly *C. skua lonnbergi*. 617–11413 was *C. maccormicki*. 647–27146 was confirmed as *C. maccormicki* from the specimen collected [*Bombay Natural History Society*, 1966]. 647–27111's identity is not certain.

TABLE 14. Recoveries of Other Species Banded in the Falkland Islands

Band Number	Resident Status Manner of Recovery	Age Breeding Status	Banding Data Recovery Data	Banding Site Recovery Site	Length of Time Banded
			Sooty Shearwater		
565–77737	*	A	4May62	Stanley, 51.45S, 57.56W, Falkland Islands	
	x		1Jun62	Barbados, 13N, 59W, **West Indies**	0–0–28
			Wattled Sheathbill		
617–52571	*	A	11Aug61	George Island, 52.20S, 59.44W, Falkland Islands	
	+		31Mar62	Signy Island, 60.43S, 45.36W, **South Orkney Islands**	0–7–20

TABLE 15. Ages* at Recovery of Wandering Albatrosses Banded as Nestlings

Recovery Location	1st Year	2nd Year	3rd Year	Total
South America	1	1	1	3
South Africa	1			1
Australia		1		1
Total recovered	2	2	1	5
Number banded per age group	1,625	1,400	400	

* Follows the U.S. Fish and Wildlife Service Bird Banding Laboratory's age classification system with modifications for the southern hemisphere breeding season. The first year extends from the mean date of hatching, March 11, to June 30 of the following year, a period of 15½ months (of which nine months are spent as a fledgling). The second and subsequent years extend from July 1 to June 30.

TABLE 16. Ages* at Recovery of Black-Browed Albatrosses Banded as Nestlings

Recovery Location	1st Year	2nd Year	3rd Year	4th Year	5th Year	Total
16A. Banded at South Georgia						
South Africa	7	74	35	3	2	121
Other African Nations	1	10				11
Australia	2	5	1			8
South America		2	1	1		4
Total recovered	10	91	37	4	2	144
Number banded per age group	18,954	18,954	16,173	8,551	3,251	
16B. Banded at Falkland Islands						
South America	15	33	7	2	1	58
Africa	2	6	1			9
Total recovered	17	39	8	2	1	67
Number banded per age group	10,793	8,793	6,799	4,805	1,805	
16C. Totals of Both Areas						
Total recovered	27	130	45	6	3	211
Number banded per age group	29,747	27,747	22,972	13,356	5,056	

* Follows the U. S. Fish and Wildlife Service Bird Banding Laboratory's age classification system with modifications for the southern hemisphere breeding season. The first year extends from the mean date of hatching, December 22, to June 30. The second and subsequent years extend from July 1 to June 30.

TABLE 17. Ages* at Recovery of Gray-Headed Albatrosses Banded as Nestlings

Recovery Location	1st Year	2nd Year	3rd Year	4th Year	Total
Africa		3	1		4
Australia	1	1	2		4
New Zealand				1	1
Total recovered	1	4	3	1	9
Number banded per age group	13,729	13,729	11,879	6,279	

* Follows the U. S. Fish and Wildlife Service Bird Banding Laboratory's age classification system with modifications for the southern hemisphere breeding season. The first year extends from the mean date of hatching, December 15, to June 30. The second and subsequent years extend from July 1 to June 30.

TABLE 18. Ages* at Recovery of Giant Petrels Banded as Nestlings

Recovery Location	1st Year	2nd Year	3rd Year	4th Year	Total
South America	1	18	2	1	22
Africa	2	12			14
Australia	23	44			67
New Zealand	3	16	1		20
Tahiti		1			1
Total recovered	29	91	3	1	124
Number banded per age group	4,825	4,279	3,732	2,932	

* Follows the U. S. Fish and Wildlife Service Bird Banding Laboratory's age classification system with modifications for the southern hemisphere breeding season. The first year extends from the mean date of hatching, December 7, to June 30. The second and subsequent years extend from July to 1 to June 30.